Lead, Mercury and Cadmium in the Aquatic Environment

Worldwide Occurrence, Fate and Toxicity

Editors

Rachel Ann Hauser-Davis
Oswaldo Cruz Foundation (Fiocruz)
Rio de Janeiro, Brazil

Natalia Soares Quinete
Department of Chemistry and Biochemistry
Florida International University
North Miami, Florida, USA

Leila Soledade Lemos
Florida International University
North Miami, Florida, USA

(CRC) **CRC Press**
Taylor & Francis Group
Boca Raton London New York

CRC Press is an imprint of the
Taylor & Francis Group, an **informa** business

A SCIENCE PUBLISHERS BOOK

First edition published 2023
by CRC Press
6000 Broken Sound Parkway NW, Suite 300, Boca Raton, FL 33487-2742

and by CRC Press
4 Park Square, Milton Park, Abingdon, Oxon, OX14 4RN

© 2023 Taylor & Francis Group, LLC

CRC Press is an imprint of Taylor & Francis Group, LLC

Library of Congress Cataloging-in-Publication Data (applied for)

ISBN: 978-1-032-03051-7 (hbk)
ISBN: 978-1-032-03054-8 (pbk)
ISBN: 978-1-003-18644-1 (ebk)

DOI: 10.1201/9781003186441

Typeset in Times New Roman
by Shubham Creation

Preface

Water is, without a doubt, one of the most important resources on Planet Earth. Human activities have, however, contaminated most water bodies worldwide with an array of environmental pollutants, resulting in many deleterious effects in exposed organisms and, of course, human populations. Metals and metalloids are among the diverse contaminants present in aquatic ecosystems, many of them extremely toxic to life. Assessments of their levels and effects are paramount towards the implementation of decision-making processes and frameworks concerning their management and future remediation. With this perspective, this book presents an integrated and holistic discussion on the toxicity of a triad of the most toxic metals, cadmium, lead and mercury, in aquatic environments, expanding general concepts on chemical speciation effects and exploring specific environmental toxicological issues, exposure routes, assessments of their intracellular deleterious effects, and conservation actions required to mitigate these impacts. Their concentrations in the aquatic environment and aquatic biota, as well as toxic effects, are discussed. We have also aimed to discuss both worldwide and regional aspects regarding the occurrence, fate, and toxicity of these metals, bringing forth new knowledge on global issues and addressing key environmental issues and health risk concerns to both humans and aquatic organisms. This book is of broader interest not only to planners and policymakers involved in water pollution control, but anyone conducting research in the fields of oceanography, geochemistry, conservation, ecotoxicology and environmental and public health, including, but not limited to, marine scientists, chemists, geochemists, toxicologists, pharmacologists, conservationists, environmental biologists, biochemists, immunologists, cell biologists, environmental analysts, molecular biologists, behaviourists, developmental biologists and risk assessors.

Editor
Rachel Ann Hauser-Davis
Natalia Soares Quinete
Leila Soledade Lemos

Contents

Subcellular Cadmium, Lead and Mercury Partitioning Assessments in Aquatic Organisms as a Tool for Assessing Actual Toxicity and Trophic Transfer

Rachel Ann Hauser-Davis

Oswaldo Cruz Foundation, Fiocruz. Email: *rachel.hauser.davis@gmail.com*

INTRODUCTION

Anthropogenic activities have wreaked havoc on the environment worldwide since the Industrial Revolution (Ahuti, 2015). Increasing industrialization and related activities, including mining, damming and farming, among many others, emit momentous volumes of different environmental contaminants, such as hydrocarbons, chlorinated pesticides, brominated and perfluorinated compounds and metals, which are then irrevocably transported to aquatic ecosystems, the ultimate environmental contamination sinks (Bashir et al., 2020).

Metals and metalloids, in particular, display high environmental persistence and stability and the ability to bioaccumulate and, in some cases, biomagnify throughout trophic food webs. These contaminants can accumulate in animal tissues at several orders of magnitude above the concentrations found in the water column, making aquatic organisms responsible for much of the dynamics of these pollutants in the aquatic environment (Espino, 2000). Metal determinations in tissues and organs of different aquatic organisms are, therefore, an important method to assess environmental metal contamination and bioavailability in aquatic ecosystems (Voets et al., 2009; Wang et al., 2011), useful in both environmental and human health risk assessments (Wallace et al., 2003; Wang and Rainbow, 2006).

However, most studies determine only total metal loads, which are inadequate to effectively assess metal toxicity (Vijver et al., 2004; Péry et al., 2008), as metals display the particular ability to undergo subcellular compartmentalization upon cell entry (Rainbow, 2002; Wallace et al., 2003), altering their bioavailability and, thus, their toxicity. Because of this, toxicity is determined by complex interactions between metal uptake, accumulation and detoxification mechanisms (Perceval et al., 2006; Rainbow, 2002; Sokolova et al., 2005; Wang et al., 2011), and not simply by total metals accumulated within tissues or organs.

In this context, this chapter aims to discuss metal subcellular compartmentalization and the need for subcellular cadmium (Cd), lead (Pb) and mercury (Hg) partitioning assessments in aquatic organisms to understand metal toxicity towards aquatic organisms and toxicological consequences throughout trophic food webs (Bragigand et al., 2004; Seebaugh and Wallace, 2004; Wallace et al., 2003).

METAL AND METALLOID CLASSIFICATION AND AQUATIC ORGANISM TOXICITY

Metals can be basically be categorized into essential, non-essential and toxic elements. This is preferable over the use of the erroneous terms such as "heavy metals" to indicate metal toxicity, which was cited over 40 years ago as being nondescript and biologically and chemically insignificant, and over 20 years ago as erroneous and misused, with more recent studies mandating its ban (Duffus, 2002; Nieboer and Richardson, 1980; Pourret and Hursthouse, 2019; Pourret et al., 2021).

Essential elements are required for adequate metabolic functioning and equilibrium. These include elements such as selenium, iron, molybdenum, manganese, copper, and zinc, among others, which are involved in different cellular processes in many taxonomic groups, such as in antioxidant defenses, thyroid hormone metabolism, oxygenation, and immune system responses, amongst many others (Janz et al., 2010; Lall, 2022). Besides being required at certain species-specific concentrations, metals are also required for correct protein and enzyme performance, either as an integral part of these molecules or transported by them. These include hemoglobin, myoglobin, transferrin and ferritin (iron), ceruloplasmin (copper), carbonic anhydrase (zinc), vitamin B12 (cobalt), superoxide dismutase (copper, zin, manganese, iron, nickel), many selenoproteins (selenium - Se) and metallothionein (MT) (selenium, zinc, copper, cadmium (Cd), silver, arsenic (As), mercury (Hg), iron and nickel, among others). Although the list is quite extensive, it does not end here. Non-essential elements, on the other hand, such as vanadium, play no known role in metabolic functioning, and exhibit potential toxicity (Ravindran and Radhakrishnan, 2020). Both essential and non-essential, elements however, become toxic if present above specific thresholds (Bansal and Asthana, 2018). Toxic elements, on the other hand, such as As, Cd, Hg and Pb, result in deleterious health effects such as cytotoxicity, genotoxicity and immunotoxicity, with the potential for teratogenic and mutagenic outcomes, at any given concentration, and thus, at extremely low thresholds (EPA, 2008; 2014; 2016; 2021). This may, in turn, lead to populational and ecosystem effects, affecting the general physical and health conditions of exposed organisms. This may lead to populational and ecosystem effects, affecting the general physical and health conditions of exposed organisms. In turn, this may reflect, for example, in decreased body conditions, and directly impacting reproductive functions and survival, in both invertebrates and vertebrates (Pantea et al., 2020; Wosnick et al., 2021).

In this regard, toxicity thresholds are often species-specific and have been determined only for certain metals and metalloids and aquatic organisms. For example, the threshold for reproductive effects concerning Se exposure in freshwater was set by Lemly in 1993 as 10 ug g^{-1} dry weight in gonads (Lemly, 1993), whereas Khadra et al. noted negative effects for yellow perch *Perca flavescens* at 10-fold lower concentrations, and other assessments indicate threshold ranges from 17 to 24 ug g^{-1} dry weight in the same organ (Janz et al., 2010). Thresholds for many metals, however, have either not been established, such as for Pb (Silberlgeld, 2004), or vary immensely among species and/or organs. For example, Hg concentrations as low as 0.008 mg kg^{-1} wet weight in muscle tissue have been indicated as causing nervous system enzymatic activity changes, while negative swimming, foraging, and reproduction effects are reported at about 0.135 mg kg^{-1} wet weight also in muscle for freshwater fish (Sandheinrich and Wiener, 2011). Arsenic, however, is considered toxic for aquatic organisms in general at over 0.09 mg kg^{-1} wet weight in any tissue (Jankong et al., 2007). In addition, these thresholds may also change according to biotic and abiotic parameters, such as reproductive efforts, salinity, temperature, dissolved oxygen levels, organic matter concentrations, geographic region, among others (Perošević et al., 2018).

Assessments regarding total metal loads and their biochemical endpoint and negative health effects in aquatic organisms in an environmental toxicology context are plentiful in literature. The first study in this regard was, in fact, carried out over 100 years ago, in 1912, by White and Thomas, who reported on copper exposure effects in a small killifish, *Fundulus heteroclitus*. This gave rise to the tissue residue model, or critical body residue model, in environmental toxicology, which aims to determine dose-response relationships for environmental contaminants and evaluate the range of tissue residues that are likely to lead to adverse effects, defining a measure of toxicity (Adams et al., 2011; Barron et al., 1997). The tissue residue model, however, has achieved only limited success, since, as stated previously, metal storage, detoxification, and elimination mechanisms, are paramount in determining metal toxicity (Adams et al., 2011). This is where the subcellular metal partitioning comes into play.

CELLULAR CADMIUM, LEAD AND MERCURY TOXICITY MECHANISMS OF ACTION

The mechanisms of action of Cd, Hg and Pb, although studied for decades, still remain somewhat elusive. All three metals, however, are known to significantly induce the production of reactive radicals, i.e., Reactive Oxygen Species (ROS), comprising singlet oxygen, hydrogen peroxide, the superoxide radical, the hydroxyl radical, and Reactive Nitrogen Species (RNS), comprising molecules derived from nitric oxide, such as peroxynitrite, and the superoxide anion (Calcerrada et al., 2011; Das and Roychoudhury, 2014). This, in turn, may result in oxidative stress, due to an imbalance between ROS production and accumulation in cells and tissues and the ability of biological systems to sequester and detoxify these reactive products (Pizzino et al., 2017). Furthermore, MT induced by exposure to these metals can be a double-edged factor, as their binding could either result in metal detoxification from the cell environment, or on the other hand, result in metal dissociation from MTs due to decreased metal binding stability, further increasing ROS production (Valko et al., 2005, 2007).

Cadmium cell uptake is directly linked to transition metal homeostasis deregulation (Droge, 2002). Once in the cell, even though it is a non-reactive metal, and, thus, unable to directly induce ROS production, it may indirectly cause oxidative stress (Valko et al., 2007). This takes place through four main mechanisms, namely displacement of redox-active metals, such as copper and iron, which then take place in the Fenton reaction, depletion of redox scavengers (mainly reduced glutathione, GSH), inhibition of anti-oxidant enzymes (mainly enzymes belonging to the glutathione system, superoxide dismutase and catalase) and inhibition of the electron transport chain resulting in mitochondrial damage, interfering with mitochondrial oxidative phosphorylation, leading to the accumulation of unstable semiubiquinones which donate electrons and create superoxide radicals, thus inhibiting basal respiration (Ercal et al., 2001; Valko et al., 2007; Young and Woodside, 2001).

Mercury in both its organic and inorganic forms, mostly acts by reacting with thiol groups present in GSH, decreasing GSH levels both due to direct metal binding to this molecule and/or enhanced thiol oxidation, which, in turn, leads to oxidative stress conditions (Elia et al., 2003; Stohs and Bagchi, 1995). Severe mitochondrial damage is also noted, due to the aforementioned GSH depletion (Nicole et al., 1998). Mercury can also bind to phosphoryl, carboxyl, amide, and amine groups, leading to generalized incapacitation of amino acids, proteins and enzymes, many of them key in cellular stress responses, protein repair, and oxidative damage prevention (Broussard et al., 2002).

Like Cd, Pb is a non-reactive metal, so it cannot readily undergo valence changes. Thus, oxidative damage is induced through direct cell membrane effects, as well as interactions with key compounds, such as hemoglobin, increasing hemoglobin auto-oxidation. Furthermore, Pb also inhibits delta-aminolevulinic acid dehydratase, which is involved in the production of heme. This takes place by altering this enzyme's quaternary structure by displacement of a zinc ion at the metal binding site, altering heme production (Warren et al., 1998). It also inhibits two other enzymes involved in the heme pathway, ferrochelatase, and coproporphyrinogen oxidase, although its effects are mostly associated to delta-aminolevulinic acid dehydratase. Glutathione reductase has also been noted, as well as selenium complexation, in turn decreasing glutathione peroxidase activity (Ercal et al., 2001).

Due to these mechanisms of action, Cd, Hg and Pb exposures have been noted as affecting a wide range of aquatic organism responses. For example, several antioxidant defenses (i.e., catalase, superoxide dismutase, xanthine oxidase, reduced glutathione, metallothionein, glutathione peroxidase and glutathione S-transferase), neurotransmitters (acetylcholinesterase), hematological parameters and ion regulation enzymes are routinely reported as altered in different marine and freshwater vertebrates and invertebrates following sublethal exposure to these three metals (Basha and Rani, 2003; Gill et al., 1991; Kataba et al., 2022; Santovito et al., 2021; Zirong and Shijun, 2007).

Toxicity assessments for these three metals range from acute to chronic exposures at both environmentally relevant and extremely high metal concentrations in laboratory settings to single and/or mixed metal exposures in field assessments concerning both invertebrates and vertebrates. Known indicator species, as well as standardized ecotoxicological model species, have been employed to this end. It is, however, important to note the lack of assessments on tropical species in this regard. In fact, it is extremely important to compare the responses of tropical and temperate species to the same contaminants, as differential responses and effects are probable, resulting in more adequate assessments concerning specific climates or regions (Kwok et al., 2007).

SUBCELLULAR METAL PARTITIONING AND ASSESSMENTS CONCERNING Cd, Hg AND Pb IN AQUATIC ORGANISMS

Organisms have evolved diverse metal sequestration and detoxification mechanisms in an attempt to mitigate deleterious metal effects. Toxicity, however, as noted previously, does not depend on total accumulated metal concentrations, but instead ensues only when metabolically bioavailable metals exceed specific threshold values, overwhelming detoxification systems (Campana et al., 2015; Rosabal et al., 2012). It is important to note that differential toxicity for the same metal among different species is a given (Mason and Jenkins, 1995; Rainbow, 2007; Rainbow, 2002), and that sex and life stage can influence bioaccumulation patterns and their consequent effects on the systemic health of exposed organisms (Vega-López et al., 2007; Wosnick et al., 2021).

In this regard, metal subcellular partitioning mechanisms may be classified into two major groups, as displayed in Figure 1.1: (i) metal binding to potentially sensitive target molecules, such as cytosolic enzymes, small peptides (i.e., glutathione), organic acids (e.g., citrates) and DNA, and organelles, like mitochondria and the endoplasmic reticulum, potentially resulting in deleterious cellular effects, due to blocked functional groups of biomolecules, displacement of essential metals from their normal sites within biomolecules and altered conformation and, consequently, activity (Campbell et al., 2008) and (ii) accumulation in detoxified metal fractions, such as heat-stable proteins, metal-rich granules, lysosomes and membrane-bound vesicles, in an attempt to minimize toxic effects (Campbell et al., 2008; Campbell and Hare, 2009; Wallace et al., 2003). These groups have, thus, been operationally defined as the metal sensitive fraction, and the detoxified fraction, and can be obtained through subcellular fractionation methods, which separate total metal contents into different functional compartments following differential centrifugations and heat treatments (Wallace et al., 2003).

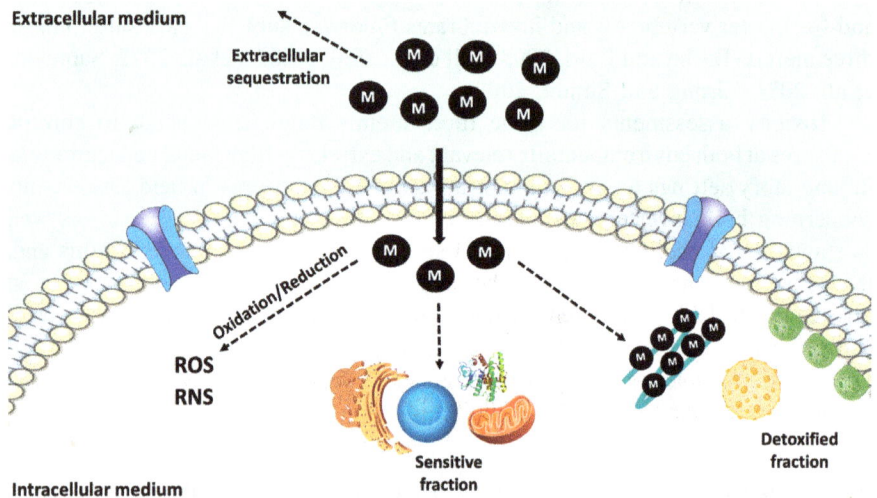

Figure 1.1 Diagram representing subcellular metal partitioning in animal cells. The sensitive fraction (middle portion of the diagram) includes binding to different organelles (endoplasmic reticulum, mitochondria), cytosolic enzymes and DNA, while the detoxified fraction (right side of the diagram) includes binding to metallothionein (indicated in teal with seven metal-binding sites), lysosomes (indicated in yellow) and membrane-bound vesicles (indicated in green). ROS – Reactive Oxygen Species; RNS – Reactive Nitrogen Species.

Some interesting subcellular partitioning assessments for Cd, Hg and Pb conducted in both laboratory and field settings for different aquatic taxonomic groups are described below.

Phytoplankton

In one study, the accumulation, subcellular distribution, and toxicity of two different Hg forms, Hg(II) and methylmercury (MeHg), were evaluated in three marine phytoplankton species, the diatom *Thalassiosira pseudonana*, the green alga *Chlorella autotrophica*, and the flagellate *Isochrysis galbana* (Wu and Wang, 2011). The authors report differential Hg(II) accumulation in both species, with a higher percentage of Hg(II) bound to the cellular debris fraction in *T. pseudonana* than in *I. galbana*. On the other hand, heat-stable proteins were a major binding pool for MeHg, indicating that speciation assessments are required in subcellular partitioning evaluations, even more so when considering that MeHg is the most toxic form of this metal and a significant concern in aquatic ecosystems (NRC, 2000).

Seaweed

One of the only assessments in the literature concerning subcellular metal partitioning in seaweed comprises an evaluation concerning the edible seaweed, *Porphyra yezoensis* exposed to different Cd concentrations for up to 96 h (Zhao et al., 2015). The authors report high cell wall Cd binding, which increased with increasing Cd concentrations and exposure times, indicating a strategy to reduce Cd toxicity in this seaweed species.

Invertebrates

One invertebrate assessment evaluated total Hg subcellular fractionation kinetics in larvae of the midge *Chironomus riparius* following laboratory exposure to Hg-spiked sediments and water (Gimbert et al., 2016). Cellular debris (including exoskeleton, gut contents, and cellular debris), granule and organelle fractions accounted only for about 10% of bioaccumulated Hg, whereas about 90% of Hg concentrations were internalized in the cytosolic compartment and one third of the Hg was found bound to metallothionein-like proteins. On the other hand, the sensitive fraction became progressively saturated with Hg, resulting in Hg excretion and physiological impairments.

One of the few studies conducted in estuarine species assessed Cd partitioning among different cytosolic fractions in the estuarine polychaete *Laeonereis acuta* (Sandrini et al., 2006). The authors report that cadmium was almost absent in the fraction lower than 3 kDa from both experimental groups, indicating that in *L. acuta*, gluta- thione or other low-molecular-weight ligands were not associated to Cd binding, and that this metal was stored mainly in the cytosolic fraction containing molecules over 10 kDa. Furthermore, high Cd concentrations were also observed in the heavy weight fraction, which might contain insoluble Cd granules.

Another study evaluated total and subcellular partitioning of Cd and Hg in juvenile eastern oysters *Crassostrea virginica* exposed to these metals for 4 weeks, both at high and low concentrations (Fitzgerald et al., 2019). Differential total body burdens for both metals were noted comparing the high and low treatments. Concerning Cd, this difference in general, mirrored at the subcellular level, although binding to heat-denatured proteins in the High Cd treatment was reduced. Mercury, on the other hand, was not appreciably partitioned to subcellular fractions.

In another bivalve assessment, the association between Cd accumulation and subcellular gill distribution and growth and mortality rates was investigated in the freshwater bivalve *Pyganodon grandis* transplanted from a clean site to four lakes along a Cd concentration gradient in a mining region in Canada (Perceval et al., 2006). The Cd distribution in gills in the various cytosolic complexes varied significantly among sites and at two of the high contamination sites, decreased survival rates were noted, alongside less cytosolic gill Cd detoxified by MT-like proteins. This demonstrates that excessive accumulation of Cd in the high molecular weight protein pool of the gill cytosol of the individual mollusks is associated to health status impairment at the population level.

Another study evaluated subcellular Cd partitioning in the freshwater bivalve *Pyganodon grandis* exposed to a known metal gradient in a mining area in northwestern Canada (Wang et al., 1999). Gill MT concentrations responded to environmental exposure to Cd in the bivalves assessed in the field, while bivalves exposed to high dissolved Cd concentrations exhibited a marked Cd increase in cytosolic gill low relative molecular mass ligand pool. The authors suggest that Cd subcellular metal partitioning is altered in bivalve populations naturally exposed to high Cd concentrations.

Subcellular partitioning profiles and metallothionein levels for different metals, including Hg and Pb, were also investigated in the digestive glands of indigenous Chinese clams *Moerella iridescens* from a metal-impacted coastal bay (Wang et al., 2016). The heat-stable protein (HSP) fraction was the dominant metal-binding compartment for Cd, but not for Pb. MT-binding was noted for both metals, although a progressive accumulation of Cd, and especially Pb due to a "spillover" effect was reported in the metal-sensitive fractions, demonstrating incomplete metal detoxification with increased toxicological risk to *M. iridescens*.

Vertebrates

One study carried out with fish verified the intracellular compartmentalization of both Cd and Pb, among other metals, in the livers of two eel species, *Anguilla rostrata* from Canada and *Anguilla etallot* from France (Rosabal et al., 2015). The authors indicate that the cytosolic heat-stable fraction was involved in the detoxification of all elements, and that granule-like structures were observed for Pb detoxification in both species. The latter were not, however, completely effective, as increasing metal concentrations in whole livers were accompanied by significant increases in metal concentrations in the sensitive subcellular fractions comprising organelles such as mitochondria, microsomes and lysosomes and heat-denatured cytosolic proteins, with mitochondria as the major binding sites for Cd and Pb. These findings, thus, indicate likely health risk for the investigated eel species.

In another study, the subcellular partitioning of Cd, Pb and Hg was determined in the livers of yelloweye rockfish (*Sebastes ruberrimus*) sampled from Alaska (Barst et al., 2018). Concerning Cd and Pb, the greatest contributions were found in the detoxified fractions, while most total Hg was detected in sensitive fractions, which the authors emphasize as indicating high health concerns for the investigated species due to subcellular level damage. Furthermore, differential metal binding was observed, as Cd > Pb > Hg.

Another study evaluated the subcellular partitioning of Cd, among other metals and metalloids, in the liver and gonads of wild white suckers *Catostomus commersonii* sampled downstream from a mining operation (Urien et al., 2018). The study investigated metal partitioning among metal-sensitive fractions (heat-denatured proteins (HDP), mitochondria and microsomes) and biologically detoxified fractions (heat-stable proteins (HSP) and metal-rich granules). The observed metal-coping strategies were similar in both liver and female gonads, but differed from male gonads, which the authors indicate as likely due to the different roles played by male and female reproductive organs. The authors also report that the HSP fraction was the most responsive to increased metal exposure, and that alterations in Cd-coping in female gonads were most evident, with Cd shifting from the metal-sensitive HDP fraction in reference fish to the metal-detoxified HSP fraction in exposed fish. The authors also state that Cd detoxification in female gonads seemed to be not fully induced in the less contaminated fish but became more effective above a threshold Cd concentration of 0.05 nmol g^{-1} dry weight.

Recently, the cytosolic distribution profile of Cd, among other metals, was evaluated for the first time in the intestinal cytosol of two fish species, brown trout (*Salmo trutta*) and Prussian carp (*Carassius gibelio*) (Mijošek et al., 2021). Some similar patterns were noted for both species, where Cd was predominantly bound to MT in the low molecular weight cytosolic fraction, although a slight association between Cd and the high molecular weight intestinal fraction was also observed in samples exhibiting the highest cytosolic Cd concentrations, which the authors postulate as potentially due to a higher susceptibility to Cd toxicity after higher bioaccumulation.

One very interesting study assessed the maternal transfer and subcellular partitioning of Hg, specifically, the more toxic MeHg, in Yellow Perch (*Perca flavescens*) from Lake Saint-Pierre (Quebec, Canada) (Khadra et al., 2019). The findings indicate that MeHg, was primarily (51%) associated to the hepatic subcellular fraction containing cytosolic enzymes, while 23% and 15% of MeHg was found in hepatic and gonadal mitochondria, respectively, suggesting inefficient Hg detoxification mechanisms in this species. A strong correlation was also observed between MeHg bioaccumulation in the liver and MeHg concentrations in gonadal mitochondria, corroborating potential risks associated to MeHg maternal transfer, as early developmental stages in aquatic biota are particularly sensitive to this toxic element.

Subcellular metal partitioning assessments in elasmobranchs (sharks and rays) are extremely rare. The only study conducted to date in this regard, to the best of our knowledge, performed by Hauser-Davis et al. (2021), investigated different metal partitioning and reported Cd, Hg and Pb concentrations, among other metals, in blue shark *Prionace glauca* muscle and liver samples. The authors report higher cytosolic metal MT-binding in liver compared to total metal loads, as expected, as this is the main detoxification organ, and indicate that the MT-metal detoxification pathway does not seem to be applicable for Hg, that Cd also exhibited low MT-binding compared to total metal burdens and that Pb exhibited the same MT-binding behavior in both tissues, indicating preferential, tissue-independent, binding, both regarding accumulation (in muscle) and subsequent excretion (liver). No other studies are, however, available for this taxonomic group concerning subcellular metal partitioning.

Scarce subcellular metal partitioning assessments are also found in the literature for marine mammals. One study in this regard determined the subcellular compartmentalization of Cd, Hg and Pb in three subcellular fractions (heat-stable, heat-labile and insoluble) in *Sotalia guianensis* Guiana dolphin kidney and liver samples from Southeastern Brazil (Hauser-Davis et al., 2020). Metallothionein detoxification was noted only for Pb, while Cd and Hg were poorly associated to MT, and mostly present in the insoluble fraction, indicating low bioavailability. Another study reported the presence of Cd in all three subcellular fractions in two other dolphin species *Stenella coeruleoalba* and *Tursiops etalloth* sampled off the coast of Italy (Decataldo et al., 2004), while another evaluation detected Cd in both the insoluble fraction and the heat-denaturable liver, kidney and muscle tissue fractions of *Steno bredanensis* dolphins sampled off the coast of Rio de Janeiro, noting MT detoxification mostly for Cd, and Pb, while Hg displayed lower cytosolic MT-associations (Monteiro et al., 2019). The differential subcellular Cd, Hg and Pb metal partitioning reported in these studies are clearly due to differential dolphin feeding habits and habitats (coastal or oceanic), further indicating that these factors should be considered in these kinds of assessments.

This list, although not all-encompassing, indicates several aspects that should be considered in subcellular Cd, Hg and Pb, assessments, such as metal speciation, species, differential organ roles and abiotic variables. Furthermore, it is very clear that Cd assessments are more numerous than Hg and Pb evaluations, indicating a knowledge gap for the latter two metals, which should be given further attention.

In addition to evaluating subcellular metal partitioning, it is also important to evaluate subcellular molar ratios. Several ecotoxicological assessments have investigated associations between essential and non-essential/toxic elements. Several ecotoxicological assessments have investigated associations between essential and non-essential/toxic elements, as protective effects by the former against the toxicity of the latter have been hypothesized and, in some cases, proven. The most well-known example in this regard is the Se:Hg association, where protective Se effects are noted against Hg toxicity in a wide variety of aquatic organisms when Se is present at a molar ratio excess over 1, although other associations, such as between the toxic element Cd and the essential elements Cu, Se and Cr and between the toxic Hg and essential Zn and Cu have also been reported for different taxonomic groups (Land et al., 2018; Ralston et al., 2007; Volpe et al., 2011). Investigations in this regard, however, are still not routine, and even less regarding subcellular metal partitioning. In this regard, we could only find one assessment concerning the protective effect of Se against Hg toxicity at the subcellular level (Khadra et al., 2019), where Se:Hg molar ratios in subcellular hepatic and gonadal fractions in yellow perch *Perca flavescens* were systematically above 1 in all tissues and subcellular fractions, above the suggested protective threshold. Studies in this regard are, however, still lacking in general.

TROPHIC TRANSFER IMPLICATIONS OF SUBCELLULAR METAL PARTITIONING

The determination of subcellular metal partitioning has also been employed to predict metal assimilation efficiencies between trophic levels from prey to predator (Wallace et al., 2003; Wang and Rainbow, 2006), as metal accumulated in prey species can be taken up by predators and accumulated according to differential, often species-specific, accumulation patterns (Cheung and Wang, 2005; Geffard et al., 2004; Rainbow, 2002; Rainbow et al., 2006; Redeker et al., 2007).

However, several aspects must be considered in this type of assessment. For example, the chemical form of the accumulated detoxified metal in prey has been shown as one of the major factors controlling metal assimilation by predators (Wang and Fisher 1999). In addition, only the soluble (cytosol) and organelle-bound metal fractions are accumulated by predators (Rainbow and Smith, 2010; Wallace and Luoma, 2003), comprising the metal subfraction of a prey which could best explain the metal assimilation efficiency in a predator, termed the trophically available metal (TAM) fraction (Wallace and Luoma, 2003), while the transfer of other fractions, such as metal-rich granules and cellular debris) are mostly dependent on the digestive capacity of a given species (Geffard et al., 2010; Seebaugh et al., 2005). Thus, the TAM may serve as a predictor of dietary metal bioavailability concerning trophic transfer between prey and predators (Dumas and Hare, 2008; Seebaugh and Wallace, 2004; Seebaugh et al., 2005; Wallace and Luoma, 2003).

The transfer of TAM, however, must not be confused with the biomagnification process, in which metals (and other environmental contaminants) may build up across food chains as a result of bioaccumulation processes, increasing concentrations with increasing trophic levels (Mackay and Fraser, 2000). The TAM fraction is simply bioavailable and transferred from prey to predator, which does not mean increased concentrations.

Subcellular metal partitioning studies, thus, clearly allow for a more complete understanding of potential trophic transfer which whole tissue burdens alone cannot offer. Figure 1.2 exhibits a representative scheme of TAM transfer to increasing tropic levels, where TAM is calculated as the sum of metals in all differentially obtained subcellular fractions (Rainbow et al., 2006).

Some interesting appraisals, conducted both in laboratory and field settings, are described below, grouped by metal.

Cadmium

Concerning Cd, one laboratory study investigated the trophic transfer of radio-labelled Cd from the polychaete worm *Nereis diversicolor* to the decapod crustacean *Palaemonetes varians* (Rainbow et al., 2006). The authors employed two worm populations containing different proportions of accumulated radiolabelled metals in different subcellular fractions obtained via sediment or a

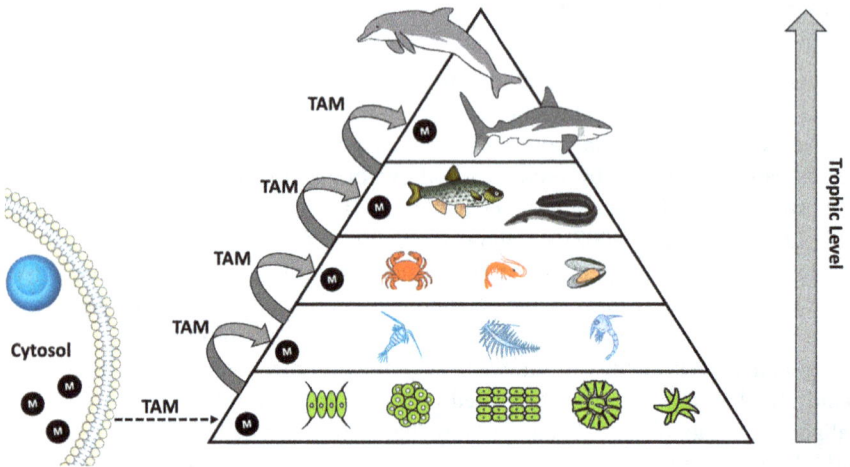

Figure 1.2 Representative scheme indicating Trophically Available Metal (TAM) transfer to increasing trophic levels from the cellular cytosol.

solution as prey and observed that metal assimilation efficiencies by *P. varians* were significantly different for the different prey categories. This confirms that accumulated metal prey origin and uptake route (sediment or solution) results in intraspecific differences concerning subsequent metal assimilation between different trophic levels.

Another laboratory assessment exposed oligochaete worms (*Limnodrilus hoffmeisteri*) to Cd at different concentrations for 1 week or 6 weeks and assessed oligochaete subcellular Cd distribution and Cd absorption in the predator grass shrimp (*Palaemonetes pugio*) (Wallace and Lopez, 1996). The authors reported a 1:1 relationship between the Cd in oligochaete cytosol and Cd adsorbed by shrimp, demonstrating that only metal bound to the soluble fraction of prey is available to higher trophic levels, and that factors influencing subcellular metal distribution in the prey will directly alter metal trophic transfer to predators.

Another study exposed rock oysters (*Saccostrea cucullata*) to cadmium for two weeks to modify subcellular Cd partitioning for two to four weeks (Cheung et al., 2006). Following predation by *Thais clavigera* gastropods, the authors reported that the predators effectively accumulated Cd from their prey, but that, interestingly, no correlation was found between Cd body concentrations in *T. clavigera* and internal metal Cd partitioning in the prey, while a significant positive correlation was found between the Cd in the TAM fraction of oysters and the Cd in the metal-sensitive fraction of *T. clavigera* and between the Cd in the TAM fraction of oysters and the metallothionein induction in *Thais clavigera*.

Concerning field studies, one assessment reported differential subcellular partitioning of Cd in two different etallothion species sampled in the field at Ria de Aveiro, a shallow estuarine ecosystem located on the northwest Atlantic coast of Portugal, the bivalve *Cerastoderma edule* and the polychaete *Diopatra neapolitana*, with higher amounts detected in the soluble fraction *of D. neapolitana*, indicating differential trophic availability potential of the bioavailable Cd fraction to predators (Freitas et al., 2012).

Lead

Concerning Pb, one laboratory assessment studied the potential trophic transfer of this element between the clam *Dosinia etallo*, which can accumulate at high concentrations of this element in the form of metal rich granules in kidneys, making Pb supposedly trophically unavailable, and a decapod consumer, the common prawn *Palaemon serratus*, compared to a mussel diet with a different subcellular Pb distribution (Sánchez-Marín and Beiras, 2017). The hypothesis that trophic transfer would be low for Pb was confirmed, as Pb was almost completely unavailable for trophic transfer and the prawns preying on highly Pb-contaminated clams displayed the same Pb accumulation as prawns fed a control diet containing much lower Pb concentrations.

Another laboratory setting study assessed Pb in the *Venerupis decussata* clam following exposure to ecologically relevant concentrations of this metal (Freitas et al., 2014). Among other investigated parameters, the findings indicate that most of the Pb accumulated in the insoluble fraction (>80%), despite the low total Pb concentration of the study, although etallothionein levels were significantly induced.

Mercury

In one study, the bioavailability of ingested Hg was investigated in a laboratory setting in the marine fish, *Terapon jarbua*, based on the mercury subcellular partitioning in different prey (brine shrimp, bivalves and small fish) and purified subcellular fractions of prey tissues (Dang and Wang, 2010). Subcellular Hg(II) differed substantially among prey types, bound mostly to cellular debris in bivalves, while MeHg distribution varied little among prey types, mostly located in the heat-stable fraction. Furthermore, the Hg(II) associated with the insoluble fraction (e.g. cellular debris) was less bioavailable than that in soluble fraction. The authors thus, indicate that subcellular prey distribution is paramount in Hg trophic transfer studies.

Again, as with subcellular partitioning assessments in general, not only focusing on trophic transfer, it becomes clear that Cd is the most assessed metal, and that Hg and Pb evaluations require further investigation.

CONCLUSIONS

The specificity of metal behavior of subcellular metal partitioning following exposure results in differential toxicity, especially for toxic elements such as Cd, Pb and Hg. These effects are more aptly assessed by evaluating metals not as total body/

organ burdens, but instead through differential centrifugation and heat treatments, in order to evaluate different subcellular fractions. This considers the fact that living organisms have developed different mechanisms to cope with negative metal effects, including extracellular sequestration, intracellular detoxification by the metallothionein route and vesicle and lysosome binding. However, when surpassing certain thresholds, which are often species-specific and may depend on different biotic and abiotic factors, these coping mechanisms may become overwhelmed. This, in turn, leads to metal binding to sensitive target molecules, such as organelles and DNA, resulting in deleterious cellular and molecular effects which may lead to higher level effects, such as populational and systemic alterations.

Subcellular molar ratios, in particular, are an interesting way to evaluate potentially protective effects against toxic elements, such as Cd, Pb and Hg, especially as the protective effect of Se against the toxic effects caused by Hg exposure are well known. This is usually postulated due to the amount of Se available for selenoprotein synthesis (Castriotta et al., 2020) and consequent alterations in biokinetic Hg processes through selenomethionine, selenocystine and selenite action (Córdoba-Tovar et al., 2022), although the mechanisms that lead to this protective effect are still not adequately understood to date and are still the subject of significant academic controversy. Furthermore, protective effects of other essential elements against Cd and Pb toxicity, as well as Hg, have also been reported in the literature. This type of assessment, however, is still lacking, requiring further investigations, although a clear potential as a tool in evaluating the toxic effects of these metals and perhaps minimizing them is noted.

Subcellular metal partitioning is also paramount in assessing trophic metal transfer, as only soluble and organelle-bound metal fractions can be accumulated by predators following prey ingestion. Studies in this regard have indicated a mostly conservative pattern among different trophic level laboratory assays, indicating that this is probably the norm regarding metal trophic transfer, becoming an invaluable tool in these kinds of evaluations.

We hope this chapter will stimulate other researchers to delve into this extremely interesting research field.

ACKNOWLEDGEMENTS

This research was funded by the Carlos Chagas Filho Foundation for Research Support of the State of Rio de Janeiro (FAPERJ) through a Jovem Cientista do Nosso Estado 2021–2024 grant (process number E-26/201.270/202) and an ARC grant (process number E-26/21.460/2019) and the Brazilian National Council of Scientific and Technological Development (CNPq), through a productivity grant. The implementation of the Projeto Pesquisa Marinha e Pesqueira is a compensatory measure established by the Conduct Adjustment Agreement under the responsibility of the PRIO company, conducted by the Federal Public Ministry—MPF/RJ.

REFERENCES

Adams, W.J., Blust, R., Borgmann, U., Brix, K.V., DeForest, D.K., Green, A.S., et al. 2011. Utility of tissue residues for predicting effects of metals on aquatic organisms. Integr. Environ. Assess. Manag. 7(1): 75–98.

Ahuti, S. 2015. Industrial growth and environmental degradation. Int. Educ. Res. J. 1(5): 5–7.

Bansal, S.L. and Asthana, S. 2018. Biologically essential and non-essential elements causing toxicity in environment. J. Environ. Anal. Toxicol. 8: 2.

Barron, M.G., Anderson, M.J., Lipton, J. and Dixon, D.G. 1997. Evaluation of critical body residue QSARs for predicting organic chemical toxicity to aquatic organisms. SAR QSAR Environ. Res. 6(1–2): 47–62.

Barst, B.D., Rosabal, M., Drevnick, P.E., Campbell, P.G.C. and Basu, N. 2018. Subcellular distributions of trace elements (Cd, Pb, As, Hg, Se) in the livers of Alaskan yelloweye rockfish (*Sebastes ruberrimus*). Environ. Pollut. 242(Pt A): 63–72.

Basha, P.S. and Rani, A.U. 2003. Cadmium-induced antioxidant defense mechanism in freshwater teleost *Oreochromis mossambicus* (Tilapia). Ecotoxicol. Environ. Saf. 56(2): 218–221.

Bashir, I., Lone, F.A., Bhat, R.A., Mir, S.A., Dar, Z.A. and Dar, S.A. 2020. Concerns and threats of contamination on aquatic ecosystems. pp. 1–26. *In*: Hakeem, K., Bhat, R. and Qadri, H. (eds). Bioremediation and Biotechnology. Bioremediation and Biotechnology. Springer, Cham.

Bragigand, V., Berthet, B., Amiard, J.C. and Rainbow, P.S. 2004. Estimates of trace metal bioavailability to humans ingesting contaminated oysters. Food Chem. Toxicol. 42(11): 1893–1902.

Broussard, L.A., Hammett-Stabler, C.A., Winecker, R.E. and Ropero-Miller, J.D. 2002. The toxicology of mercury. Lab. Med. 33(8): 614–625.

Calcerrada, P, Peluffo, G. and Radi, R. 2011. Nitric oxide-derived oxidants with a focus on peroxynitrite: molecular targets, cellular responses and therapeutic implications. Curr. Pharm. Des. 17(35): 3905–3932. Doi: 10.2174/138161211798357719.

Campana, O., Taylor, A.M., Blasco, J., Maher, W.A. and Simpson, S.L. 2015. Importance of subcellular metal partitioning and kinetics to predicting sublethal effects of copper in two deposit-feeding organisms. Environ. Sci. Technol. 49(3): 1806–1814.

Campbell, P.G. and Hare, L. 2009. Metal detoxification in freshwater animals. Roles of metallothioneins. pp. 239–277. *In*: Sigel, A., Sigel, H. and Sigel, R.K.O. (eds). Metallothioneins and Related Chelators. Royal Society of Chemistry, Cambridge, UK.

Campbell, P.G.C., Kraemer, L.D., Giguere, A., Hare, L. and Hontela, A. 2008. Subcellular Distribution of Cadmium and Nickel in Chronically Exposed Wild Fish: Inferences Regarding Metal Detoxification Strategies and Implications for Setting Water Quality Guidelines for Dissolved Metals. HERA. 14(2): 290–316.

Cheung, M.S. and Wang, W.X. 2005. Influence of subcellular metal compartmentalization in different prey on the transfer of metals to a predatory gastropod. Mar. Ecol. Prog. Ser. 286: 155–166.

Cheung, M.S., Fok, E.M., Ng, T.Y., Yen, Y.F. and Wang, W.X. 2006. Subcellular cadmium distribution, accumulation, and toxicity in a predatory gastropod, thais Clavigera, fed different prey. Environ. Toxicol. Chem. 25(1): 174–181.

Castriotta, L., Rosolen, V., Biggeri, A., Ronfani, L., Catelan, D., Mariuz M., et al. 2020. The role of mercury, selenium and the Se-Hg antagonism on cognitive neurodevelopment: A 40-month follow-up of the Italian mother-child PHIME cohort. Int. J. Hyg. Environ. Health. 230: 113604.

Córdoba-Tovar, L., Marrugo-Negrete, J., Barón, P.R. and Díez, S. 2022. Drivers of biomagnification of Hg, As and Se in aquatic food webs: A review. Environ. Res. 204(Pt C): 112226.

Croiziera, G.L., Lacroix, C., Artigauda, S., Le Floch, S., Munaron, J.-M., Raffray, J., et al. 2019. Metal subcellular partitioning determines excretion pathways and sensitivity to cadmium toxicity in two marine fish species. Chemosphere. 217: 754–762.

Dang, F. and Wang, W.X. 2010. Subcellular controls of mercury trophic transfer to a marine fish. Aquat. Toxicol. 99(4): 500–506.

Das, K. and Roychoudhury, A. 2014. Reactive oxygen species (ROS) and response of antioxidants as ROS-scavengers during environmental stress in plants. Front. Environ. Sci. 2: 53. Doi: 10.3389/fenvs.2014.00053.

Decataldo, A., Di Leo, A., Giandomenico, S. and Cadelliccio, N. 2004. Association of metals (mercury, cadmium and zinc) with metallothionein-like proteins in storage organs of stranded dolphins from the Mediterranean Sea (Southern Italy). J. Environ. Monitor. 6: 361–367.

Depledge, M.H. and Rainbow, P.S. 1990. Models of regulation and accumulations of trace metals in marine invertebrates: a mini-review. Comp. Biochem. Physiol. 97C: 1–7.

Droge W. 2002. Free radicals in the physiological control of cell function. Physiol. Rev. 82: 47–95. Doi: 10.1152/physrev.00018.2001.

Duffus, J.H. 2022. Heavy Metals—A Meaningless Term? Pure Appl. Chem. 74(5): 793–807.

Dumas, J. and Hare, L. 2008. The internal distribution of nickel and thallium in two freshwater invertebrates and its relevance to trophic transfer. Environ. Sci. Technol. 42(14): 5144-5149.

Elia, A.C., Galarini, R., Taticchi, M.I., Dorr, A.J.M. and Mantilacci, L. 2003. Antioxidant responses and bioaccumulation in Ictalurus melas under mercury exposure. Ecotoxicol. Environ. Saf. 55: 162–167.

EPA, 2008. National Ambient Air Quality Standards for Lead. Available at: https://www. govinfo.gov/content/pkg/FR-2008-11-12/pdf/E8-25654.pdf

EPA, 2014. Arsenic. Available at: https://www.epa.gov/sites/default/files/2014-03/documents /arsenic_toxfaqs_3v.pdf

EPA, 2016. Cadmium compounds. Available at: https://www.epa.gov/sites/default/files/ 2016-09/documents/cadmium-compounds.pdf

EPA, 2021. Health Effects of Exposures to Mercury. Available at: https://www.epa.gov/ mercury/health-effects-exposures-mercury

Ercal, N, Gurer-Orhan, H. and Aykin-Burns, N. 2001. Toxic metals and oxidative stress part I: Mechanisms involved in metal induced oxidative damage. Curr. Top. Med. Chem. 1: 529–539.

Espino, G.L. 2000. Organismo indicadores de la calidade del agua y de la contaminación (bioindicadores). Plaza y Valdes Editores, Mexico.

Fang, T., Lu, W., Cui, K., Li, J., Yang, K., Zhao, X., et al. 2019. Distribution bioaccumulation and trophic transfer of trace metals in the food web of Chaohu Lake, Anhui, China. Chemosphere. 218: 1122–1130.

Fitzgerald, A.M., Zarnoch, C.B. and Wallace, W.G. 2019. Examining the relationship between metal exposure (Cd and Hg), subcellular accumulation, and physiology of juvenile Crassostrea virginica. Environ. Sci. Pollut. Res. Int. 26: 25958–25968. Doi: 10.1007/s11356-019-05860-1.

Freitas, R., Pires, A., Quintino, V., Rodrigues, A.M. and Figueira, E. 2012. Subcellular partitioning of elements and availability for trophic transfer: comparison between the bivalve *Cerastoderma edule* and the Polychaete *Diopatra neapolitana*. Estuarine Coastal Shelf Sci. 99: 21–30.

Freitas, R., Martins, R., Antunes, S., Velez, C., Moreira, A., Cardoso, P., et al. 2014. Venerupis decussata under environmentally relevant lead concentrations: Bioconcentration, tolerance, and biochemical alterations. Environ. Toxicol. Chem. 33(12): 2786-2794.

Geffard, A., Jeantet, A.Y., Amiard, J.C., Le Pennec, M., Ballan-Dufrançais, C. and Amiard-Triquet, C. 2004. Comparative study of metal handling strategies in bivalves *Mytilus edulis* and *Crassostrea gigas*: a multidisciplinary approach. J. Mar. Biol. Assoc. U.K. 84: 641–650.

Geffard, A., Sartelet, H., Garric, J., Biagianti-Risbourg, S., Delahaut, L. and Geffard, O. 2010. Subcellular compartmentalization of cadmium, nickel, and lead in Gammarus fossarum: Comparison of methods. Chemosphere. 78(7): 822–829. Doi: 10.1016/j.chemosphere.2009.11.051. Epub 2009 Dec. 31.

George, S.G. 1984. Intracellular control of Cd concentrations in marine mussels. Mar. Environ Res. 14: 465–468.

Gill, S., Tewari, H. and Pande, J. 1991. In vivo and in vitro effects of cadmium on selected enzymes in different organs of the fish *Barbus conchonius* ham. (Rosy Barb). Comp. Biochem. Physiol. C Comp. Pharmacol. 100(3): 501–505.

Gimbert, F., Geffard, A., Guédron, S., Dominik, J. and Ferrari, B.J. 2016. Mercury tissue residue approach in *Chironomus riparius*: Involvement of toxicokinetics and comparison of subcellular fractionation methods. Aquat. Toxicol. 171: 1-8.

Hauser-Davis, R.A., Figueiredo, L., Lemos, L., de Moura, J.F., Rocha, R.C.C., Saint'Pierre, T., et al. 2020. Subcellular cadmium, lead and mercury compartmentalization in guiana dolphins (*Sotalia guianensis*) from Southeastern Brazil. Front. Mar. Sci. 7: 584195. doi: 10.3389/fmars.2020.584195.

Hauser-Davis, R.A., Rocha, R.C.C., Saint'Pierre, T.D. and Adams D.H. 2021. Metal concentrations and metallothionein metal detoxification in blue sharks, *Prionace glauca* L. from the Western North Atlantic Ocean. J. Trace. Elem. Med. Biol. 68: 126813. Doi: 10.1016/j.jtemb.2021.126813.

Hawkins, C.A. and Sokolova, M. 2017. Effects of elevated CO_2 levels on subcellular distribution of trace metals (Cd and Cu) in marine bivalves. Aquat. Toxicol. 92: 251–264.

Jankong, P., Chalhoub, C., Kienzi, N., Goessler, W., Francesconi, K.A. and Visoottiviseth, P. 2007. Arsenic accumulation and speciation in freshwater fish living in arsenic-contaminated waters, Environ. Chem. 4: 11–17.

Janz, D.M., DeForest, D.K., Brooks, M.L., Chapman, P.M., Gilron, G., Hoff, D., et al. 2010. Selenium toxicity to aquatic organisms. pp. 141–231. *In*: Chapman, P.M., Adams, W.J., Brooks, M.L., Delos, C.G., Luoma, S.N., Maher, et al. (eds.), Ecological Assessment of Selenium in the Aquatic Environment. Society of Environmental Toxicology and Chemistry (SETAC) and CRC Press, Pensacola, Florida.

Janz, D.M. 2011. Selenium. pp. 327-374. *In*: Wood, C.M. Farrell, A.P. and Brauner, C.J. (eds). Fish Physiology, Volume 31, Part A. Academic Press.

Khadra, M., Planas, D., Brodeur, P. and Amyot, M. 2019. Mercury and selenium distribution in key tissues and early life stages of Yellow Perch (*Perca flavescens*). Environ. Pollut. 254(Pt A): 112963.

Kataba, A., Botha, T.L., Nakayama, S.M.M., Yohannes, Y.B., Ikenaka, Y, Wepener, V., et al. 2022. Environmentally relevant lead (Pb) water concentration induce toxicity in zebrafish (*Danio rerio*) larvae. Comp. Biochem. Physiol. C Toxicol. Pharmacol. 252: 109215.

Kraemer, L.D., Campbell, P.G.C. and Hare, L. 2006. Seasonal variations in hepatic Cd and Cu concentrations and in the sub-cellular distribution of these metals in juvenile yellow perch (*Perca flavescens*). Environ. Pollut. 142: 313–325.

Krasnići, N., Dragun, Z., Erk, M. and Raspor, B. 2014. Distribution of Co, Cu, Fe, Mn, Se, Zn, and Cd among cytosolic proteins of different molecular masses in gills of European chub (*Squalius cephalus* L.). Environ. Sci. Pollut. Res. Int. 21(23): 13512–13521.

Kwok, K.W.H., Leung, K.M.Y., Lui, G.S., Chu, V.K.H., Lam, P.K.S., Morritt, D., et al. 2007. Comparison of tropical and temperate freshwater animal species' acute sensitivities to chemicals: implications for deriving safe extrapolation factors. Integr. Environ. Assess. Manag. 3: 49–67.

Lall, S.P. 2022. The minerals. pp. 469–554. *In*: Hardy, R.W. and Kaushik, S.J. (eds). Fish Nutrition (Fourth Edition). Academic Press, London, UK.

Land, S.N., Rocha, R.C.C., Bordon, I.C., Saint'Pierre, T.D., Ziolli, R.L. and Hauser-Davis, R.A. 2018. Biliary and hepatic metallothionein, metals and trace elements in environmentally exposed neotropical cichlids *Geophagus brasiliensis*. J. Trace. Elem. Med. Biol. 50: 347–355.

Lee, B.G., Wallace, W.G. and Luoma, S.N. 1998. Uptake and loss kinetics of Cd, Cr and Zn in the bivalves Potamocorbula amurensis and Macoma balthica: effects of size and salinity. Mar. Ecol. Prog. Ser. 175: 177–189.

Lehtonen, K.K. and Schiedek, D. 2006. Monitoring biological effects of pollution in the Baltic Sea: neglected—but still wanted? Mar. Pollut. Bull. 53: 377–386.

Lemly, A.D., 1993. Guidelines for evaluating selenium data from aquatic monitoring and assessment studies. Environ. Monit. Assess. 28: 83–100.

Mackay, D. and Fraser, A. 2000. Bioaccumulation of persistent organic chemicals: mechanisms and models. Environ. Pollut. 110(3): 375–391.

Mason, A.Z. and Jenkins, K.D. 1995. Metal detoxification in aquatic organisms. pp. 479-608. *In*: Tessier, A. and Turner, D.R. (eds). Metal Speciation and Bioavailability in Aquatic Systems. John Wiley & Sons Ltd, New York, NY.

Mijošek, T., Filipović Marijić, V., Dragun, Z., Krasnići, N., Ivanković, D., Redžović, Z., et al. 2021. First insight in trace element distribution in the intestinal cytosol of two freshwater fish species challenged with moderate environmental contamination. Sci. Total Environ. 798: 149274.

Monteiro, F., Lemos, L.S., de Moura, J.F., Rocha, R.C.C., Moreira, I., Di Beneditto, A.P., et al. 2019. Subcellular metal distributions and metallothionein associations in rough-toothed dolphins (*Steno bredanensis*) from Southeastern Brazil. Mar. Pollut. Bull. 146: 263–273.

National Research Council (US) Committee on the toxicological effects of methylmercury. toxicological effects of methylmercury. Washington (DC): National Academies Press (US); 2000. Available at: https://www.ncbi.nlm.nih.gov/books/NBK225769/

Nicole, A., Santiard-Baron, D. and Ceballos-Picot, I. 1998. Direct evidence for glutathione as mediator of apoptosis in neuronal cells. Biomed. Pharmacother. 52: 349–355.

Nieboer, E. and Richardson, D.H.S. 1980. The replacement of the nondescript term 'heavy metals' by a biologically and chemically significant classification of metal ions. Environ. Pollut. Ser. B. 1: 3–26.

Pantea, E.-D., Oros, A., Roşioru, D.M. and Roşoiu, N. 2020. Condition Index of Mussel *Mytilus galloprovincialis* (Lamarck, 1819) as a Physiological Indicator of Heavy Metals contamination. Academy of Romanian Scientists. Annals—Series on Biological Sciences. (9): 20–36.

Perceval, O., Couillard, Y., Pinel-Alloul, B. and Campbell, P.G. 2006. Linking changes in subcellular cadmium distribution to growth and mortality rates in transplanted freshwater bivalves (*Pyganodon grandis*). Aquat. Toxicol. 79(1): 87–98.

Perošević, A., Pezo, L., Joksimović, D., Đurović, D., Milašević, I., Radomirović, M. and Stanković, S. 2018. The impacts of seawater physicochemical parameters and sediment metal contents on trace metal concentrations in mussels—a chemometric approach. Environ. Sci. Pollut. Res. 25: 28248–28263.

Péry, A.R.R., Geffard, A., Conrad, A., Mons, R. and Garric J. 2008. Assessing the risk of metal mixtures in contaminated sediments on *Chironomus riparius* based on cytosolic accumulation. Ecotox. Environ. Safe. 71: 869–873.

Pizzino, G., Irrera, N., Cucinotta, M., Pallio, G., Mannino, F., Arcoraci, V., et al. 2017. Oxidative Stress: Harms and Benefits for Human Health. Oxid. Med. Cell. Longev. 2017: 8416763.

Pourret, O. and Hursthouse, A. 2019. It's time to replace the term "Heavy Metals" with "Potentially Toxic Elements" when reporting environmental research. Int. J. Environ. Res. Public Health. 16(22): 4446. https://doi.org/10.3390/ijerph16224446.

Pourret, O., Bollinger, J.-C. and Hursthouse, A. 2021. Heavy metal: A misused term? Acta Geochim. 40: 466–471.

Rainbow, P.S. 2002. Trace metal concentrations in aquatic invertebrates: why and so what? Environ. Pollut. 120: 497–507.

Rainbow, P.S., Poirier, L., Smith, B.D., Brix, K.V. and Luoma, S.N. 2006. Trophic transfer of trace metals: subcellular compartmentalization in a polychaete and assimilation by a decapod crustacean. Mar. Ecol. Prog. Ser. 308: 91–100.

Rainbow, P.S. 2007. Trace metal bioaccumulation: models, metabolic availability and toxicity. Environ. Int. 33: 576–582.

Rainbow, P.S. and Smith, B.D. 2010. Trophic transfer of trace metals: subcellular compartmentalization in bivalve prey and comparative assimilation efficiencies of two invertebrate predators. J. Exp. Mar. Biol. Ecol. 390: 143–148.

Ralston, N.V.C., Lloyd Blackwell, J. and Raymond, L.J. 2007. Importance of molar ratios in selenium-dependent protection against methylmercury toxicity, Biol. Trace Elem. Res. 119: 255–268.

Ravindran, A. and Radhakrishnan, M.V. 2020. Bioaccumulation of vanadium in selected organs of the freshwater fish *Heteropneustes fossilis* (Bloch). Nat. Env. & Poll. Tech. 9(3): 1149–1153.

Redeker, E.S., van Campenhout, K., Bervoets, L., Reijnders, H. and Blust, R. 2007. Subcellular distribution of Cd in the aquatic oligochaete *Tubifex tubifex*, implications for trophic availability and toxicity. Environ. Poll. 148: 166–175.

Rosabal, M., Hare, L., Campbell, P.G. 2012. Subcellular metal partitioning in larvae of the insect *Chaoborus* collected along an environmental metal exposure gradient (Cd, Cu, Ni and Zn). Aquat. Toxicol. 120-121: 67–78.

Rosabal, M., Pierron, F., Couture, P., Baudrimont, M., Hare, L. and Campbell, P.G. 2015. Subcellular partitioning of non-essential trace metals (Ag, As, Cd, Ni, Pb, and Tl) in livers of American (Anguilla rostrata) and European (*Anguilla anguilla*) yellow eels. Aquat. Toxicol. 160: 128–41. doi: 10.1016/j.aquatox.2015.01.011.

Sánchez-Marín, P. and Beiras, R. 2017. Subcellular distribution and trophic transfer of Pb from bivalves to the common prawn *Palaemon serratus*. Ecotoxicol. Environ. Saf. 138: 253–259.

Sandheinrich, M.B. and Wiener, J.G. 2011. Methylmercury freshwater fish: Recent advances in assessing toxicity and environmentally relevant exposures. pp. 169–191. *In*: Beyer, W.N. and Meador, J.P. (eds). Environmental Contaminants in Biota. Boca Raton: CRC Press. DOI: http://dx.doi.org/10.1201/B10598-6.

Sandrini, J.Z., Regoli, F., Fattorini, D., Notti, A., Inácio, A.F., Linde-Arias, A.R., et al. 2006. Short-term responses to cadmium exposure in the estuarine polychaete *Laeonereis acuta* (polychaeta, Nereididae): subcellular distribution and oxidative stress generation. Environ. Toxicol. Chem. 25(5): 1337–1344. doi: 10.1897/05-275r.1.

Santovito, G., Trentin, E., Gobbi, I., Bisaccia, P., Tallandini, L. and Irato, P. 2021. Non-enzymatic antioxidant responses of *Mytilus galloprovincialis*: Insights into the physiological role against metal-induced oxidative stress. Comp. Biochem. Physiol. C Toxicol. Pharmacol. 240: 108909.

Saper, R.B. and Rash, R. 2009. Zinc: an essential micronutrient. Am. Fam. Physician. 79(9): 768–772.

Seebaugh, D.R. and Wallace, W.G. 2004. Importance of metal-binding proteins in the partitioning of Cd and Zn as trophically available metal (TAM) in the brine shrimp *Artemia franciscana*. Mar. Ecol. Prog. Ser. 272: 215–230.

Seebaugh, D.R., Goto, D. and Wallace, W.G. 2005. Bioenhancement of cadmium transfer along a multi-level chain. Mar. Environ. Res. 59: 473–491.

Silberlgeld, E.K. 2004. Testimony of Professor Ellen K Silbergeld. Bloomberg School of Public Health, Johns Hopkins University, Baltimore, MD. Lead Contamination in the District of Columbia Water Supply. Oversight Committee on Government Reform, House of Representatives, U.S. Congress.

Sokolova, I.M., Ringwood, A.H. and Johnson, C. 2005. Tissue-specific accumulation of cadmium in subcellular compartments of eastern oysters *Crassostrea virginica* Gmelin (Bivalvia: Ostreidae). Aquat. Toxicol. 74: 218–228.

Stohs, S.J. and Bagchi, D. 1995. Oxidative mechanisms in the toxicity of metals ions. Free Radical Biol. Med. 2: 321–336.

Urien, N., Cooper, S., Caron, A., Sonnenberg, H., Rozon-Ramilo, L., Campbell, P.G.C., et al. 2018. Subcellular partitioning of metals and metalloids (As, Cd, Cu, Se and Zn) in liver and gonads of wild white suckers (*Catostomus commersonii*) collected downstream from a mining operation. Aquat. Toxicol. 202: 105–116. doi: 10.1016/j. aquatox.2018.07.001.

Valko, M., Leibfritz, D., Moncola, J., Cronin, M.D., Mazur, M. and Telser, J. 2007. Free radicals and antioxidants in normal physiological functions and human disease. Int. J. Biochem. Cell. Biol. 39: 44–84. doi: 10.1016/j.biocel.2006.07.001.

Valko, M., Morris, H. and Cronin, M.T.D. 2005. Metals, toxicity and oxidative stress. Curr. Med. Chem. 12: 1161–1208.

Vega-López, A., Galar-Martínez, M., Jiménez-Orozco, F.A., García-Latorre, E. and Domínguez-López, M.L. 2007. Gender related differences in the oxidative stress response to PCB exposure in an endangered goodeid fish (*Girardinichthys viviparus*). Comp. Biochem. Physiol. Part A. Mol. Integr. Physiol. 146(4): 672–678.

Viarengo, A. 1989. Heavy metals in marine invertebrates: mechanisms of regulation and toxicity at the cellular level. Rev. Aquat. Sci. 1: 295–317.

Vijver, M.G., van Gestel, C.A., Lanno, R.P., van Straalen, N.M. and Peijnenburg, W.J. 2004. Internal metal sequestration and its ecotoxicological relevance: a review. Environ. Sci. Technol. 38(18): 4705–4712. doi: 10.1021/es040354g.

Voets, J., Redeker, E.S., Blust, R. and Bervoets, L. 2009. Differences in metal sequestration between zebra mussels from clean and polluted field locations. Aquat. Toxicol. 93: 53–60.

Volpe, A.R., Cesare, P., Aimola, P., Boscolo, M., Valle, G. and Carmignani, M. 2011. Zinc opposes genotoxicity of cadmium and vanadium but not of lead. J. Biol. Regul. Homeost. Agents. 25(4): 589–601.

Wallace, W.G. and Lopez, G.R. 1996. Relationship between subcellular cadmium distribution in prey and cadmium trophic transfer to a predator. Estuaries. 19: 923–930.

Wallace, W.G. and Luoma, S.N. 2003. Subcellular compartmentalization of Cd and Zn in two bivalves. II. Significance of trophically available metal (TAM). Mar. Ecol. Prog. Ser. 257: 125-137.

Wallace, W.G., Lee, B.G. and Luoma, S.N. 2003. Subcellular compartmentalization of Cd and Zn in two bivalves. I. Significance of metal-sensitive fractions (MSF) and biologically detoxified metal (BDM). Mar. Ecol. Prog. Ser. 249: 183–197.

Wang, W.X. and Fisher, N.S. 1999. Assimilation efficiencies of chemical contaminants in aquatic invertebrates: a synthesis. Environ. Toxicol. Chem. 18: 2034–2045.

Wang, D., Couillard, Y., Campbell, P.G.C. and Jolicoeur, P. 1999. Changes in subcellular metal partitioning in the gills of freshwater bivalves (*Pyganodon grandis*) living along an environmental cadmium gradient. Can. J. Fish. Aquat. Sci. 56: 774–784.

Wang, W.X. 2002. Interactions of trace metals and different marine food chains. Mar. Ecol. Prog. Ser. 243: 295–309.

Wang, W.X. and Rainbow, P.S. 2006. Subcellular partitioning and the prediction of cadmium toxicity to aquatic organisms. Environ. Chem. 3: 395–99.

Wang, Z.S., Yan, C.Z., Pan, Q.K. and Yan, Y.J. 2011. Concentrations of some heavy metals in water, suspended solids, and biota species from Maluan Bay, China and their environmental significance. Environ. Monit. Assess. 175: 239–249.

Wang, Z., Feng, C., Ye, C., Wang, Y., Yan, C., Li. R., et al. 2016. Subcellular partitioning profiles and metallothionein levels in indigenous clams Moerella iridescens from a metal-impacted coastal bay. Aquat. Toxicol. 176: 10-23.

Warren, M.J., Cooper, J.B., Wood, S.P. and Shoolingin-Jordan, P.M. 1998. Lead poisoning, haem synthesis and 5-aminolaevulinic acid dehydratase. Trends Biochem. Sci. 23: 217–221.

Wosnick, N., Niella, Y., Hammerschlag, N., Chaves, A.P., Hauser-Davis, R.A., da Rocha, RCC, et al. 2021. Negative metal bioaccumulation impacts on systemic shark health and homeostatic balance. Mar. Pollut. Bull. 168: 112398.

Wu, Y. and Wang, W.X. 2011. Accumulation, subcellular distribution and toxicity of inorganic mercury and methylmercury in marine phytoplankton. Environ. Pollut. 159(10): 3097–3105.

Young, I. and Woodside, J. 2001. Antioxidants in health and disease. J. Clin. Pathol. 54: 176–186.

Zhao, Y., Wu, J., Shang, D., Ning, J., Zhai, Y., Sheng, X., et al. 2015. Subcellular distribution and chemical forms of cadmium in the edible seaweed, *Porphyra yezoensis*. Food Chem. 168: 48–54.

Zirong, X. and Shijun, B. 2007. Effects of waterborne Cd exposure on glutathione metabolism in Nile tilapia (*Oreochromis niloticus*) liver. Ecotoxicol. Environ. Saf. 67(1): 89–94.

Chapter 2

Assessment of Lead, Cadmium and Mercury in Coastal Aquatic Environments in South Florida and Abroad: Identification of Anthropogenic and Natural Sources in Surface Waters

Natalia Quinete*[1,2], Yang Ding[2], Rob Menzies[3] and Douglas Seba[3]

[1]Department of Chemistry and Biochemistry, Florida International University, Biscayne Bay Campus and University Park, North Miami, FL 33181 & Miami, FL 33199

[2]Institute of Environment, Florida International University, University Park, Miami, FL 33199

[3]Academy of Marine Sciences, Inc. Fort Lauderdale, FL USA

INTRODUCTION

Water, one of the most precious and essential resources for life, has been increasingly negatively impacted over the years. Estuarine and coastal environments, in particular, have been constantly suffering with increased urban development, agricultural activities, and wastewater generation, which constitute important sources of anthropogenic pollution. Metals and metalloids are natural constituents of the marine environment, and some are essential or beneficial for

*Corresponding authors: nsoaresq@fiu.edu

life, generally found at very low concentrations (Ansari et al., 2004). Others, however, can be highly toxic to living organisms, even at low concentrations, such as Lead (Pb), Cadmium (Cd) and Mercury (Hg). These metals are being increasingly discharged into aquatic systems by human activities, such as mining, oil and gas exploration, industrial production (pharmaceuticals, electronics, plastic stabilizers, paints and fertilizers, among others), agriculture run-off, and climate change, in turn altering metal solubility and resulting in acid rain (Lim et al., 2012; Nriagu and Pacyna, 1988). In fact, these metals are regarded as global environmental contaminants of great concern due to their toxicity, persistence and bioaccumulation potential, found in surface and drinking waters, sediments and biota in several regions worldwide (Fisher et al., 2021; Hauser-Davis et al., 2020; HELCOM, 2010; Lim et al., 2012; Schaefer et al., 2015; Zhou et al., 2020).

Cadmium (Cd) is a natural element in the earth's crust and a non-essential metal with no biological function in aquatic life, which can enter the environment through natural processes, such as forestfires, rock weathering and volcanic emissions, as well as from human sources due to its industrial use in the manufacturing of batteries, plastic stabilizers, pigments, coatings, plating, electronics and more recently, as nanoparticles for solar cells (Hutton, 1983; U.S. EPA, 2015). Therefore, several Cd sources, such as domestic and industrial discharge, urban run-off and the deterioration of galvanized pipes, can lead to dissolved and readily available Cd in waterbodies. The exposure of aquatic organisms, including fish, marine mammals and seabirds, to this element has been linked to adverse effects concerning development, behavior, reproduction, immune and endocrine systems, including, specifically, kidney damage, disturbed calcium metabolism and bone loss (WHO, 2003a).

Similarly to Cd, anthropogenic lead (Pb) sources, such as ammunition discharges, leaded fuel and paints, coal and wood combustion, mining activities, water pipe corrosion and industrial uses in metal production and manufacturing processes, offset natural sources (e.g., sulfide ore weathering, forest fires, volcanoes), leading to significant Pb inputs into the aquatic environment through atmospheric routes (precipitation, lead dust and road dust fall-out), street runoff and industrial and municipal wastewater discharges (ARMCANZ and ANZECC, 2000; Lee et al., 2019; U.S. EPA, 1984). Metal speciation and bioavailability have a significant influence on its toxicity. In seawater, Pb chlorides and carbonate are abundant and readily available to accumulate in sediments and organisms, causing a wide range of toxic behavioral, physiological, and biochemical effects in animals, including kidney and liver, cardiovascular, nervous, and immune system damage, as well as oxidative stress (Hsu and Guo, 2002; Kim and Kang, 2016; Lee et al., 2019).

Mercury (Hg) is one of most toxic metals found in the aquatic system. Natural sources of this element comprise volcanic eruptions, geologic deposits, and ocean emissions, with human-caused emissions including release from fuels or raw materials, mining, coal combustion, waste incineration, and industrial processes (Siddiqi, 2018). In the aquatic environment, Hg speciation, uptake, bioavailability and toxicity depends on several environmental parameters, such as pH, salinity, hardness and potential interactions with biotic and abiotic ligands (Erickson et al., 2008). Mercury can be transformed into the highly neurotoxic methylmercury (MeHg), which undergoes increased accumulation and biomagnification in the aquatic biota (Morcillo et al., 2017). Exposure to this metal leads to irreversible toxic effects in fish at the biochemical, histological, genetic and physiological levels (Morcillo et al., 2017). Mercury has, in fact, been one of the contaminants of significant concern in Florida for several years (Kannan et al., 1998).

The metals Pb, Cd and Hg are listed as the most hazardous inorganic contaminants in the Unites States Environmental Protection Agency (U.S.EPA) Hazardous Substance Priority List and are regularly monitored in surface waters for the protection of sensitive aquatic species. The recommended water quality criteria by the U.S.EPA (U.S.EPA, 2015, 2004, 1984), European Commission (EC) (EC, 2013), Canadian Council of Ministers of the Environment (CCME) (CCME, 2008) and Australian and New Zealand Environment and Conservation Council (ANZECC) (ANZECC, 2000) for the marine environment considering both acute and chronic scenarios are presented in Table 2.1.

Table 2.1 Summary of water quality standards for cadmium, lead and mercury in estuarine/marine environments established by several regulating agencies for both acute and chronic exposures

	Acute ($\mu g\ L^{-1}$)		Chronic ($\mu g\ L^{-1}$)			
Element	EPA Saltwater CMC[1]	EC MAC-EQS for Seawater	EPA Saltwater CCC[2]	EC AA-EQS for Seawater	ANZECC trigger values for marine water (95% protection)	CCME Marine Long-term concentration
Cd	33	1.5	7.9	0.2	5.5	0.12
Pb	210	14	8.1	1.3	4.4	No data
Hg	1.8	0.07	0.94	No data	0.4	0.016

[1]CMC: criterion maximum concentration, [2]MAC-EQS: maximum allowable concentration-environmental quality standards,[3]CCC: criterion continuous concentration, [4]AA-EQS: annual average-environmental quality standards.

In Biscayne Bay and adjacent canals and in the Everglades, both in South Florida, urban and agricultural runoffs have been mostly associated with water quality degradation (Castro et al., 2013; Miller et al., 2004) and can contribute to anthropogenic metal loads in the aquatic environment. In this regard, this study aims to assess the spatial variations of Pb, Cd and Hg alongside other relevant metals in surface waters from selected coastal marine environment of importance for biodiversity and conservation efforts in South Florida, and abroad (where opportunistic sampling was conducted by RM). The identification of potential sources in South Florida, contextualization of the detected levels within a global scenario, and toxicity considerations are also explored.

MATERIAL AND METHODS

Chemicals

Concentrated nitric acid (HNO_3) and 30% v/v hydrogen peroxide were Optima™ grade purchased from Fisher Scientific (Fairlawn, NJ, USA). Multi-elemental ICP-MS certified calibration standards (10 mg L^{-1} in 5% HNO_3) containing aluminum (Al), antimony (Sb), arsenic (As), beryllium (Be), cadmium (Cd), chromium (Cr), cobalt (Co), copper (Cu), lead (Pb) and manganese (Mn) (ICP-MS-200.8-CAL1-1) and barium (Ba) and silver (Ag) (ICP-MS-200.8-CAL2-1) were purchased from Accustandard (New Haven, CT, USA). The mercury (Hg) (ICP, 34-N-1; 1000 mg L^{-1} in 10% HNO_3) and tin (Sn) standards (ICP-MS-63N-0.01x-1, 100 mg L^{-1} in 2-5% HNO_3) were also obtained from Accustandard. A second multi-elemental certified standard solution (QC Standard 20) containing silver Ag, Al, As, Ba, beryllium (Be), Cd, Co, Cr, Cu, Mn, molybdenum (Mo), nickel (Ni), Pb, Sb, selenium (Se), thorium (Th), titanium (TI), uranium (U), vanadium (V), zinc (Zn); 10 mg L^{-1} in 5% HNO_3) was purchased from Crescent Chemical (Islandia, NY, USA). The internal standard (IS) mixture (Claritas PPT™ grade; 10 mg L^{-1} in 5% HNO_3) containing Ge, Y, In, Bi, Sc, Tb, and ^{6}Li, was purchased from Spex Certiprep (Metuchen, NJ, USA). A second source of Hg and Sn standards was obtained from Spex Certiprep and Crescent Chemical, respectively. The ICP-MS tuning solution (^{6}Li, Y, Ce, Tl) was also purchased from Accustandard. Ultrahigh purity liquid argon was purchased from Airgas (Kennesaw, GA, USA). Artificial seawater (3.5% w/v) was prepared using the commercially available Instant Ocean® salt diluted in ultrapure water.

METHODOLOGY

Sampling and Study Area

Surface water samples were collected in October 2017, July 2018 and January 2019 from Biscayne Bay and adjacent canals, in Miami (N=41); between March and April 2019 from Port Everglades, Fort Lauderdale (N=23) and from Clear Lake, West Palm Beach (N=1); in July 2016 and June 2018 near the Everglades area (N=4); and in September 2018 from Key Largo (N=1), Key West (N=12) and Dry Tortuga (N=5), South Florida, USA. In addition, a total of 16 water samples were collected between June and October 2018 near shore coastal marine locations and oceanic islands in Turkey (N=3), Dominican Republic (N=1), and Greece (N=12). A map indicating the South Florida sampling sites is shown in Figure 2.1. Locations in Greece included the islands Symi, Nisyros, Rhode's harbor, Livadia Tilos Port, Lakki, in Leros, Patmos, Lipsi, Agios Kirykos and Pythagoras (not identified in Figure 2.1). In Turkey, samples were collected along Bozburun Peninsula, in Marmaris (not identified in Figure 2.1).

Our main study area was Biscayne Bay, a semi-tropical lagoon extending throughout most of the length of the populated Miami-Dade County and classified as habitat focus by the NOAA habitat blueprint program, which address the challenge of coastal and marine habitat loss and degradation by identifying areas for habitat protection and restauration (NOAA, 2019). Biscayne Bay is a coastal marine environment that plays a critical role in the function and dynamics of the larger Florida Keys and coral reef ecosystem (Ault et al., 2001). This region was chosen due the importance of its adjacent coral reef and pelagic habitats for many sensitive species, recreational and tourism activities, and for being heavily impacted by urbanization and, thus, wastewater intrusion (Ng et al., 2021), which could contribute to anthropogenic metal input in this environment.

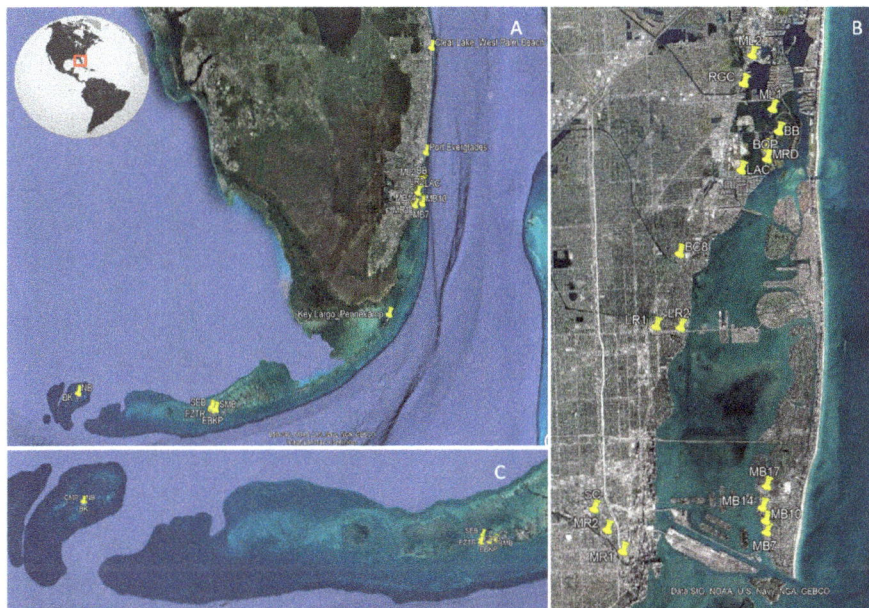

Figure 2.1 Sampling stations of surface water collection for the determination of metals in South Florida, USA, (A) Overview of the South Florida locations, USA, (B) Miami: Maule Lake Site 1 (ML1), Maule Lake Site 2 (ML2), Royal Galdes Canal (RGC), Biscayne Bay (BB), Main Road (MRD), Bridge Oleta Park (BOP), Little Arch Creek (LAC), Biscayne Canal 8 (BC8), Little River Site 1 (LR1), Little River Site 2 (LR2), Seybold Canal (SC), Miami River Site 1 (MR1), Miami River Site 2 (MR2), Miami Beach 7th Street (MB7), Miami Beach 10th Street (MB10), Miami Beach 14th Street (MB14), Miami Beach 17th Street (MB17), (C) Key West: Edward B Knight Pier (EBKP), Smathers Beach (SMB), Ft. Zachary Taylor Beach (FZTB), Ft. Zachary Taylor Ramp (FZTR), Seaport Bight (SEB); and Dry Tortugas: Bush Key (BK), North Beach (NB) and Camp (CMP).

Sample Preparation and Analysis

Surface water samples (N=76) were sampled using collection poles at >30 cm depth to avoid sampling surface microlayers using new, disposable 500 mL polyethylene terephthalate bottles (PET), and transferred to a 50 mL certified clean, disposable digestion vessel for filtration using a 0.45 μm polyvinylidene difluoride/ polytetrafluoroethylene polymer (PVDF/PTFE membrane) filter system (Environmental Express, Charleston, SC, USA). An aliquot (1 mL) of each sample was taken, amended with 100 μL of IS mixture (1000 μg L^{-1} containing gold at 10,000 μg L^{-1}), then diluted to a final volume of 10 mL with 3% nitric acid and homogenized using a vortex before analysis. The analytical method was based on the EPA 200.8 method (U.S.EPA, 1994) for the determination of trace elements by inductively coupled plasma-mass spectrometry (ICP-MS). Besides lead (Pb), cadmium (Cd) and mercury (Hg), the following metals were also determined on an ICP-triple-quadrupole MS instrument (Thermo ICAP ICP-QqQ-MS, Thermo Scientific): Beryllium (Be), Barium (Ba), Silver (Ag), Cobalt (Co), Copper (Cu), Chromium (Cr), Aluminum (Al), Arsenic (As), Nickel (Ni), Selenium (Se), Manganese (Mn), Molybdenum (Mo), Vanadium (V), and Zinc (Zn). Blank samples consisted of ultrapure water (16 MΩcm^{-1}) obtained from a Nanopure Infinity Ultrapure Water system. Method detection limits (MDLs) were evaluated by spiking eight replicates of artificial seawater (3.5% w/v) at a final volume of 1 μg L^{-1} and were statistically estimated by multiplying the standard deviation of the analysis by the one-side Student's t value at a 99% confidence interval. The MDLs ranged from 0.08 to 0.13 ng mL^{-1} for Hg, Cd and Hg and from 0.06 to 1.06 ng mL^{-1} for the other metals (Table 2.2).

The software package used for the statistical analyses was Origin version 2021(b) and Google charts were used to plot the bar graphs and concentration maps.

RESULTS AND DISCUSSION

Levels of Pb, Cd and Hg and other Metals in Coastal Marine Environments in South Florida and Selected Locations Abroad

The occurrence and concentrations of selected metals and metalloids, including Pb, Cd and Hg in surface waters from South Florida, are shown in Table 2.2, while the spatial distribution of frequently detected toxic metals in South Florida aquatic environments is displayed in Figure 2.2. Metal concentrations in surface water samples from abroad are presented in Table 2.3 and a comparison map is illustrated in Figure 2.3.

Mercury was below the MDL in all analyzed surface water samples. Contamination by Hg has been previously reported for several taxonomic groups in Florida (Kannan et al., 1998; Schaefer et al., 2015), but levels in surface water are rarely found to exceed state guidelines (Castro et al., 2013). A previous study in canals and creeks in South Florida reported levels ranging from 0.003–0.0074 μg L^{-1}, which is within the range of 0.002 to 0.015 μg L^{-1} normally reported for coastal estuarine waters (Kannan et al., 1998) and below the levels of our current MDL (0.08 μg L^{-1}).

Cadmium was found in 9% of all samples at concentrations ranging from <MDL to 0.3 μg L^{-1} at Port Everglades and in Greece, and Turkey samples. In

the Biscayne Bay and tributaries of Clear Lake (West Palm Beach), Key Largo (Pennekamp), Key West and Dry Tortuga, surface water Cd concentrations were <MDL.

Lead was detected in 12% of the samples, with concentrations ranging from <MDL to 1.35 μg L^{-1}. Other measured anthropogenic metals of known toxicity, such as Cr (detected in 77% of the samples), Zn (27%), Cu (73%) and As (100%) were more frequently found in the surface water samples at levels from <MDL to 1.66 μg L^{-1}, <MDL to 168 μg L^{-1}, <MDL to 31.5 μg L^{-1} and 0.42 to 6.08 μg L^{-1}, respectively. In addition, common earth crust and naturally occurring metals such as Al (33%), V (96%), Mo (93%) and Sb (33%) exhibited concentrations up to 37.7 μg L^{-1} (for Al), while Co, Se, and Sn were not detected or not reported in the samples (due to analytical issues). Berilium and Ni were rarely detected in Florida but were present in >50% of the samples obtained abroad, in up to 5.59 μg L^{-1} (for Ni) in Greece (Kirykos).

Different processes such as dissolution, precipitation, sorption and complexation play a role in metal speciation in the aquatic environment, affecting their behavior and bioavailability (Islam et al., 2015). In general, metals such as Cd, Hg, and Pb tend to partition to the sediment rather than the water column (Bhuyan et al., 2017) or bioaccumulate and biomagnify throughout the food chain (Bawuro et al., 2018; Moiseenko and Gashkina, 2020; Rajeshkumar and Li, 2018), which can explain their low occurrence in the surface water. As observed in Figure 2.2, arranged from North to South sites, increasing concentrations of selected metals, such as Cu and Zn, are more evident in the southernmost locations. The extremely high Zn level observed in the Everglades area (collected from the airboat trail) is likely associated with airboat activities such as boat paint and diesel burning during operations (Duan, 2012).

Molybdenum is an essential trace metal for human and animal health, occurring naturally in trace amounts in most rocks and soils, which leads to concentrations of less than 10 μg L^{-1} in seawaters, which is consistent with the levels observed herein, whereas the highest concentration was 12.5 μg L^{-1} (Smedley and Kinniburgh, 2017). Arsenic was one of the most prevalent metals in surface waters with an average concentration of about 1–2 μg L^{-1}, two orders of magnitude higher than the U.S.EPA human health criterion value (regarding fish consumption) of 0.0175 μg L^{-1} (Neff, 1997) and is, thus, of significant environmental concern. Copper is ubiquitous in the environment, occurring both naturally, constituting 60 mg kg^{-1} of the Earth's crust, and anthropogenically, and can be present at 0.25 μg L^{-1} in seawater (Lewis, 1995). High Cu levels (up to 18.3 μg L^{-1}) found in Dry Tortugas can be attributed to historical backgrounds, such as shipwrecks and ammunitions (Geisler and Schmidt, 1991). Chromium concentrations in open ocean waters range from 0.10 to 0.26 μg L^{-1} and typical sources include mineral weathering processes and riverine and atmospheric input. Nevertheless, anthropogenic sources such as electroplating, leather tanning, and textile industries have been shown to discharge relatively high amounts of Cr to surface waters (Geisler and Schmidt, 1991). The levels observed herein are within the typical range, except for the LAC site, with levels up to 1.7 μg L^{-1}, suggesting impacts from human activities. The natural occurrence of V and Al in the earth crust makes their release into the surface water take place primarily as a result of rock weathering and soil erosion. The conservative behavior of V leads to concentrations between 1.5–2 μg L^{-1} in oceans (Wu et al., 2019), consistent with the results observed herein, with average concentrations ranging from 0.41 to 2.22 μg L^{-1}.

In general, levels of selected metals (Al, Cu, Zn, and Pb) found in Greece were remarkably higher than in South Florida locations. The highest average Cd concentration (0.24 μg L^{-1}) was also observed abroad, in Turkey, but still similar to average levels detected in Miami (0.21 μg L^{-1}).

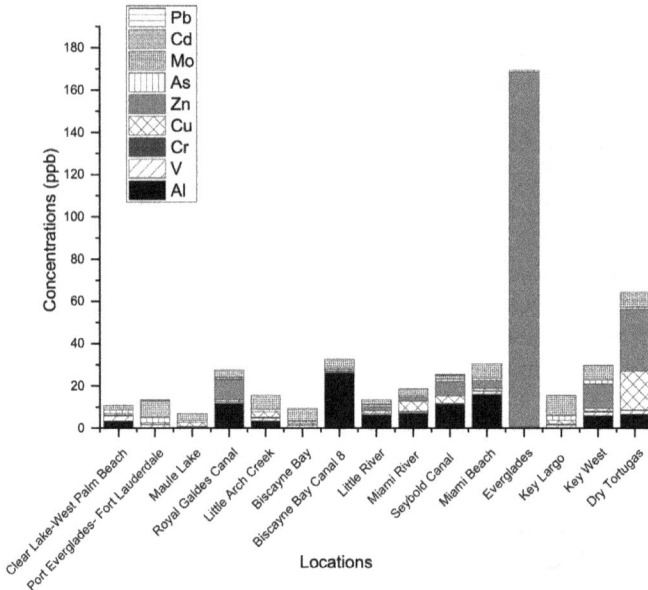

Figure 2.2   Spatial distribution of metals detected in the South Florida's surface waters.

Comparison of Marine Pb, Cd and Hg Levels

Aquatic metal contamination has become a major global concern, especially in industrialized countries, owing to the abundance, persistence, and toxicity of these contaminants. Metals are normally found bound to organic matter, clay, sulfides and Fe/Mn oxides in sediments, which may constitute an important metal reservoir (Lim et al., 2012). Nevertheless, metals are increasingly and ubiquitously detected in surface waters in rivers, lakes and other water bodies due to their continuous discharge into the environment and their non-degradable character. Cadmium, Pb and Hg have been routinely reported in many studies conducted across the continents (Alves et al., 2014; Ayandiran et al., 2018; Edokpayi et al., 2016; Gabrielyan et al., 2018; Islam et al., 2015; Krishna et al., 2009; Kumar et al., 2020; Mahfooz et al., 2019; Rautenberg et al., 2015; Sadeghi et al., 2020; Vilavert et al., 2015;

Table 2.2 Average (Min–Max) concentrations (µg L^{-1}) of metals and metalloids in surface water samples from South Florida

Elements (monitored isotopes)	MDL (µg L^{-1})	DF (%)	CL (n=1)	PE (n=23)	ML (n=4)	RGC (n=4)	LAC (n=5)	BB (n=2)	BC8 (n=3)	LR (n=6)	MR (n=6)	SC (n=3)	MB (n=8)	EVER (n=4)	KL (n=1)	KW (n=12)	DT (n=5)
9Be	0.78	6	<MDL	0.24 (<MDL–0.34)	<MDL	<MDL	<MDL	<MDL	<MDL	<MDL	<MDL	<MDL	<MDL	<MDL	<MDL	(<MDL –1.47)	(<MDL–1.04)
27Al	1.06	32	3.41	<MDL	<MDL	(<MDL–11.7)	3.44 (<MDL–6.01)	<MDL	(<MDL–26.2)	6.38 (<MDL–13.3)	7.24 (<MDL–10.6)	11.4 (<MDL–17.0)	16.0 (<MDL–37.7)	<MDL	<MDL	6.01 (<MDL–23.9)	6.71 (<MDL–9.39)
51V	0.09	95	2.54	1.70 (1.14–2.06)	0.68 (0.31–1.27)	0.74 (0.26–1.22)	1.28 (0.99–1.41)	1.36 (1.30–1.42)	0.58 (<MDL–1.06)	0.69 (<MDL–1.01)	0.57 (<MDL–1.06)	0.41 (<MDL–0.67)	1.51 (1.30–1.69)	(<MDL–0.14)	1.57	1.69 (1.34–2.22)	1.79 (1.64–1.75)
52Cr	0.09	75	0.16	0.16 (0.09–0.24)	(<MDL–0.27)	0.10 (<MDL–0.14)	0.46 (0.11–1.66)	0.22 (0.13–0.31)	0.18 (<MDL–0.21)	<MDL	0.22 (<MDL–0.28)	<MDL	0.27 (0.13–0.37)	<MDL	0.44	0.39 (0.12–0.50)	0.32 (0.23–0.41)
55Mn	0.57	0	NR	<MDL	<MDL	<MDL	<MDL	<MDL	<MDL	<MDL	<MDL	<MDL	N/A	<MDL	<MDL	NR	NR
59Co	0.06	0	<MDL	<MDL	<MDL	<MDL	<MDL	<MDL	<MDL	<MDL	<MDL	<MDL	<MDL	<MDL	<MDL	<MDL	<MDL
60Ni	0.25	0	<MDL	<MDL	<MDL	<MDL	<MDL	<MDL	<MDL	<MDL	<MDL	<MDL	<MDL	<MDL	<MDL	<MDL	<MDL
63Cu	0.11	71	0.95	0.80 (0.11–2.34)	1.88 (<MDL–2.45)	0.96 (<MDL–1.53)	2.62 (0.41–5.58)	(<MDL–0.79)	0.93 (<MDL–1.55)	1.36 (<MDL–3.47)	4.85 (<MDL–10.5)	3.57 (<MDL–4.18)	1.19 (<MDL–1.50)	0.55 (<MDL–1.21)	1.83	1.32 (<MDL–2.86)	18.3 (<MDL–18.3)
66Zn	0.25	23	<MDL	<MDL	<MDL	(<MDL–9.63)	<MDL	<MDL	<MDL	(<MDL–1.78)	(<MDL–2.10)	6.69 (<MDL–11.3)	3.64 (<MDL–7.33)	(<MDL–168)	<MDL	11.5 (<MDL–22.3)	29.2 (<MDL–59.2)
75As	0.11	100	1.73	2.46 (1.24–4.83)	0.97 (0.77–1.09)	0.81 (0.68–0.95)	1.20 (1.01–1.42)	1.07 (0.97–1.17)	1.00 (<MDL–1.22)	0.99 (<MDL–1.40)	0.82 (0.75–0.90)	1.15 (<MDL–1.21)	1.08 (0.85–1.42)	0.62 (0.42–0.82)	2.25	1.96 (1.22–5.32)	0.98 (0.86–1.10)
80Se	0.11	0	NR	<MDL	<MDL	<MDL	<MDL	<MDL	<MDL	<MDL	<MDL	<MDL	<MDL	<MDL	<MDL	NR	NR

(*Contd.*)

Table 2.2   Average (Min–Max) concentrations ($\mu g/L^{-1}$) of metals and metalloids in surface water samples from South Florida (*Contd.*)

Elements (monitored isotopes)	MDL ($\mu g\ L^{-1}$)	DF (%)	CL (n=1)	PE (n=23)	ML (n=4)	RGC (n=4)	LAC (n=5)	BB (n=2)	BC8 (n=3)	LR (n=6)	MR (n=6)	SC (n=3)	MB (n=8)	EVER (n=4)	KL (n=1)	KW (n=12)	DT (n=5)
98Mo	0.24	92	2.07	8.02 (7.15–9.86)	3.19 (0.02–7.08)	3.65 (0.21–7.55)	6.61 (3.32–8.03)	5.92 (5.65–6.19)	3.78 (<MDL–7.37)	2.20 (<MDL–4.07)	2.73 (<MDL–6.02)	1.93 (<MDL–4.41)	6.86 (2.33–8.99)	<MDL	9.48	6.78 (5.93–8.69)	7.03 (6.39–7.53)
107Ag	0.07	0	<MDL	<MDL	<MDL	<MDL	<MDL	<MDL	<MDL	<MDL	<MDL	<MDL	<MDL	<MDL	<MDL	<MDL	<MDL
111Cd	0.13	9	<MDL	0.21 (<MDL–0.30)	<MDL	<MDL	<MDL	<MDL	<MDL	<MDL	<MDL	<MDL	<MDL	<MDL	<MDL	<MDL	<MDL
118Sn	0.11	0	<MDL	NR	<MDL	<MDL	<MDL	<MDL	<MDL	<MDL	<MDL	<MDL	<MDL	<MDL	<MDL	<MDL	<MDL
121Sb	0.11	33	0.05	(<MDL–0.17)	0.36 (<MDL–0.45)	(<MDL–0.27)	0.27 (<MDL–0.63)	(<MDL–0.29)	(<MDL–0.37)	0.34 (<MDL–0.53)	0.38 (<MDL–0.53)	0.49 (<MDL–0.62)	0.20 (<MDL–0.22)	0.27 (0.13–0.43)	<MDL	0.43 (<MDL–0.45)	<MDL
138Ba	1.88	0	NR	<MDL	<MDL	<MDL	<MDL	<MDL	<MDL	<MDL	<MDL	<MDL	<MDL	<MDL	<MDL	NR	NR
202Hg	0.08	0	<MDL	<MDL	<MDL	<MDL	<MDL	<MDL	<MDL	<MDL	<MDL	<MDL	<MDL	<MDL	<MDL	<MDL	<MDL
207Pb	0.08	7	<MDL	(<MDL–0.10)	<MDL	<MDL	<MDL	<MDL	<MDL	<MDL	(<MDL–0.18)	(<MDL–0.34)	<MDL	<MDL	0.08	(<MDL–0.14)	<MDL

MDL: Method detection limit; DF: Detection frequency; CL: Clear Lake, West Palm Beach; PE: Port Everglades, Fort Lauderdale; ML: Maule Lake; RGC: Royal Galdes Canal; LAC: Little Arch Creek; BB: Biscayne Bay; BC8: Biscayne Canal 8; LR: Little River; MR: Miami River; SC: Seybold Canal; MB: Miami Beach; EVER: Everglades; KL: Key Largo_Pennekamp; KY: Key West; DT: Dry Tortugas; NR: Not reported, n: number of samples.

Table 2.3 Average (Min-Max) concentrations ($\mu g\ L^{-1}$) of metals and metalloids in surface water samples from locations abroad (Dominic Republic, Greece, and Turkey)

Elements (monitored isotopes)	MDL ($\mu g\ L^{-1}$)	DF (%)	Dominican Republic (n=1)	Greece (n=12)	Turkey (n=3)
9Be	0.78	56	<MDL	0.46 (<MDL–1.32)	<MDL
27Al	1.06	63	<MDL	15.5 (<MDL–25.1)	(<MDL–1.73)
51V	0.09	100	1.04	2.04 (1.41–2.82)	1.58 (1.37–1.75)
52Cr	0.09	75	0.12	0.42 (<MDL–0.65)	0.47 (0.35–0.56)
55Mn	0.57	0	<MDL	NR	NR
59Co	0.06	19	<MDL	0.08 (<MDL–0.12)	<MDL
60Ni	0.25	69	<MDL	2.56 (<MDL–5.59)	<MDL–2.01
63Cu	0.11	94	0.11	13.7 (<MDL–32.8)	6.58 (1.78–13.7)
66Zn	0.25	75	<MDL	33.5 (<MDL–95.7)	10.4 (<MDL–18.8)
75As	0.11	100	1.57	1.43 (0.99–2.20)	0.96 (0.83–1.04)
80Se	0.11	0	<MDL	NR	NR
98Mo	0.24	100	7.85	10.7 (8.51–12.5)	9.55 (9.19–10.1)
107Ag	0.07	6	<MDL	(<MDL–0.14)	<MDL
111Cd	0.13	25	<MDL	0.13 (<MDL–0.13)	<MDL–0.25
118Sn	0.11	6	<MDL	(<MDL–0.12)	<MDL
121Sb	0.11	0	<MDL	<MDL	<MDL
138Ba	1.88	0	<MDL	NR	NR
202Hg	0.08	0	<MDL	<MDL	<MDL
207Pb	0.08	75	<MDL	0.81 (<MDL–1.51)	0.16 (<MDL–0.19)

MDL: Method detection limit; DF: Detection frequency; NR: Not reported.

Withanachchi et al., 2018; Zhang et al., 2009, 2010). Figure 2.3 illustrates Cd, Pb and Hg concentrations in worldwide locations (South and North China, India, Pakistan, Bangladesh, Iran, South Africa, Nigeria, Spain, Armenia, Georgia, Brazil, and Argentina), including values reported in the current study (Florida, U.S., Greece, Turkey, and Dominican Republic). Table 2.4 presents the average concentrations in surface waters in South Florida locations and abroad, comparing metal levels worldwide within five continents. As observed in Figure 2.3, the highest Cd and Pb levels are reported for the Oluwa River, South-Western Nigeria, Africa (Ayandiran et al., 2018), with average concentrations of 100 $\mu g\ L^{-1}$ and 360 $\mu g\ L^{-1}$, respectively, 3 orders of magnitude higher than the levels found in our study. Mercury was either not reported or had <MDL values, also consistent with our study. High Hg values, however, were detected in China (0.59 $\mu g\ L^{-1}$) and Argentina (0.33 $\mu g\ L^{-1}$), at levels above European and Australian guidelines for the protection of aquatic species (Rautenberg et al., 2015; Zhang et al., 2009). Regarding other metals, As, Cu, Cr, and Zn showed average concentrations up to 54.7, 1407, 21.3 and 11,968 $\mu g\ L^{-1}$, respectively, in the Mashavera River, Georgia (Withanachchi et al., 2018), at least 50-fold higher than the levels detected in both

South Florida and selected locations abroad. The higher levels in the Mashavera River basin are likely linked to direct discharges of industrial wastewater and untreated sewage in the Bolnisi and the Dmanisi regions (Withanachchi et al., 2018). Overall, metal levels observed in South Florida were similar to concentrations found in South America, for example in Rio Pardo, Brazil and the Suquía River in Argentina, as well as in Dominican Republic and Europe (Ebro River, Spain and Voghji River Basin, Armenia, Greece and Marmaris, Turkey). Particularly regarding Pb, average concentrations observed in North and South China, Georgia, Bangladesh, Brazil, and Argentina are usually 10–100-fold higher than in our study. In general, excluding the values reported from Georgia, average metal concentrations increased across the five continents in the following order: North America<South America<Europe<Asia<Africa (Table 2.4), even though profiles might vary depending on the metal. For example, the average calculated Cd level in South America (0.02 µg L^{-1}) was lower than in North America (0.04 µg L^{-1}), while for Pb, average concentrations in South America (1.30 µg L^{-1}) were higher than in Europe (0.34 µg L^{-1}).

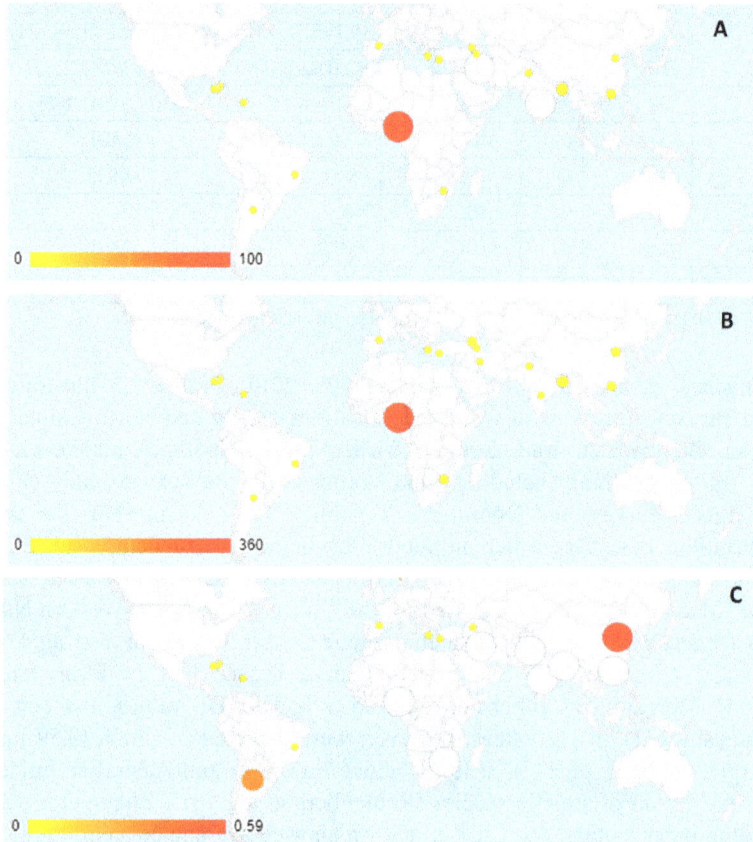

Figure 2.3   Worldwide levels (in µg L^{-1}) of Cd (A), Pb (B) and Hg (C) in surface waters. Increasing circle sizes and colors represent higher concentration, while white circles represent non-available data (N/A).

Table 2.4 Average metal concentrations (µg L^{-1}) in surface waters from global rivers across five continents

Continents	Cr	Cu	Zn	As	Cd	Pb	Hg	References
Africa								
Mvudi River, South Africa	357	185	261	N/A	0.3	14	N/A	Edokpayi et al. (2016)
Oluwa River, Nigeria	600	130	2900	N/A	100	360	N/A	Ayandiran et al. (2018)
Asia								
Hengshuihu Wetland, North China	<MDL	0.77	0.00	2.88	<MDL	10.5	0.59	Zhang et al. (2009)
Pearl River Estuary, South China	8.50	1.60	8.90	N/A	2.90	12.8	N/A	Zhang et al. (2010)
Nakkavagu stream, Hyderabad, India	16.8	N/A	98.6	29.2	N/A	2.1	N/A	Krishna et al. (2009)
Faisalabad, Pakistan	5	220	900	30	2	1	N/A	Mahfooz et al. (2019)
Korotoa River, Bangladesh	83	73	N/A	46	11	35	N/A	Islam et al. (2015)
Ardabil, Iran	5	6.1	34.8	34.7	<MDL	3	N/A	Sadeghi et al. (2020)
Europe								
Greece	0.42	13.70	33.50	1.43	0.13	0.81	<MDL	This study
Turkey	0.47	6.58	10.37	0.96	0.25	0.16	<MDL	This study
Ebro River, Spain	6.97	6.63	N/A	0.83	0.00	0.52	<MDL	Vilavert et al., (2015)
Mashavera River, Georgia	21.3	1407	11,968	54.7	0.4	11.5	<MDL	Withanachchi et al. (2018)
Voghji River Basin, Armenia	0.395	3.07	3.21	0.938	0.045	0.039	N/A	Gabrielyan et al. (2018)
North America								
Miami, U.S.	0.23	1.70	32.0	1.16	0.21	0.21	<MDL	This study
Key Largo, U.S.	0.44	1.83	0.00	2.25	<MDL	0.08	<MDL	This study
Key West, U.S.	0.39	1.32	11.5	1.96	<MDL	0.14	<MDL	This study
Dry Tortugas, U.S.	0.32	18.3	29.2	0.98	<MDL	<MDL	<MDL	This study
Punta Cana, Dominican Republic	0.12	0.11	<MDL	1.57	<MDL	<MDL	<MDL	This study
South America								
Rio Pardo, Brazil	0.52	3.28	13.3	2.14	0.05	1.8	<MDL	Alves et al. (2014)
Suquía River, Argentina	4.47	3.23	38.3	1.91	<MDL	2.11	0.33	Rautenberg et al. (2015)

MDL: method detection limit; N/A: not available or not assessed.

Identification of Pb, Cd and Hg Sources in South Florida

Large quantities of hazardous chemicals, including toxic metals, have been released into the ocean waters worldwide through several anthropogenic activities, associated to rapid global population growth, intensive domestic activities, and the expansion of industrial and agricultural production, as well as natural processes such as weathering (Garrett, 2000; Moiseenko and Gashkina, 2020; Tarras-Wahlberg et al., 2001).

Although some metals, such as Al, mainly originate from natural sources, others such as Cd, Hg, Pb, As, and Cr are mainly associated with human-related sources, which includes industrial activities, agriculture, atmospheric pollution and others. Crustal elements, especially Al, have been previously identified as of geological material contribution tracers, such as mineral dust from North Africa (African dust events) and marine aerosol (Ramirez et al., 2022).

Cadmium has been routinely employed in the manufacturing of batteries, plating, pigments and plastics worldwide (Hutton, 1983; WHO, 2003b). In South Florida, Cd levels may be related to shipping activities, electricity plants and agricultural sources (Ramirez et al., 2022). In fact, the only location where Cd had been previously detected above the MDL was in Port Everglades, which is home to three terminals for large, diesel-powered vessels (cargo ships, oil tankers and cruise ships) (Ramirez et al., 2022).

The use of inorganic fertilizers (containing Cd, Cr, Ni, Pb and Zn) can also contribute to the levels of these contaminants by agricultural run-off to Biscayne Bay and its adjacent canals, the Miami river and the Everglades (Duan, 2012). Therefore, Cd and Pb deposition near agricultural areas due to rainwater and aerosols is also a possibility and may constitute another source. Pb was used as a fuel additive in gasoline for many years in piston aircraft in Florida before it was phased-out (Florea and Büsselberg, 2006). In addition, it was also used in paint formulations, batteries, glass and soldering (Florea and Büsselberg, 2006; WHO, 2011). Emissions from motor vehicles and coal combustion power plants, wastewater discharges from domestic and industrials treatment plants, urban run-off and landfill leachate could aggravate the levels of Cd, Pb and Zn in South Florida aquatic environments, including preserved Everglades areas (Duan, 2012). In addition, one of the major sources of Hg in Florida's water comprises atmospheric deposition from fossil fuel burning, especially coal and fertilizers used in nearby agricultural areas (Duan, 2012; Sillman et al., 2007).

Other metals contained in fungicides, fertilizers and bactericides used in the agriculture are Cu, and As, commonly used for citrus and vegetable crop production in Florida, may also contribute to the observed levels through agricultural run-off and atmospheric deposition in nearby water bodies in South Florida, specially near the Everglades area (Yang et al., 2009). Ni has been widely used in the metallurgical industry in the stainless-steel and alloys processing (Florea and Büsselberg, 2006), also associated to the coal, diesel oil and fuel oil combustion.

Toxicity Considerations on Pb, Cd and Hg and other Toxic Metals in South Florida

Surface water metal concentrations may be considered as metal contamination indicators and potential environmental risks to aquatic marine organisms. Even though some metals are essential to living organisms, when present in excess of water quality concentrations thresholds, they can exert toxic adverse effects to a variety of aquatic species (Jaishankar et al., 2014). In this regard, metal bioaccumulation and biomagnification processes at higher trophic food web levels are of significant concern, as they may exacerbate toxicological effects and represent potential risks to human health (Ali and Khan, 2019; Bawuro et al., 2018; Moiseenko and Gashkina, 2020).

In South Florida, recreational saltwater fisheries are a multi-billion-dollar industry (Ohs et al., 2018), contributing significantly to tourism and fishing tackle businesses in Florida. In fact, Florida is considered the "fishing capital of the world", home to one of the highest number of recreational anglers in the U.S. (NOAA, 2018). Therefore, the presence of toxic metals at particularly high concentrations in the surface waters is regarded not only as a threat to fish heath and the biota in general, but also a potential human health risk.

In this regard, Cd is a non-essential metal toxic to aquatic life, affecting biota growth, reproduction, development, behavior and immune and endocrine systems, leading to increased mortality of aquatic organisms (U.S. EPA, 2015). Its effects may be exacerbated due to the presence of certain essential elements, such as Zn, which has shown to increase Cd toxicity in aquatic invertebrates (WHO, 2003b). Mercury, although not encountered in this study due to analytical capabilities, has been previously reported at high concentrations in aquatic systems throughout Florida (Kannan et al., 1998; Royals and Lange, 1990). The lower Hg concentrations (values <MDL) compared to previous studies might be related to the 86% reduction of point-source emissions in Florida from 1994 to more recent years (FLDEP, 2013). In the marine environment, exposure to Hg results in reduced biota metabolism and liver function, altered behavior, deformity, impaired reproduction, gills damage and increased mortality (Morcillo et al., 2017; Zheng et al., 2019). Lead is a highly toxic biologically non-essential metal known to affect several organs, especially the central nervous system (neurotoxicity), inducing synaptic damage and neurotransmitter malfunctions in aquatic organisms, also acting as an immune-toxicant (Lee et al., 2019).

Other metals displaying toxicity concerns, such as Cr, Zn, Ni, Cu and As, may be present in surface waters at concentrations above aquatic life protective thresholds, causing adverse effects in fish and other marine species, such as acute hypoxia, hyperglycemia, depletion of enzymatic activities and severe damage to the gills, liver and kidneys, and various acute and chronic toxicity, and immune system dysfunctions have been reported (Javed and Usmani, 2013; Kumari et al., 2017).

In addition, there are some growing evidences showing that environmental metal exposures (including Cu, Mo, As, Mn, Pb, Cd, Hg and Zn) can cause epigenetic modifications, which may lead to heritable alterations in gene function and expression (mutations) and disease susceptibility and development, e.g., can be implicated in the pathophysiology of cancer and endocrine/metabolic diseases (Bitto et al., 2014; Cai et al., 2022).

Herein, we estimated the ecological risk index (ERI) to evaluate the overall environmental toxicity and ecological risks of toxic metals in the coastal marine environment as follows:

$$ERI = \frac{\sum_{i}^{n}\left(Tr \times \frac{MEC}{PNEC}\right)}{\sum_{i}^{n} Tr}$$

where Tr represents the toxic-response factor of the investigated metal, MEC is the average measured environmental concentration in surface water, and PNEC is the predicted no-effect concentration, calculated by dividing the U.S.EPA chronic CCC by a safety factor of 5 (values obtained from Kumar et al., 2020; Luo et al., 2022). It should be noted that to determine the overall risk posed by the simultaneous presence of multiple metals, accounting for their different toxicity degrees, as noted in the equation, ERI is normalized by the toxicity response factor (Tr). The Tr values for Cd, Pb, As, Cr, Cu, Ni, and Zn are 30, 5, 10, 2, 5, 5, and 1, respectively (Kumar et al., 2020; Luo et al., 2022). The estimated PNEC values for Cd, Pb, As, Cr, Cu, Ni, and Zn are 1.58, 1.62, 7.20, 10.00, 0.62, 1.64, and 16.2, respectively (U.S.EPA, 2004).

The calculated $Er\left(\text{defined as } Tr \times \frac{MEC}{PNEC}\right)$, which refers to the potential ecological risk factor for each element, and the ERI are presented in Table 2.5 for all sampling locations in South Florida and abroad. The ERI values can be classified into the following categories, according to their degree of ecological toxicity and risks: negligible ecological risk (ERI ≤ 1), low ecological risk (1 < ERI ≤ 5), moderate ecological risk (5 < ERI ≤ 10), considerable ecological risk (10 < ERI ≤ 15) and high ecological risk (ERI > 15) (Luo et al., 2022). Furthermore, based on the calculated Er values for each location, these values can be categorized as: Er < 40 (low ecological risk for the water body), 40 ≤ Er < 80 (moderate ecological risk for the water body), 80 ≤ Er < 160 (considerable ecological risk for the water body), 160 ≤ Er < 320 (very high ecological risk for the water body) and Er > 320 (extremely high ecological risk for the water body) (Protano et al., 2014).

The ecological risk assessment revealed negligible risks (ERI values < 1) for all sampling locations in Miami as well as Key Largo, Key West, and Punta Cana. ERI exceeding 1 (but below 5) were observed for the Dry Tortugas, Greece and Turkey sites, indicating a low ecological risk due to the presence of some toxic metals at relatively high concentrations, such as Cu, Cd and Pb levels. Nevertheless, based on the Er values, it is evident that Cu levels pose moderate to considerable ecological risks in Turkey, Dry Tortugas (U.S.), and Greece. Taking into consideration, the continuous input from these toxic metals, such as Cd and Pb, through numerous anthropogenic sources, metal aquatic environment screening is warranted to obtain reliable and up-to-date information about the occurrence, concentrations, and distribution of these environmental pollutants, assisting managers and decision makers in treating and, hopefully, reducing their concentrations in water systems.

Table 2.5 Ecological risk index (ERI) for toxic metals in surface water samples from South Florida and abroad

Locations	ErCr	ErNi	ErCu	ErZn	ErAs	ErCd	ErPb	∑ERI
CL	0.03	0.00	7.66	0.00	2.40	0.00	0.00	0.17
PE	0.03	0.00	6.45	0.00	3.42	3.99	0.31	0.24
RGC	0.02	0.00	7.74	0.59	1.13	0.00	0.00	0.16
ML	0.05	0.00	15.16	0.00	1.35	0.00	0.00	0.29
LAC	0.09	0.00	21.13	0.00	1.67	0.00	0.00	0.39
BB	0.04	0.00	6.37	0.00	1.49	0.00	0.00	0.14
BC8	0.04	0.00	7.50	0.00	1.39	0.00	0.00	0.15
LR	0.00	0.00	10.97	0.11	1.38	0.00	0.00	0.21
MR	0.04	0.00	39.11	0.13	1.14	0.00	0.56	0.71
SC	0.00	0.00	28.79	0.41	1.60	0.00	1.05	0.55
MB	0.05	0.00	9.60	0.22	1.50	0.00	0.00	0.20
EVER	0.00	0.00	4.44	10.40	0.89	0.00	0.00	0.27
KL	0.09	0.00	14.79	0.00	3.13	0.00	0.26	0.31
KW	0.08	0.00	10.65	0.71	2.72	0.00	0.43	0.25
DT	0.06	0.00	147.18	1.80	1.36	0.00	0.00	2.59
Greece	0.08	7.80	110.40	2.07	1.99	2.47	2.50	2.20
Turkey	0.09	6.13	53.06	0.64	1.33	4.75	0.49	1.15
Punta Cana	0.02	0.00	0.89	0.00	2.18	0.00	0.00	0.05

CONCLUSIONS

Lead, cadmium, mercury, as well as other metals of significant importance were assessed in the coastal aquatic environments in South Florida and abroad. Mercury was not found in any of the samples, possibly due to the high MDL of the technique or sediment speciation and adsorption. Cadmium and Pb exhibited low occurrence in South Florida (detection frequency < 10%) but were more frequently detected (25–75%) in Greece and Turkey. Cadmium was found at concentrations ranging from < MDL to 0.3 µg L^{-1} at Port Everglades (Florida) and in Turkey, while Pb was detected in 12% of the samples, concentrations ranging from < MDL to 1.35 µg L^{-1}. Cadmium levels are likely related to shipping activities at Port Everglades (Florida), Greece and Turkey. In Florida, agricultural sources are a major contributor to increased Cd, Cr, Ni, Pb and Zn concentrations due to agricultural run-off into Biscayne Bay and its adjacent canals, the Miami river and the Everglades. The major source of Pb in Florida could be from leaded paints and gasoline used for many years before their ban. Other sources, such as emissions from motor vehicles and coal combustion power plants, domestic and industrial wastewater discharges, urban run-off, and landfill leachate could aggravate Cd, Pb, and Zn levels in South Florida aquatic environments. The Pb concentrations observed in the present study were at least 10–100 times lower than in North and South China, Georgia, Bangladesh, Brazil and Argentina and three orders of magnitude lower than in Africa. Average metal levels observed in South Florida were at the same order of magnitude than studies from South America, Punta Cana and Europe. Overall, metal concentrations observed in South Florida's surface waters pose a low ecological risk to aquatic life, as indicated by an estimated ecological risk index (ERI) below 5. Nevertheless, the ecological risk for Cu (Er > 50–150) suggests moderate to considerable ecological risks to aquatic organisms in Dry Tortugas (Florida), Greece, and Turkey, and further monitoring is warranted.

REFERENCES

Ali, H. and Khan, E. 2019. Trophic transfer, bioaccumulation, and biomagnification of non-essential hazardous heavy metals and metalloids in food chains/webs—Concepts and implications for wildlife and human health. Hum. Ecol. Risk Assess. An Int. J. 25: 1353–1376. https://doi.org/10.1080/10807039.2018.1469398

Alves, R.I.S., Sampaio, C.F., Nadal, M., Schuhmacher, M., Domingo, J.L. and Segura-Muñoz, S.I. 2014. Metal concentrations in surface water and sediments from Pardo River, Brazil: Human health risks. Environ. Res. 133: 149–155. https://doi.org/https://doi.org/10.1016/j.envres.2014.05.012

Ansari, T.M., Marr, I.L. and Tariq, N. 2004. Heavy metals in marine pollution perspective–A mini review. J. Appl. Sci. 4: 1–20. https://doi.org/10.3923/jas.2004.1.20

ANZECC, 2000. Australian and New Zealand guidelines for fresh and marine water quality. Australian and New Zealand Environment and Conservation Council, Canberra, ACT.

ARMCANZ and ANZECC, 2000. Australian and New Zealand guidelines for fresh and marine water quality. Canberra.

Ault, J.S., Smith, S.G., Meester, G.A., Luo, J. and Bohnsack, J.A. 2001. Site characterization for biscayne national park: assessment of fisheries resources and habitats. NOAA Technical Memorandum NMFS-SEFSC-468, 156 p.

Ayandiran, T.A., Fawole, O.O. and Dahunsi, S.O. 2018. Water quality assessment of bitumen polluted Oluwa River, South-Western Nigeria. Water Resour. Ind. 19: 13–24. https://doi.org/https://doi.org/10.1016/j.wri.2017.12.002

Bawuro, A.A., Voegborlo, R.B. and Adimado, A.A. 2018. Bioaccumulation of heavy metals in some tissues of fish in Lake Geriyo, Adamawa State, Nigeria. J. Environ. Public Health. 2018: Article ID 1854892. https://doi.org/10.1155/2018/1854892

Bhuyan, M.S., Bakar, M.A., Akhtar, A., Hossain, M.B., Ali, M.M. and Islam, M.S. 2017. Heavy metal contamination in surface water and sediment of the Meghna River,

Bangladesh. Environ. Nanotechnol. Monit. Manage. 8: 273–279. https://doi.org/10.1016/j.enmm.2017.10.003

Bitto, A., Pizzino, G., Irrera, N., Galfo, F. and Squadrito, F. 2014. Epigenetic modifications due to heavy metals exposure in children living in polluted areas. Curr. Genomics 15: 464–468. https://doi.org/10.2174/1389202915066150106153336

Cai, G., Yu, X., Hutchins, D. and McDermott, S. 2022. A pilot study that provides evidence of epigenetic changes among mother–child pairs living proximal to mining in the US. Environ. Geochem. Health. (In Press). https://doi.org/10.1007/s10653-022-01217-9

Castro, J.E., Fernandez, A.M., Gonzalez-Caccia, V. and Gardinali, P.R. 2013. Concentration of trace metals in sediments and soils from protected lands in south Florida: Background levels and risk evaluation. Environ. Monit. Assess. 185: 6311–6332. https://doi.org/10.1007/s10661-012-3027-9

CCME, 2008. Canadian Water Quality Guidelines. Ottawa, Ontario.

Duan, Z. 2012. The Distribution of Toxic and Essential Metals in the Florida Everglades. Florida International University.

EC, 2013. Directive 2013/39/EU of the European Parliament and of the Council of 12 August 2013 amending Directives 2000/60/EC and 2008/105/EC as regards priority substances in the field of water policy. Brussels, Belgium.

Edokpayi, J.N., Odiyo, J.O., Popoola, O.E. and Msagati, T.A.M. 2016. Assessment of trace metals contamination of surface water and sediment: A case study of Mvudi River, South Africa. Sustainability 8(2): 135. https://doi.org/10.3390/su8020135

Erickson, J.D., Nichols, J.W., Cook, P.M. and Ankley, G.T. 2008. Bioavailability of chemical contaminants in aquatic systems. pp. 9–45. *In:* Di Giulio, R. and Hinton, D. (eds). The Toxicology of Fishes. CRC Press, Florida, USA.

Fisher, M.B., Guo, A.Z., Tracy, J.W., Prasad, S.K., Cronk, R.D., Browning, E.G., et al. 2021. Occurrence of lead and other toxic metals derived from drinking-water systems in three West African countries. Environ. Health Perspect. 129: 47012. https://doi.org/10.1289/EHP7804

FLDEP, 2013. Mercury TMDL for the state of Florida. Florida Department of Environmental Protection (FLDEP), Division of Environmental Assessment and Restoration.

Florea, A.-M. and Büsselberg, D. 2006. Occurrence, use and potential toxic effects of metals and metal compounds. BioMetals 19: 419–427. https://doi.org/10.1007/s10534-005-4451-x

Gabrielyan, A.V., Shahnazaryan, G.A. and Minasyan, S.H. 2018. Distribution and identification of sources of heavy metals in the Voghji River basin impacted by mining activities (Armenia). J. Chem. 2018: 7172426. https://doi.org/10.1155/2018/7172426

Garrett, R.G. 2000. Natural sources of metals to the environment. Hum. Ecol. Risk Assess. 6(6): 945–996. https://doi.org/10.1080/10807030091124383

Geisler, C.-D. and Schmidt, D. 1991. An overview of chromium in the marine environment. Dtsch. Hydrogr. Zeitschrift 44: 185–196. https://doi.org/10.1007/BF02226462

Hauser-Davis, R.A., Figueiredo, L., Lemos, L., de Moura, J.F., Rocha, R.C.C., Saint'Pierre, T., et al. 2020. Subcellular cadmium, lead and mercury compartmentalization in Guiana dolphins (*Sotalia guianensis*) from southeastern Brazil. Front. Mar. Sci. 7: 584195. https://doi.org/10.3389/fmars.2020.584195

HELCOM, 2010. Hazardous Substances in the Baltic Sea—An integrated thematic assessment of Hazardous Substances in the Baltic Sea. Baltic Sea Environment Proceedings (120B). Helsinki Commission, Helsinki.

Hsu, P.-C. and Guo, Y.L. 2002. Antioxidant nutrients and lead toxicity. Toxicology 180: 33–44. https://doi.org/10.1016/S0300-483X(02)00380-3

Hutton, M. 1983. Sources of cadmium in the environment. Ecotoxicol. Environ. Saf. 7: 9–24. https://doi.org/10.1016/0147-6513(83)90044-1

Islam, M.S., Ahmed, M.K., Raknuzzaman, M., Habibullah-Al-Mamun, M. and Islam, M.K. 2015. Heavy metal pollution in surface water and sediment: A preliminary assessment of an urban river in a developing country. Ecol. Indic. 48: 282–291. https://doi.org/10.1016/j.ecolind.2014.08.016

Jaishankar, M., Tseten, T., Anbalagan, N., Mathew, B.B. and Beeregowda, K.N. 2014. Toxicity, mechanism and health effects of some heavy metals. Interdiscip. Toxicol. 7: 60–72. https://doi.org/10.2478/intox-2014-0009

Javed, M. and Usmani, N. 2013. Assessment of heavy metal (Cu, Ni, Fe, Co, Mn, Cr, Zn) pollution in effluent dominated rivulet water and their effect on glycogen metabolism and histology of Mastacembelus armatus. Springerplus 2: 390. https://doi.org/10.1186/2193-1801-2-390

Kannan, K., Smith, Jr., R.G., Lee, R.F., Windom, H.L., Heitmuller, P.T., Macauley, J.M., et al. 1998. Distribution of total mercury and methyl mercury in water, sediment, and fish from South Florida estuaries. Arch. Environ. Contam. Toxicol. 34: 109–118. https://doi.org/10.1007/s002449900294

Kim, J.-H. and Kang, J.-C. 2016. The toxic effects on the stress and immune responses in juvenile rockfish, *Sebastes schlegelii* exposed to hexavalent chromium. Environ. Toxicol. Pharmacol. 43: 128–133. https://doi.org/10.1016/j.etap.2016.03.008

Krishna, A.K., Satyanarayanan, M. and Govil, P.K. 2009. Assessment of heavy metal pollution in water using multivariate statistical techniques in an industrial area: A case study from Patancheru, Medak District, Andhra Pradesh, India. J. Hazard. Mater. 167: 366–373. https://doi.org/10.1016/j.jhazmat.2008.12.131

Kumar, S.B., Padhi, R.K., Mohanty, A.K. and Satpathy, K.K. 2020. Distribution and ecological- and health-risk assessment of heavy metals in the seawater of the southeast coast of India. Mar. Pollut. Bull. 161 Pt A: 111712.

Kumari, B., Kumar, V., Sinha, A.K., Ahsan, J., Ghosh, A.K., Wang, H., et al. 2017. Toxicology of arsenic in fish and aquatic systems. Environ. Chem. Lett. 15: 43–64. https://doi.org/10.1007/s10311-016-0588-9

Lee, J.-W., Choi, H., Hwang, U.-K., Kang, J.-C., Kang, Y.J., Kim, K. Il., et al. 2019. Toxic effects of lead exposure on bioaccumulation, oxidative stress, neurotoxicity, and immune responses in fish: A review. Environ. Toxicol. Pharmacol. 68: 101–108. https://doi.org/10.1016/j.etap.2019.03.010

Lewis, A.G. 1995. Copper in Water and Aquatic Environments. American Chemet Corporation.

Lim, W.Y., Aris, A.Z. and Zakaria, M.P. 2012. Spatial variability of metals in surface water and sediment in the Langat River and geochemical factors that influence their water-sediment interactions. Sci. World J. 2012: 652150. https://doi.org/10.1100/2012/652150

Luo, M., Zhang, Y., Li, H., Hu, W., Xiao, K., Yu, S., et al. 2022. Pollution assessment and sources of dissolved heavy metals in coastal water of a highly urbanized coastal area: The role of groundwater discharge. Sci. Total Environ. 807: 151070. https://doi.org/https://doi.org/10.1016/j.scitotenv.2021.151070

Mahfooz, Y., Yasar, A., Sohail, M.T., Tabinda, A.B., Rasheed, R., Irshad, S. et al. 2019. Investigating the drinking and surface water quality and associated health risks in a semi-arid multi-industrial metropolis (Faisalabad), Pakistan. Environ. Sci. Pollut. Res. Int. 26: 20853–20865. https://doi.org/10.1007/s11356-019-05367-9

Miller, R.L., McPherson, B.F., Sobczak, R. and Clark, C. 2004. Water quality in big cypress national preserve and everglades national park—trends and spatial characteristics of selected constituents. Water-Resources Investigations Report 2003-4249. Reston, VA. https://doi.org/10.3133/wri034249

Moiseenko, T.I. and Gashkina, N.A. 2020. Distribution and bioaccumulation of heavy metals (Hg, Cd and Pb) in fish: Influence of the aquatic environment and climate. Environ. Res. Lett. 15: 115013.

Morcillo, P., Esteban, M.A. and Cuesta, A. 2017. Mercury and its toxic effects on fish. AIMS Environ. Sci. 4: 386–402. https://doi.org/10.3934/environsci.2017.3.386

Neff, J.M. 1997. Ecotoxicology of arsenic in the marine environment. Environ. Toxicol. Chem. 16: 917–927. https://doi.org/10.1002/etc.5620160511

Ng, B., Quinete, N., Maldonado, S., Lugo, K., Purrinos, J., Briceño, H., et al. 2021. Understanding the occurrence and distribution of emerging pollutants and endocrine disruptors in sensitive coastal South Florida Ecosystems. Sci. Total Environ. 757: 143720. https://doi.org/10.1016/j.scitotenv.2020.143720

NOAA, 2018. Fisheries economics of the United States, 2016. NOAA Technical Memorandum NMFS-F/SPO-187A0, U.S. Department of Commerce, Silver Spring, MD.

NOAA, 2019. NOAA Habitat Blueprint. Biscayne Bay Habitat Focus Area Implementation Plan (2015–2020) [WWW Document]. URL https://www.habitatblueprint.noaa.gov/habitat-focus-areas/biscayne-bay-florida/

Nriagu, J.O. and Pacyna, J.M. 1988. Quantitative assessment of worldwide contamination of air, water and soils by trace metals. Nature 333: 134–139. https://doi.org/10.1038/333134a0

Ohs, C.L., DiMaggio, M.A. and Beany, A.H. 2018. Preferences for and perception of cultured marine baitfish by recreational saltwater anglers in Florida. Aquac. Econ. and Manag. 22: 264–278. https://doi.org/10.1080/13657305.2017.1298007

Protano, C., Zinnà, L., Giampaoli, S., Spica, V.R., Chiavarini, S. and Vitali, M. 2014. Heavy metal pollution and potential ecological risks in rivers: A case study from southern Italy. Bull. Environ. Contam. Toxicol. 92: 75–80. https://doi.org/10.1007/s00128-013-1150-0

Rajeshkumar, S. and Li, X. 2018. Bioaccumulation of heavy metals in fish species from the Meiliang Bay, Taihu Lake, China. Toxicol. Reports 5: 288–295. https://doi.org/10.1016/j.toxrep.2018.01.007

Ramirez, C.E., Quinete, N., Rojas de Astudillo, L., Arroyo-Mora, L.E., Seba, D. and Gardinali, P. 2022. Elemental composition of airborne particulate matter from coastal South Florida area influenced by African dust events. Aeolian Res. 54: 100774. https://doi.org/10.1016/j.aeolia.2022.100774

Rautenberg, G.E., Amé, M.V., Monferrán, M.V., Bonansea, R.I. and Hued, A.C. 2015. A multi-level approach using Gambusia affinis as a bioindicator of environmental pollution in the middle-lower basin of Suquía River. Ecol. Indic. 48: 706–720. https://doi.org/10.1016/j.ecolind.2014.09.025

Royals, H. and Lange, T. 1990. Mercury in Florida fish and wildlife. Florida Wildl. 44: 3–6.

Sadeghi, H., Fazlzadeh, M., Zarei, A., Mahvi, A.H. and Nazmara, S. 2022. Spatial distribution and contamination of heavy metals in surface water, groundwater and topsoil surrounding Moghan's tannery site in Ardabil, Iran. Int. J. Environ. Anal. Chem. 102(5): 1049–1059. https://doi.org/10.1080/03067319.2020.1730342

Schaefer, A.M., Titcomb, E.M., Fair, P.A., Stavros, H.-C.W., Mazzoil, M., Bossart, G.D., et al. 2015. Mercury concentrations in Atlantic bottlenose dolphins (*Tursiops truncatus*) inhabiting the Indian River Lagoon, Florida: Patterns of spatial and temporal distribution. Mar. Pollut. Bull. 97: 544–547. https://doi.org/10.1016/j.marpolbul.2015.05.007

Siddiqi, Z.M. 2018. Transport and fate of mercury (Hg) in the environment: Need for continuous monitoring. pp. 1–20. *In*: Hussain, C.M. (ed.). Handbook of Environmental Materials Management. Springer International Publishing, Cham. https://doi.org/10.1007/978-3-319-58538-3_56-1

Sillman, S., Marsik, F.J., Al-Wali, K.I., Keeler, G.J. and Landis, M.S. 2007. Reactive mercury in the troposphere: Model formation and results for Florida, the northeastern United States, and the Atlantic Ocean. J. Geophys. Res. 112: D23305. https://doi:10.1029/2006JD008227

Smedley, P.L. and Kinniburgh, D.G., 2017. Molybdenum in natural waters: A review of occurrence, distributions and controls. Appl. Geochemistry 84: 387–432. https://doi.org/10.1016/j.apgeochem.2017.05.008

Tarras-Wahlberg, N.H., Flachier, A., Lane, S.N. and Sangfors, O. 2001. Environmental impacts and metal exposure of aquatic ecosystems in rivers contaminated by small scale gold mining: the Puyango River basin, southern Ecuador. Sci. Total Environ. 278: 239–261. https://doi.org/10.1016/S0048-9697(01)00655-6

U.S.EPA, 1984. Ambient Water Quality Criteria for Lead-1984.

U.S.EPA, 1994. Method 200.8: Determination of Trace Elements in Waters and Wastes by Inductively Coupled Plasma-Mass Spectrometry.

U.S.EPA, 2004. National Recommended Water Quality Criteria [WWW Document]. URL https://www.epa.ghttps//www.epa.gov/wqc/national-recommended-water-quality-criteria-aquaticlife-criteria-table

U.S.EPA, 2015. Recommended Aquatic Life Ambient Water Quality Criteria for Cadmium-2016. EPA-HQ-OW-2015-0753.

Vilavert, L., Sisteré, C., Schuhmacher, M., Nadal, M. and Domingo, J.L. 2015. Environmental concentrations of metals in the catalan stretch of the ebro river, Spain: assessment of temporal trends. Biol. Trace Elem. Res. 163: 48–57. https://doi.org/10.1007/s12011-014-0140-3.

WHO, 2003a. Chromium in Drinking-water, Background document for development of WHO Guidelines for Drinking-water Quality. https://doi.org/WHO/SDE/WSH/03.04/05

WHO, 2003b. Cadmium Review.

WHO, 2011. Lead in Drinking-water—Background document for development of WHO Guidelines for Drinking-water Quality.

Withanachchi, S.S., Ghambashidze, G., Kunchulia, I., Urushadze, T. and Ploeger, A. 2018. Water quality in surface water: A preliminary assessment of heavy metal contamination of the Mashavera River, Georgia. Int. J. Environ. Res. Public Health 15(4): 621. https://doi.org/10.3390/ijerph15040621

Wu, F., Owens, J.D., Huang, T., Sarafian, A., Huang, K.-F., Sen, I.S., et al. 2019. Vanadium isotope composition of seawater. Geochim. Cosmochim. Acta. 244: 403–415. https://doi.org/10.1016/j.gca.2018.10.010

Yang, Y., He, Z., Lin, Y., Phlips, E.J., Stoffella, P.J. amd Powell, C.A., 2009. Temporal and spatial variations of copper, cadmium, lead, and zinc in Ten Mile Creek in South Florida, USA. Water Environ. Res. a Res. Publ. Water Environ. Fed. 81: 40–50. https://doi.org/10.2175/106143008x296479

Zhang, M., Cui, L., Sheng, L. and Wang, Y. 2009. Distribution and enrichment of heavy metals among sediments, water body and plants in Hengshuihu Wetland of Northern China. Ecol. Eng. 35: 563–569. https://doi.org/10.1016/j.ecoleng.2008.05.012

Zhang, H., Cui, B., Xiao, R. and Zhao, H., 2010. Heavy metals in water, soils and plants in riparian wetlands in the Pearl River Estuary, South China. Procedia Environ. Sci. 2: 1344–1354. https://doi.org/10.1016/j.proenv.2010.10.145

Zheng, N., Wang, S., Dong, W., Hua, X., Li, Y., Song, X., et al. 2019. The toxicological effects of mercury exposure in marine fish. Bull. Environ. Contam. Toxicol. 102: 714–720. https://doi.org/10.1007/s00128-019-02593-2

Zhou, Q., Yang, N., Li, Y., Ren, B., Ding, X., Bian, H., et al. 2020. Total concentrations and sources of heavy metal pollution in global river and lake water bodies from 1972 to 2017. Glob. Ecol. Conserv. 22: e00925. https://doi.org/10.1016/j.gecco.2020.e00925

Chapter 3

Reviewing Cd, Hg and Pb Assessments and Effects in Elasmobranchs

Natascha Wosnick[1]* Aline Cristina Prado[1], Mariana Martins[2],
Liza Merly[3], Ana Paula Chaves[4], Neil Hammerschlag[5],
Oliver Shipley[6] and Rachel Ann Hauser-Davis[7]

[1]Programa de Pós-graduação em Zoologia, Universidade Federal do Paraná, Brazil.

[2]Programa de Pós-graduação em Ciências Fisiológicas, Universidade Federal do Rio Grande, Brazil.

[3]Department of Marine Biology and Ecology, Rosenstiel School of Marine and Atmospheric Science, University of Miami, USA.

[4]Programa de Pós-graduação em Toxicologia, Universidade de São Paulo, Brazil.

[5]Department of Environmental Research and Policy, Rosenstiel School of Marine and Atmospheric Science, University of Miami, USA.

[6]Biology Department, University of New Mexico, USA.

[7]Laboratório de Avaliação e Promoção da Saúde Ambiental, Instituto Oswaldo Cruz (Fiocruz), Brazil.

INTRODUCTION

Since the beginning of the industrial revolution, anthropogenic activities such as fossil fuel combustion and associated atmospheric deposition, mining, and

*Corresponding authors: n.wosnick@gmail.com and rachel.hauser.davis@gmail.com

smelting have dramatically increased the quantities of metals and metalloids entering marine environments (Förstner and Wittmann, 2012; Karimi et al., 2012). Due to their essential role in cellular homeostasis, many metals (e.g., Cu, Zn, Cd) persist at low concentrations in marine organisms, however, they can become toxic when they exceed threshold concentrations. Metals generally enter marine food-webs via direct assimilation by marine primary producers (i.e., phytoplankton and cyanobacteria), which can biomagnify concentrations by several orders of magnitude relative to the ambient water column (Lee and Fisher, 2016, 2017). Biomagnification and bioaccumulation continue to occur with subsequent trophic transfers, the rate at which is driven by a suite of environmental (e.g., temperature) and biological processes (e.g., zooplankton allometry) (Wu et al., 2020).

In marine organisms, relative cadmium (Cd), mercury (Hg), and lead (Pb) concentrations are of particular interest, owing to their long-term persistence and chronic toxicological effects on higher consumers (including humans) at relatively low threshold concentrations (Chunhabundit, 2016). Cadmium is a non-essential heavy metal that is most commonly associated with renal and hepatic dysfunction, though high exposure has been linked to a suite of cancers among other degenerative diseases (Genchi et al., 2020). In marine consumers, mercury (Hg) occurs mostly in methylated form of monomethylmercury ($MeHg^+$), which is highly lipophilic and cannot be readily metabolized by most taxa (Bloom, 1992; Harris et al., 2003). Exposure to high levels of Hg has been associated with chronic neurological effects such as sensory and motor dysfunction (Mergler et al., 2007), in addition to cardiovascular disease (Choi and Grandjean, 2008). Finally, Pb is a divalent cation that associates largely with proteins and typically mimics or competes with calcium (Morales et al., 2011). High Pb exposure can strongly impact the central nervous system and kidneys while preventing cellular respiration (Goyer, 1993; Pain et al., 2019; Papanikolaou et al., 2005).

In recent decades there has been a rapid proliferation of scientific studies describing the relative concentrations of Cd, Hg, and Pb in the tissues of higher marine consumers, especially those that comprise a substantial proportion of the global seafood trade (Fraser et al., 2013). Concentrations are often highest in large bodied, slow-growing species that exhibit high relative trophic positions, such as tunas (Kumar, 2018; Lee et al., 2016), swordfish (Barone et al., 2018), and elasmobranchs (Matulik et al., 2017; Rumbold et al., 2014; Shipley et al., 2021; Tiktak et al., 2020). Elasmobranchs (sharks, skates, and rays) are a speciose and globally distributed group of taxa (Carrier et al., 2012). Though commercial fishing has had detrimental impacts on the conservation status of many elasmobranch species (Dulvy et al., 2021), a second major threat relates to the fact that they also exhibit some of the highest tissue concentrations of Cd, Hg, and Pb throughout the oceans (Shipley et al., 2021). This is largely attributed to their slow rate of growth, longevity, and meso-to-apex trophic positions in marine food-webs (Cortes, 1999; Hussey et al., 2014). Many elasmobranch species can therefore serve as important sentinels for establishing regional (for resident species), basin-scale (for highly migratory species), and temporal trends in Cd, Hg, and Pb bioaccumulation through marine food-webs (Alves et al., 2016).

Given that the persistent consumption of elasmobranch meat can have detrimental impacts for human consumers, much of the existing literature has focused on describing the relative concentrations of Cd, Hg, and Pb in various elasmobranch species and tissues in relation to recommended exposure thresholds (e.g., Matulik et al., 2017; Merly et al., 2019; Shipley et al., 2021). However, the actual health impacts of high Cd, Hg, and Pb concentrations in elasmobranchs have not yet been well investigated, nor the physiological mechanisms to counteract high tissue concentrations considered highly toxic. A thorough mechanistic understanding of Cd, Hg, and Pb dynamics in elasmobranchs is therefore critical, given that elevated concentrations of these toxicants could potentially impact longer-term fitness (i.e., survival and reproductive success).

This chapter provides a global perspective on the current knowledge of Cd, Hg, and Pb dynamics in elasmobranchs. First, we synthesize current knowledge of the physiological (i.e., oxidative stress, homeostatic balance, systemic health, immune function, and gut microbiome composition) and potential behavioral and ecological consequences of high Cd, Hg, and Pb concentrations based on wild and experimental studies. Finally, we outline several avenues for future work that will provide greater understanding on how increasing concentrations of Cd, Hg, and Pb may impact the long-term fitness in this threatened group of animals as well as the implications of toxic metal exposure for elasmobranch conservation and human consumption.

PHYSIOLOGICAL OUTCOMES OF Cd, Hg, AND Pb EXPOSURE

Contaminant exposure has been linked to several biochemical alterations, comprising key biomarkers in assessing early exposure effects (Van der Oost et al., 2003). These minor alterations may lead to metabolic and physiologic imbalances, compromising individual fitness (Mearns et al., 2015). In this context, such measurements are paramount for predicting higher hierarchical level physiological effects in exposed populations, serving as biomonitoring tools (Monserrat et al., 2003). Till date, most studies on the physiological effects of Cd, Hg, and Pb contamination were performed with sharks, and data for batoids (i.e., rays, skates, guitarfish, and sawfish) are surprisingly scarce. This is of particular concern, as batoids represent the largest portion of Chondrichthyes diversity. Batoids are neglected in toxicological studies, as well as for most areas of research, where data on sharks is significantly higher. Moreover, batoids are more threatened (36%) than sharks (31.2%) or chimeras (7.7%) (Dulvy et al., 2021), indicating the urgent need to generate data for their management and conservation.

Concerning oxidative stress responses, most of the studies were carried out with the blue shark (*Prionace glauca*) possibly due to its high consumption by humans, which could explain the greater interest for toxicological investigations. While most physiological investigations on the negative effects of Cd, Hg and Pb on free-ranging sharks were performed with Carcharhiniformes (Alves et al., 2016; Wosnick et al., 2021a; Norris et al., 2021), studies under controlled conditions/exposure experiments were performed with Squaliformes, more specifically the spiny dogfish (*Squalus acanthias*) and the small-spotted catshark (*Scyliorhinuscanicula*).

Due to logistical challenges, most toxicological studies with wild populations are performed opportunistically, that is, from commercial fisheries (Alves et al., 2016; Wosnick et al., 2021a,b) (i.e. blue sharks, tiger sharks—*Galeocerdo cuvier*, nurse sharks—*Ginglymostoma cirratum*) or during scientific fishing campaigns (Merly et al., 2019) (i.e., great white sharks—*Carcharodon carcharias*). Such approach, although less controlled, allows the assessment of physiological effects under more realistic scenarios, besides the opportunity to investigate large-sized species, such as the great white shark, the tiger shark, and the bull shark (*Carcharhinus leucas*) (Merly et al., 2019; Wosnick et al., 2021a). However, there are some study limitations that need to be considered, as the complexity of natural environments and synergetic effects of the exposure to several contaminants makes real understanding of the physiological effects of a given element very challenging. In contrast, studies performed under controlled conditions (i.e., exposure to a given element in a set of concentrations) allow a more robust analysis of the physiological effects of exposure. However, exposure experiments can only be carried out with small-sized species (e.g., *Squalus* spp. and *Scyliorhinus* spp.) due to logistical limitations (Bernal and Lowe, 2015). Furthermore, lethal studies are categorically discouraged for both top predators (Hammerschlag and Sulikowski, 2011) (e.g., great white sharks, tiger sharks, and bull sharks) and threatened species (e.g., hammerhead sharks—*Sphyrna* spp., limiting the applications of controlled toxicological studies to inform elasmobranch conservation.

OXIDATIVE STRESS

Oxidative stress is the most assessed topic among the different biochemical effects of pollution on elasmobranchs, being largely applied to different species and associated with several contaminant classes, including toxic and non-essential metals (Alves et al., 2016; Barrera-Garcia et al., 2012, 2013; Hauser-Davis et al., 2021; Somerville et al., 2020; Vélez-Alavez et al., 2013). In this regard, inorganic contaminants are able to induce reactive oxygen species (ROS) formation through the Fenton Reaction or through depletion of non-enzymatic antioxidants and protein bonding (Valko et al., 2007). Superoxide dismutase (SOD) and catalase (CAT), for example, play a direct role in battling reactive oxygen species (ROS), whereas other enzymes such as glutathione peroxidase (GPx) and glutathione reductase (GR) are associated with non-enzymatic scavenger intracellular glutathione levels (Halliwell and Gutteridge, 2007). Proteins, such as metallothionein and vitamins, among others, also play important roles in antioxidant defenses. However, the antioxidant system, although efficient, may not be able to counteract extra ROS formation driven by environmental contamination, resulting in high ROS concentrations, which, in turn, lead to homeostatic imbalances, including molecular oxidation and signaling pathway disruption, termed oxidative stress (Jones, 2006). This may, in turn, damage the cell membrane, cellular constituents and, finally, DNA, consequently leading to cell death (Dittman et al., 2005).

Transcriptional regulation of superoxide dismutase, for example, has been reported as influenced by Cd exposure in the torazame catshark (*Scyliorhinustorazame*) (Cho et al., 2005a), whereas post-translational regulation seems to be modulated by different non-essential metals. Altered enzymatic antioxidant activities, on the other hand, have been directly associated with Cd, Hg and Pb exposure in different shark species. These elements, however, result in different effects depending on the assessed species and enzymes. For example, muscle GPx activity in the shortfin mako shark (*Isurus oxyrinchus*) has been negatively associated with Cd levels, whereas positively related to Pb (Vélez-Alavez et al., 2013). Kidney SOD and GPx activities were not modulated by any element. Barrera-Garcia et al. (2013) and Somerville et al. (2020) also observed a positive relationship between Cd and Pb concentrations and GPx and SOD activities in blue shark kidney samples and in the Atlantic sharpnose shark (*Rhizoprionodon terranovae*) muscle, respectively. In another assessment, however, liver Cd and Hg concentrations were shown to negatively influence GPx and SOD in Blue sharks (Alves et al., 2016), demonstrating variable inter-species and tissue enzymatic responses to non-essential metal exposure.

Non-enzymatic antioxidant content is considerably high in elasmobranchs, remarkably in sharks. Glutathione is an essential oxyradical scavenger, although other thiol-rich proteins (hemoglobin) and antioxidants (vitamin E) also buffer ROS generation (Lopéz-Cruz et al., 2011). Despite this, most studies evaluating metal exposure to glutathione depletion did not detect any significant correlations, suggesting that GSH may not play such an important role in metal-induced oxidative stress in this taxonomic group. For instance, glutathione levels were not correlated to Hg levels in the Atlantic sharpnose shark, nor to several non-essential elements in Atlantic nurse sharks (*Ginglymostoma cirratum*) (Wosnick et al., 2021b) and blue sharks (Alves et al., 2016; Hauser-Davis et al., 2021). Glutathione also seems to have little effect on Hg-induced toxicity in little skate (*Leucoraja erinacea*) isolated hepatocytes (Ballatori et al., 1988; Ballatori and Boyer, 1996).

Differential macromolecular oxidative damage markers, also varying among species and tissues, have been reported in elasmobranch ecotoxicological assessments, mostly regarding lipid and protein damage. Barrera-Garcia et al. (2012), for example, observed sexual differences for protein carbonyl groups in the muscle of blue sharks, whereas antioxidant enzymatic activity was similar between sexes and maturity stages. Barrera-Garcia et al. (2013), however, reported similar damage among sexes and ontogenic stages, but significantly higher in hepatic tissues. Despite this, the authors did not correlate damage with contaminant levels. Also in blue sharks, DNA damage was positively related to a set of elements, which was related to inhibition of GPx and SOD activities (Alves et al., 2016). Shortfin mako sharks also had a more conspicuous lipid and protein oxidation in liver, whereas macromolecules damage was more subtle in muscle (Vélez-Alavez et al., 2013). Lipid peroxidation in neuronal tissue was also associated with Hg levels for the velvet belly lanternshark (*Etmopterus spinax*) (Rodrigues et al., 2022). Finally, hepatic damage represented by a decrease in lipid deposition and higher melanomacrophages was suggested as a result from lipid peroxidation in blacktip sharks (*Carcharhinus limbatus*) (Norris et al., 2021), but such assumptions are speculative due to the lack of oxidative damage analysis.

Metallothionein (MT) are low molecular weight cysteine-rich cytosolic proteins associated with metal exposure due to their role in homeostatic regulation and metal detoxification (Kägi, 1991). Other functions, however, such as free radical scavenging are also associated with these proteins (Coyle et al., 2001; Livingstone, 1993). This metalloprotein is commonly employed as a specific biomarker for metal exposure (Van der Oost et al., 2003). However, the role of MT against metal toxicity in elasmobranchs is still controversial. Some studies, for example, indicate that overall MT detoxification mechanisms concerning toxic metals do not seem to be predominant in sharks (Company et al., 2010; De Boeck et al., 2010; Vas et al., 1993; Walker et al., 2014; Wosnick et al., 2021b). Other assessments, however, have reported direct MT detoxification for several metals. Examples includethe reports of MT associations in a dose-dependent manner following Cd exposure in torazame catshark (*Scyliorhinus torazame*) liver and kidney samples, observed in the form of temporary transcriptional MT induction in kidney and constant in liver (Cho et al., 2005b). The same was reported for small-spotted catshark (*Scyliorhinus canicula*) liver (Hidalgo et al., 1985; Hidalgo and Flos, 1986), and testis (Betka and Callard, 1999). Metallithioneins were also associated to Cd, Hg and Pb concentrations in liver and Cd and Pb in muscle tissue in blue sharks (Hauser-Davis et al., 2021). Furthermore, some assessments, although not specifically analyzing this metalloprotein, have noted imbalances in essential metals with increasing toxic metal burdens, such as Cd, during the growth stage of the tiger shark (*Galeocerdo cuvier*), postulating that this may be due to the induction of MT synthesis due to high hepatic Cd burdens and the subsequent binding of these metals to metallothionein (Endo et al., 2015). A summary of molecular damage and main enzymes affected by ROS is provided in Figure 3.1.

Biomarkers of neuronal damage and energetic metabolism have also been applied to elasmobranchs ecotoxicological assessments. Alves et al. (2016) analyzed lactate dehydrogenase (LDH) and isocitrate dehydrogenase (IDH) in muscle samples of blue sharks but did not observe direct relationships between these enzymes function and Cd, Hg and Pb. Additionally, muscle and brain cholinesterase (ChE) activity were measured but were also not related to contaminant levels, questioning their suitability as biomarkers. Rodrigues et al. (2022) also assessed acetylcholinesterase (AChE) responses as a function of brain Hg levels in the velvet belly lanternshark (*Etmopterus spinax*) and found potential neurotoxic effects related to decreased AChE activity. The same authors also assessed the aerobic metabolism through the electron transport system, which was negatively affected by Hg, thus explaining the observed oxidative imbalance found for this species. Nam et al. (2010) found no correlation between Hg levels and neurochemical features, especially ChE activity and transcription in lemon sharks

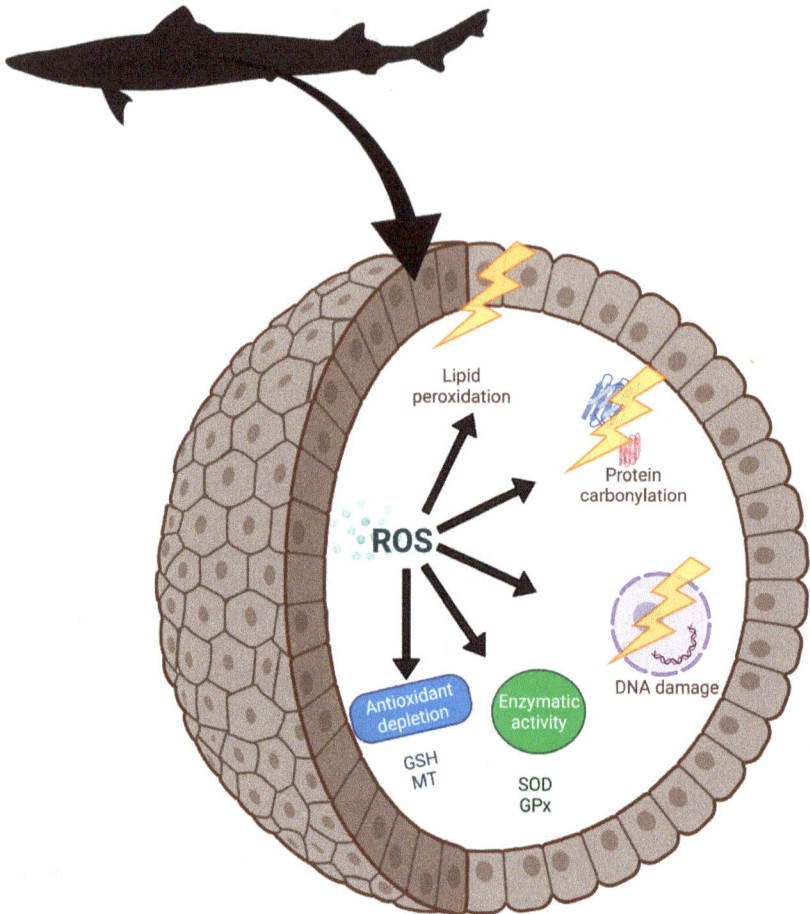

Figure 3.1 Oxidative stress outcomes reported for metal exposure in elasmobranchs. Main biomarkers affected by reactive oxygen species driven by metal exposure in elasmobranchs. Created with BioRender.com

(*Negaprion brevirostris*). Cholinesterases activity should, however, be carefully considered as a biomarker since muscular ChE activity was not considered a good biomarker in sharks (Solé et al., 2008). Besides, shark ChE is targeted by specific contaminants, such as organophosphate pesticides (Alves et al., 2015). Nevertheless, more studies would better elucidate the effects of metals in neuronal activity since Hg is considered a neurotoxic element (Zheng et al., 2019).

HOMEOSTATIC BALANCE AND SYSTEMIC HEALTH

For aquatic organisms, osmo-ionic and acid-base pathways are the main allostatic components involved in the maintenance of the homeostatic balance (Cameron and Iwama, 1989; Edwards and Marshall, 2012). In the presence of abiotic and/or biotic stressors, physiological stability can be affected, and compensatory mechanisms are needed to re-establish the homeostatic balance. In the case of elasmobranchs, some distinct mechanisms of osmo-ionic regulation are observed. More specifically, sharks and batoids are ureotelic and rely on urea (along with Trimethylamine N-oxide; TMAO) as the main osmolyte for osmoregulatory processes (Ballantyne and Fraser, 2012; Yancey and Somero, 1979). Unlike bony fish, NaCl plays a secondary role in the osmotic balance of this taxonomic group, as well as the gills (Pang et al., 1977; Wosnick and Freire, 2013). In elasmobranchs, the rectal gland is the main structure involved in osmo-ionic regulation, which makes this organ particularly vulnerable to exogenous substances with toxic potential (Wosnick et al., 2021a). In the molecular level, elasmobranch osmo-ionic control is performed by a complex of transmembrane proteins, transporters/cotransporters, and ion channels, highlighting the Na^+/K^+-ATPase (NAK), the Na-K-2Cl cotransporter (NKCC), and chloride channels (CFTR) (Ballantyne and Robinson, 2010). As for the acid-base balance, the gills are the main structures involved in regulation, through mechanisms directly linked to ionic regulation, i.e., the absorption of Cl^- is coupled with the excretion of HCO^{3-} through the Cl^-/HCO^3 exchanger, and the absorption of Na^+ coupled with the excretion of H^+ through the Na^+/H^+ exchanger (Evans, 1984). Carbonic anhydrase also plays an important role in pH homeostasis, through the catalysis of carbon dioxide and water into carbonic acid (H_2CO_3), which dissociates into hydrogen (H^+) and bicarbonate (HCO^{3-}) ions, serving as substrates for biochemical regulatory processes in elasmobranch gills (Ballantyne and Robinson, 2010; Evans, 1984).

Till date, most toxicological studies concerning elasmobranch health have focused on the physiological processes cited above, highlighting the negative impacts that toxic metals have on the homeostatic balance of exposed animals. The most commonly used markers are components of plasma associated with osmo-ionic regulation, coupled with assessments of Cd, Hg, and Pb in the muscle, liver, kidneys, gills and rectal gland (Eyckmans et al., 2013; Jacoby et al., 1999; Kinne-Saffran and Kinne, 2001; Silva et al., 1992; Wosnick et al., 2021a). Recently, more attention has been given to other metrics of systemic health, more focused on the impacts that metal accumulation may have on proper organ functioning (Norris et al., 2021; Wosnick et al., 2021a). Understanding the systemic effects of pollution is imperative, as this holistic approach helps to assess the impacts of contamination on a macroscale (Lehnert et al., 2018; Maceda-Veiga et al., 2015; Moore, 2004). Even through the use of biochemical markers, it is possible to elucidate the effects of contamination at the systemic level, and further studies with this approach are needed. In fact, biochemical assessments such as analysis of ions, proteins (e.g., albumin), metabolites, and enzyme activity (i.e., alanine aminotransferase ALT, aspartate aminotransferase AST, Na^+-K^+-ATPase), bring important information on acid-base balance, osmo-ionic and hydro-electrolytic regulation, impairment in both liver and kidneys, as well as respiratory function (Monserrat et al., 2007; Maceda-Veiga et al., 2015). Furthermore, these markers can also indicate chronic stress due to contamination at the secondary level (Skomal and Mandelman, 2012).

Most studies focused on the negative impacts of exposure on osmoregulation processes. In the small-spotted catshark (*Scyliorhinus canicula*), exposure to Cd alters plasma ion dynamics, causing an increase in potassium and calcium, both predominantly intracellular ions, which may indicate cell membrane rupture (De Boeck et al., 2010). In the spiny dogfish (*Squalus acanthias*), HgCl$_2$ and Hg$_2^+$ have the potential to inhibit the activity of the NKCC in the rectal gland (Kinne-Saffran and Kinne, 2001; Silva et al., 1992), as well as chloride secretion, and subsequent impairment of homeostatic balance (Jacoby et al., 1999; Ratner et al., 2006). In the little skate hepatocytes, Hg was proven to be more deleterious than Cd, compromising membrane permeability (Ballatori et al., 1988). Inhibition of L-alanine transporters in the little skate exposed to HgCl$_2$ is possibly driven by either direct interaction with the proteins—possibly with reduced sulfhydryl groups, or Na$^+$ permeability, which in turn has a negative effect in these transporters (Seelinger and Kalka, 1991). Taurine channels modulation was also attributed to HgCl$_2$ exposure, through direct inhibition of the channels or cellular ATP depletion (Ballatory and Boyer, 1996), leading to cell swelling and inhibition of osmotic regulation. Since cellular volume signals other functions, Hg might impact other regulatory pathways in elasmobranchs.

In free-ranging coastal sharks, Hg has also the potential to affect phosphorus regulation in the rectal gland, leading to increased concentrations in serum compatible with cell membrane disruption (Wosnick et al., 2021a). Besides, Hg accumulation in the gills of free-ranging sharks increases serum lactate concentrations, causing loss of the acid-base balance and leading to respiratory acidosis (Wosnick et al., 2021a). Although less marked, alterations in osmoregulatory processes were also observed upon Pb accumulation in experimentally exposed spiny dogfish, with transient loss of acid-base balance and impaired Na$^+$/K$^+$-ATPase activity in the gills, and decrease in NaCl concentrations in the plasma coupled with high rates of urea excretion and short-term increase in blood pH (Eyckmans et al., 2013). Unlike Hg, no activity inhibition was observed in the rectal gland transporters, indicating that the negative effects of Pb exposure are more evident in the gills.

There is also evidence of cardiovascular impairment due to metal exposure. In the spiny dogfish, Cd can lead to vasoconstriction of the ventral aorta, while Hg exhibits a marginally constrictive effect, and Pb has no effect on the species vascular tension (Evans and Weingarten, 1990). However, as stated by the authors, it is possible that the marginal effect of Hg and lack of effect of Pb may be due to tissue isolation, with no exposure performed in the vascular endothelium. Cd has also the capacity to impair heart functioning in sharks, as also reported for the spiny dogfish. More specifically, this toxic metal alters ventricular developed pressure (VDP), and ventricular depolarization (QRS complex) (Wang et al., 2010). Under controlled conditions, such effects are reversible. The alterations detailed above have the potential to affect free-ranging individuals for which intervention protocols are not feasible. Yet, the consequences at both individual and population-level are still unknown and deserve further attention in future investigations.

The toxic effects of non-essential metal exposure can also impair the systemic health of elasmobranchs. Concerning metabolism, for the spiny dogfish, exposure to Cd may lead to increases in plasma glucose and lactate with no recovery to basal levels, indicating energy mobilization through gluconeogenesis and metabolic acidosis (Tort and Torres, 1988). At the biochemical level, metals also impair the enzymatic kinetics related to energy metabolism that can aid oxidative protection against environmental toxicity. The metals Cd_2^+ and Hg_2^+, for example, have inhibitory effects and alter the activity of paraoxonase in the small-spotted catshark (Sayin et al., 2012), leading to metabolic impairment upon exposure. Non-essential metal accumulation also has the potential to affect proper organ functioning. In free-ranging coastal sharks, increased concentrations of hepatic Hg are related to increased activity of alkaline phosphatase (AK), indicating impairment on liver functioning (Wosnick et al., 2021a). Although no correlation was observed between liver Hg concentrations and alanine aminotransferase (ALT) activity in the blacktip shark, increased activity was positively correlated to histological alterations in the liver, considered compatible with high hepatic Hg concentrations, indicating that increased ALT in the species may be an indirect effect of Hg bioaccumulation (Norris et al., 2021). Physiological effects of exposure to Cd, Hg and Pb are summarized in Figure 3.2.

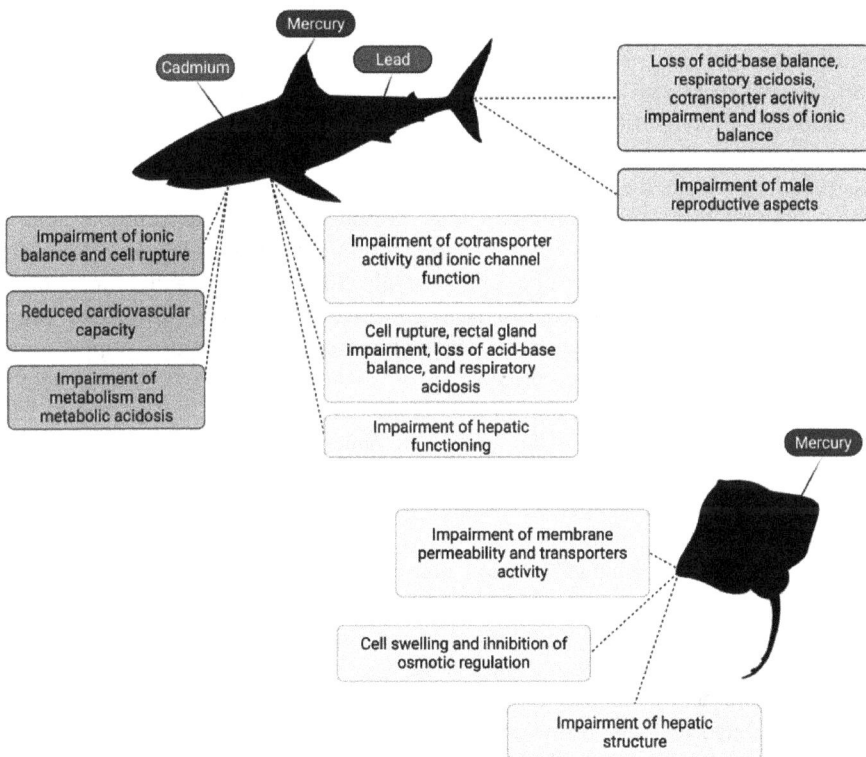

Figure 3.2 Negative impacts on physiological pathways reported for elasmobranchs. A summary is provided of the reported impacts that Cd, Hg, and Pb exposure have on sharks and rays' physiology. Created with BioRender.com

Despite the importance of reproduction for species recruitment and subsequent conservation, the impacts of toxicology on elasmobranch reproductive physiology remain unknown. In fact, so far, reproductive effects associated with toxic metals were only observed in experimentally exposed male spiny dogfish (*Squalus acanthias*) to Cd. Time-dependent cell death was observed in post-meiotic germinative cells, suggesting that Cd induces apoptosis during spermiogenesis. Blood-testis barrier was also negatively affected, potentially due to tight-junctions disruption (McClusky, 2006). Metallothionein levels were also higher in immature germinative cells of the same species (Betka and Callard, 1999). Thus, future investigations should focus on unraveling the effects of toxic metals in reproductive aspects of sharks and rays, aiming to elucidate not only which physiological pathways can be harmed, but also be associated with histology studies that indicate damage at the cellular and tissue level.

IMMUNE FUNCTION

Elasmobranchs are dependent on their immune systems to protect them against pathogens and diseases as well as to maintain their homeostatic state (Janeway and Medzhitov, 2002; Smith et al., 2014). In wild populations, immune function can be used to assess the health of elasmobranchs and relate those assessments to population dynamics and environmental variables (Keller et al., 2000). Environmental disturbances such as capture stress, climate change, habitat degradation, diminished water quality, light/noise pollution, and exposure to heavy metals and other pollutants may present acute and chronic stressors to which elasmobranch populations are responding (Consales and Marsili, 2021; King et al., 2011; Pereira Santos et al., 2021; Skomal and Mandelman, 2012; Van Rijn and Reina, 2010). Heavy metals in particular have long been studied as contaminants that have substantial impacts on immunosuppression and immunoregulation which in turn could make fish living in heavy metal-contaminated waters more susceptible to disease (Zelikoff, 1993). The metals Cd, Hg, and Pb are all metal ions that have been shown to affect vertebrate immune system functioning and changes in immunocompetence (Lehmann et al., 2011). Measures of immunocompetence among individual elasmobranchs may reveal trade-offs that exist within populations (Luster et al., 1992; Lochmiller and Deerenberg, 2000). Immune function may vary within and across species and seasonally, even within a single population, but reference values can be established to identify biomarkers of compromised health status (i.e., a decrease in optimal function), brought on by both short-term and long-term exposure to organic pollutants and heavy metals (Atallah-Benson et al., 2020). Health can be defined as being related to both intrinsic and extrinsic factors, where immune function is a primary driver of health outcomes for both individuals and populations (Figure 3.3).

A central question to examining the relationship between environmental metal exposure and immune status is developing methods that allow sampling from live wild animals. Multiple challenges persist in finding useful biomarkers for immune function in response to metallic stress in elasmobranchs. For example, melanomacrophage centers (MMCs) are pigmented aggregates of macrophages and lymphocytes that are common in the head kidney, spleen, and other immunological organs in bony fish (Agius and Roberts, 2003). They are used routinely in toxicology studies to evaluate the immune reactivity to metal stressors (Steinel and Bolnick, 2017). However, unlike bony fish, elasmobranchs have diffused MMCs restricted primarily to the liver and their role in immune function and reactivity to environmental stressors is not well understood (Borucinska et al., 2009; Steinel and Bolnick, 2017). The current goal should be to apply evidence gathered from the comparative study of elasmobranch immunobiology, as model organisms and from a phylogenetic perspective, to the development of assessment methods that can address eco immunological research questions about elasmobranch health.

Elasmobranch health is dependent upon a robust set of both innate and adaptive immune responses that are similar to that observed in higher vertebrates, such as mammals, and that provide elasmobranchs substantial protection from disease (Smith et al., 2019). Innate immune responses are those that are considered broadly non-specific and that the animal is capable of mounting initially each time it encounters a foreign invader such as a microbe (Akira et al., 2006). Adaptive responses are those that are initiated once a pathogen (microbe that causes disease) has been recognized and targeted by the immune system and it involves a highly specific response to individual foreign molecules and provides the host animal with long-term protection against that pathogen in the future (Flajnik and Rumfelt, 2000).

An elasmobranch's first line of defense is their tough skin, covered in dermal denticles, which presents a significant physical barrier to infection (Raschi and Tabit, 1992). These placoid scales have specialized, micropatterned arrangements that decrease the incidence of bacterial invasion and have been used as a template in engineering antibacterial surfaces (Mann et al., 2014). Elasmobranch skin also contains a robust layer of mucus, consisting of lysins, agglutinins, and antimicrobial factors (Alexander and Ingram, 1992; Hinds Vaughan and Smith, 2013). These molecules along with the architecture of elasmobranch epithelial surfaces are believed to minimize the rate of infection by various pathogens present in the marine environment. It is unknown if elasmobranch skin may also contain molecular and physical factors that protect them against heavy metal exposure. In freshwater fish, mucus and scales have been shown to chelate heavy metals like lead and mercury and remove them from water. The addition of scales to toxicology experiments where fish were exposed to lead and mercury made several species tolerant to doses considered lethal (Coello and Khan, 1996). The level of mucus production in elasmobranch epithelium is highly variable between species and has not been as well characterized as that found in bony fish. However, many of the same glycoproteins, enzymes, and immune factors are present (Reverter et al., 2018).

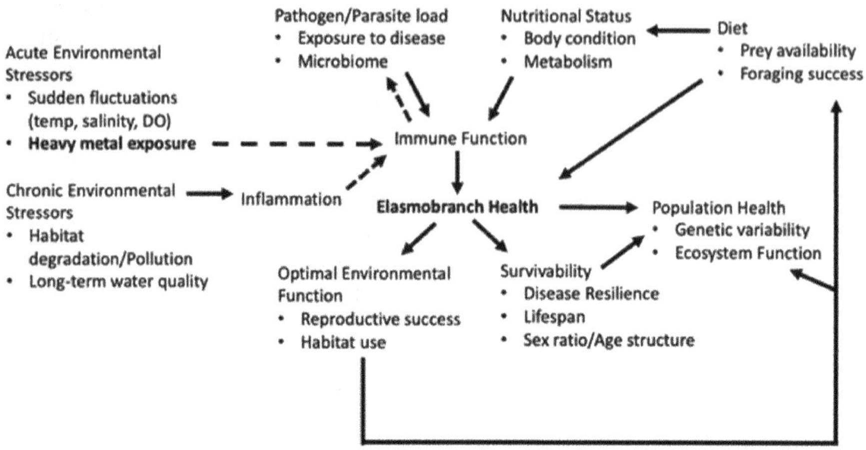

Figure 3.3 Impacts of environmental stressors on immune function and elasmobranch health. A summary is provided of those factors believed to be critical in an integrated approach to assessing the impacts of heavy metal exposure and other environmental stressors on immune function in elasmobranch populations. Dotted lines (--->) indicate down-regulation whereas solid lines (→) indicate up-regulation.

Another biological component of defense can include an animal's microbiome and studies of elasmobranch microbiome composition have increased substantially, specifically those characterizing the microbiota in epidermal surfaces (Caballero et al., 2020; Doane et al., 2017). Recent studies have also investigated the role of various commensal bacteria in providing protection by degrading organic pollutants and heavy metals such as mercury (Mahbub et al., 2016). A recent study found that one such commensal bacteria, Pseudoxanthomonas, was prevalent in the mouth and skin of bull sharks. Bull sharks inhabit near-shore and estuarine waters and are capable of moving into freshwater for extended periods of time. This wider range of habitat preferences may leave them susceptible to areas that have more pollutants, and the microbiome may play a role in providing protection (Black et al., 2021). Further investigation into the role of the elasmobranch microbiome in defense against exposure to heavy metals is needed.

Immune responses in elasmobranchs are initiated when a microbial pathogen or homeostatic disruption is detected within an animal's tissues. Cellular signaling induces recruitment of innate immune cells and inflammation at sites of infection, injury, or stress (Akira, 2008; Hanada and Yoshimura, 2002). Chemokines (receptor-mediated chemical substances that attract immune cells) are released by responding cells that help recruit neutrophils, monocytes, and other immune responsive cells responsible for cell-mediated responses, including phagocytosis (Sokol and Luster, 2015). Responding cells express receptors that recognize microbial ligands and are induced to produce additional signaling molecules like cytokines. Cytokines are cell to cell messenger molecules that up-regulate or down-regulate the activity of immune cells in response to infection. Several cytokine and chemokine genes and gene families have been identified in elasmobranchs and are believed to serve a similar regulatory role in the orchestration of immune responses and the maintenance of homeostasis in these animals as they do in other vertebrates (Li et al., 2015; Redmond et al., 2019).

Homeostasis and immune function are tightly linked and can be evaluated using hematological assays for soluble and cellular components of the blood and their responses to extrinsic factors. They are routinely used in clinical evaluations to assess physiological status in an organism (Grant, 2015; Schmaier, 2019). Hematocrit is used to measure the amount of blood cells relative to plasma and indicates how successfully the animal can produce erythrocytes (Van Beaumont, 1972). Erythrocytes play a critical role in oxygenation of the body's tissues because of their hemoglobin content as hemoglobin binds and transports oxygen. It is, therefore, important to assess the number of erythrocytes and their physiological function in the form of hemoglobin level (Van Kampen and Zijilstra, 1966). Low levels of hemoglobin could be related to low erythrocyte counts, dysfunctional erythrocytes, or genetically altered hemoglobin and may point to decreased health or a medical condition, such as anemia. Similarly, total blood counts are used clinically to assess immune status and health. Evaluating blood cell parameters is likely an important target tissue for determining the impacts of heavy metals on immune status as metals accumulate in the internal organs of elasmobranchs through the blood after dietary uptake (Pethybridge et al., 2010).

In heavy metal studies, hematology has often been used as the outcome being measured in response to exposure and/or concentrations of metals found in the tissues because blood collection is minimally invasive and can provide substantial information on overall well-being. Hematological parameters have been shown to be useful biomarkers of physiological stress in response to metals in fish species as there are substantial variations in blood water content with metal exposure (Dhanapakiam and Ramasamy, 2001; Vinodhini and Narayanan, 2009). Tort and Torres (1988) conducted a study in the small-spotted catshark, (*Scyliorhinus canicula*), to measure hematological responses to sublethal Cd levels. Decreases in both hematocrit (proportion of erythrocytes in whole blood) and leukocrit (volume percentage of leukocytes in whole blood) were observed in fish exposed to Cd for 24 and 96 hours while hemoglobin concentration was increased after 24 hours and then returned to low levels after the 4-day exposure. An important conclusion from this study was that the proportion of erythrocytes in the blood was lower, but that did not appear to be the result of hemolysis since total counts remained elevated (Tort and Torres, 1988). This indicates that the overall blood volume was altered by the metallic stress. The number of leukocytes were also decreased which could indicate an immunosuppressive effect of Cd. However, differential leukocyte counts were not performed so it is not clear what subset of immune cells were impacted in this experiment (Tort and Torres, 1988).

Total blood leukocyte counts and differential counts serve as useful indicators of immune status in wild populations being sampled, where the only available tissue sample is blood. Total leukocyte counts were assessed in great white sharks (*Carcharodon carcharias*) in South Africa and related to exposure levels to various heavy metals including Cd, Pb, and Hg. Differential counts were used to determine the granulocyte to lymphocyte ratio (GLR). The results suggested no immunotoxic effects in white sharks as there was no correlation between counts and heavy metal blood concentrations. It is difficult to assess immunotoxicity from single measures, however, and in this study, counts were only available for a subset of sharks (Merly et al., 2019). In mammals, the neutrophil to lymphocyte (NLR) ratio is used to indicate immune status as shifts outside the normal range are indicative of infection or compromised health. In elasmobranchs, there are at least three different prominent granulocytic cells present in peripheral blood, so using a granulocyte to lymphocyte ratio is more appropriate than counting neutrophils alone. Reference intervals for total blood counts, differential counts, and GLR are now being established for various species of elasmobranch as there is substantial interspecies variation in the morphology and cell types in peripheral blood (Atallah-Benson et al., 2020). This will allow future investigators to apply these measures to ecotoxicology studies in specific species.

In other vertebrates, Cd is known to accumulate in immune cells entering via calcium channels and binding to intracellular components (Marchetti, 2013). Its impact on lymphocytes and other immune cells is variable as it has been shown to induce apoptosis in a dose-dependent fashion by upregulating caspase enzymes. It can also induce inflammatory responses by activating a number of signaling pathways, including the nuclear factor kappa B (NF-kB) and mitogen-activated protein kinase (MAPK) pathways, that are central to immunological regulation and the production of inflammatory mediators such as cytokines and chemokines (Desforges et al., 2016; Wang et al., 2021). Cadmium has been shown to induce the production of inflammatory cytokines and chemokines in vertebrate immune cells as well as recruiting neutrophils and macrophages to tissue sites which can cause inflammation and tissue damage (Hossein-Khannazer et al., 2020).

Lead can also specifically block and/or bind to calcium channels and impact calcium-dependent proteins. Both Cd and Pb can mimic calcium and zinc at their specific sites in mammalian cells (Marchetti, 2013). In elasmobranchs, several key immunoregulatory molecules and innate immune mediators are calcium dependent. In particular, shark complement defenses are dependent on macromolecular structures that require calcium ions. Complement is a specialized defense system important to immune function (Nonaka and Smith, 2000). It is a system of proteins that are sequentially activated in a cascade and bind to specific receptors on immune cells. A complement cascade can be activated in several ways once foreign substances are recognized (Müller-Eberhard, 1988). Complement cascades release bioactive peptides that can bind to microbial targets making them more susceptible to immune reactivity and more likely to undergo phagocytosis. Complement can also directly clear infections because activated proteins can assemble onto target bacterial membranes (membrane attack complexes or MAC) and induce the formation of pores, effectively lysing and killing bacteria (Müller-Ederhard, 1988; Smith, 1998; Takeda et al., 2003; Zou and Secombes, 2016). Complement is also considered a bridge between innate and adaptive immune defenses (Dunkelberger and Song, 2010). Metals that disrupt calcium availability may significantly impact complement activation and regulation. Further study on complement reactivity in elasmobranchs exposed to metals is necessary.

Mercury toxicity in fish has not been as well studied as in other vertebrate groups, but its role in inducing oxidative stress can have downstream effects including immune cell damage and apoptosis (Morcillo et al., 2017). It has also been shown to disrupt enzymatic function which could severely compromise immune regulation. Most studies in bony fish have focused on immunosuppressive effects, but mercury immunotoxicity may include diverse mechanisms. In some fish, induction of inflammatory cytokine production has been observed as well as enhanced lysozyme activity (Begam and Sengupta, 2015; Kong et al., 2012; Sanchez-Dardan et al., 1999). Exposure to some metals may have immuno-stimulatory effects that could have significant health consequences for long-lived vertebrates such as elasmobranchs. For example, Hg has been shown to induce increases in immune reactivity that leads to autoimmunity and hypersensitivity in mammals (Crowe et al., 2016). To date, there is very little information about the immunotoxic effects of mercury exposure in elasmobranchs.

As long-lived vertebrates that occupy mid to high trophic levels in marine food webs, the effect of Cd, Pb, and Hg on elasmobranch immunocompetence may be complex and multifaceted. The immunobiology of elasmobranchs is similar to other vertebrates in that it develops slowly and, therefore, there are significant differences in the immune reactivity in juveniles and adults of most species studied to date (Haines et al., 2005; Rumfelt et al., 2002). Furthermore, the few studies that have investigated the impacts of heavy metals on elasmobranch health have revealed that there may be substantial variation among species in their ability to mitigate high levels of metal stress, which is not unexpected given the millions of years of divergence between various elasmobranch lineages. Recent genomic studies have confirmed substantial variation in the genomic size and composition among the evaluated species (the brownbanded bambooshark—*Chiloscylium punctatum,* the whale shark—*Rhincodon typus*, and the torazame catshark—*Scyliorhinus torazame*) (Hara et al., 2018; Kuraku, 2021). There are significant gaps in our understanding of the impacts of heavy metals and other pollutants on elasmobranch immune status and health. Further inquiry is required given that for most species studied thus far, heavy metal concentrations are typically very high. Sub-lethal impacts of metals on immune function should be considered from a conservation point of view as they may have consequences at the community and ecosystem level.

POTENTIAL BEHAVIORAL AND ECOLOGICAL OUTCOMES OF CD, HG, AND PB EXPOSURE

To our knowledge, no studies have investigated the effects of Hg, Cd, and Pb exposure on the behavior of elasmobranchs. We suspect this is likely due to the logistical and ethical challenges of conducting such studies on relatively large, mobile, and threatened species, which would require controlled experimentation in laboratory settings.

Studies evaluating the effects Hg, Cd, and Pb on the behavior of teleosts, however, provide some insights on possible negative impacts of these metals on elasmobranch behavior. In a variety of studies, exposure of freshwater and saltwater fishes to Hg, Cd, and Pb has been found to impact several behaviors, including those associated with predator avoidance, reproduction, foraging, and sociality (reviewed in Scott and Sloman, 2004); some notable examples are as follows. Exposure of rainbow trout *Oncorhynchus mykiss* to Cd has been found to eliminate their normal antipredatory responses to chemical predation cues (Scott et al., 2003), while dietary intake of MeHg has been found to impair schooling behaviors of golden shiners (*Notemigonus crysoleucas*) exposed to visual predator cues (Webber and Haines, 2003). Exposure of larval mummichogs (*Fundulus heteroclitus*) to MeHg has been found to increase collision with conspecifics within their schools as well as increase unnecessary parallel swimming (Ososkov and Weis, 1996). Social behaviors, such as agnostic interactions and the formation of dominance hierarchies, have been found to be impaired due to Cd exposure (Almeida et al., 2009; Sloman et al., 2003). Lead exposure has been found to reduce spawning frequency in male fathead minnows (*Pimephales promelas*) (Weber, 1993) and dietary intake of MeHg has been shown to delay their onset of spawning (Hammerschmidt et al., 2002). Experiments with yellowfin bream (*Acanthopagrus australis*) fed inorganic Hg, exhibited increased feeding attempts, but lower success rate than controls over short time periods (Harayashiki et al., 2018). Following single and co-exposure of Cd and Hg, adult zebrafish (*Danio rerio)* have been found to decrease their exploratory behavior (Patel et al., 2021). Both Hg (Vieira et al., 2009) and Cd exposure (Eissa et al., 2010) have shown to impair swimming performance in several teleost species (Hg: *Pomatoschistus microps*; Cd: *Cyprinus carpio, Australoheros facetum, Astyanax fasciatu*).

Based on the well-documented impacts of heavy metal toxicity on the behaviors of teleost fishes, it is plausible that such effects could also occur in elasmobranchs. The potential negative effects of Hg, Cd, and Pb behaviors associated with predator avoidance, foraging, and reproduction, is particularly concerning as these impacts could have demographic and ecosystem consequences. For example, in the case of lower trophic level elasmobranchs, the inability to respond appropriately to predators (most likely larger elasmobranchs) could increase their vulnerability to predation, thus increasing natural mortality rates. In turn, this increased predation of contaminated elasmobranchs by higher trophic level species could lead to increased bioaccumulation and biomagnification rates and the associated negative consequences. Given a total paucity of research to date on the potential effects of Hg, Cd, and Pb on elasmobranch behavior, this is a clear research priority.

Since elasmobranchs are often upper-level predators, they can impact a diverse range of ecosystem functions and services through top-down effects, including regulating food webs, cycling nutrients, engineering habitats, transmitting diseases/parasites, mediating ecological invasions, affecting climate, supporting fisheries, generating tourism, and providing bioinspiration (See review by Hammerschlag et al., 2019). As such, mortality or sublethal impacts from Hg, Cd, and Pb toxicity could alter these ecosystem functions and services. For example, rays support biogeochemical cycling by excavating the seafloor feed on buried infauna that in turn enables oxygen and organic matter to penetrate deeper into sediments (Lohrer et al., 2004, O'Shea et al., 2012). Declines in ray populations or impairments to their physiology, behavior or health from exposure to Hg, Cd, and Pb, could thus impact biogeochemical cycling processes. While overfishing represents the greatest extinction risk to elasmobranchs globally (Dulvy et al., 2021), any mortality or fitness loss attributed to Hg, Cd, and Pb exposure would exacerbate the threats faced by elasmobranch populations, especially coastal species that may be disproportionately exposed to these heavy metals due to proximity to human activities.

IMPLICATIONS FOR ELASMOBRANCH CONSERVATION AND FUTURE DIRECTIONS

The International Union for Conservation of Nature (IUCN) red list recognizes the role that contaminants play in population size reduction, addressing this specific threat in the criteria A (subcriteria e) when assessing the risk of extinction of a species (IUCN, 2012). Till date, pollution has proved to be a significant threat for only a small portion (6.9%) of threatened elasmobranchs. However, some anthropic activities related to metal pollution (i.e., urbanization and mining) contribute to habitat loss and degradation, a threat that affects 18.7% of assessed species (Dulvy et al., 2021). So, it is possible that other species are suffering from the indirect effects of metal contamination, but data is not available to confirm such a statement.

Conservation physiology is a research field that aims to investigate physiological responses of species to anthropogenic disturbances that might contribute to population size reduction (Wikelski et al., 2006). Till date, most studies on elasmobranchs focus on the physiological effects of capture or climate change, but little attention has been given to pollution and its potential to negatively affect sharks and rays. Within the conservation physiology field, pollution was recently identified as a priority research theme to inform conservation policy and practice (Cooke et al., 2021). According to the authors, it is necessary to elucidate the physiological responses to heavy metals and other environmental toxicants. In fact, such a question is of great relevance, as data on the physiological effects of the most traditional toxic metals (Cd, Hg, and Pb) is scarce, as highlighted by this chapter. According to the authors, it is also necessary that the effects of pollution on population dynamics (e.g., growth, reproduction and survival) be elucidated, to inform policy makers on mitigation interventions (Cooke et al., 2021).

That said, it is now necessary that toxicology be recognized as a research field relevant for elasmobranch conservation, and toxicological investigations propose not only to describe concentrations of toxic non-essential metals in several tissues, but also to advance in the understanding of the harmful effects of contamination on survival, health and reproductive potential, focusing not only on threatened species, but also on those that are residents (endemic or not) of heavily polluted regions, as it is possible that the negative effects of pollution are already occurring, even at a regional population level. It is also necessary that the effects of metal contamination on elasmobranch ability to cope with additional stressors (e.g., overfishing, predation, habitat loss, and climate change) be evaluated, as some pollutants have the potential to attenuate acute stress response in cartilaginous fish (i.e., the round stingray (*Urobatis helleri*); for more information see Lyons et al., 2019).

The human component should also be considered, as elasmobranch fin and meat consumption are a global issue (Dent and Clarke, 2015). As Cd, Hg and Pb are known for their significant toxic effects on humans, the consumption of fins and meat containing high concentrations of these metals can represent a risk to human health and a food safety issue (Hammerschlag et al., 2016), resulting in significant public health concerns. Thus, an opportunity to address elasmobranch conservation from this perspective through consumer awareness is also noted, as well as through the establishment of governmental regulations for the safe consumption of elasmobranch-derived products, which can indirectly assist in the recovery of overexploited and commonly traded species and therefore, their conservation.

Another point of concern is that no toxicological thresholds for any contaminants have been yet established for elasmobranchs, due to the inherent difficulties of working with this taxonomic group. That said, future studies should focus on establishing thresholds for many species as possible, aiming to advance in this regard. Future studies should also focus on elucidating the effects of Cd, Hg and Pb on elasmobranchs' reproductive traits, as for most aquatic vertebrates, fertility impairment due to metal exposure is well-documented. Also, more studies on the effects of contamination on systemic health are necessary, to elucidate how toxic metals can compromise species' survival and resilience. Finally, although assessments on the behavioral effects of metals are impractical, the use of biomarkers for assessing neurotoxicity of metals could contribute to the inferences on potential neurological and, consequently, behavioral effects.

Finally, it is important to note that in an ever-changing world, multidisciplinary approaches are extremely valuable and must be implemented to achieve the goal of sustainably managing the oceans and achieving the 2030 Agenda for Sustainable Development (UN, 2020). Sustainable Development Goal (SDG) 14, in particular, "Conserve and sustainably use the oceans, seas and marine resources for sustainable development", in the form of its sub-items 14-2 (protect and restore marine ecosystems), 14-5 (conserving coastal and marine areas), 14-A (increasing scientific knowledge, research and technology for ocean health), clearly indicates the importance of monitoring the oceans in a context where pressures such as overfishing and overexploitation of marine resources, pollution, invasive alien species, habitat destruction and climate change (UNDESA, 2014; OECD, 2017) are increasingly compromising the ability of the oceans to provide economic, social and environmental benefits, collectively referred to as ecosystem services (Virto, 2018). In addition, we can also cite contributions to ODS 3-9 (substantially reduce the number of deaths and illnesses from hazardous chemicals, contamination and pollution of air and soil water), ODS 6-6 (protect and restore aquatic ecosystems) and, finally, SDG 15-5 (protecting biodiversity and natural habitats) that are also directly affected by elasmobranch contamination.

ACKNOWLEDGEMENTS

This study was financed in part by the Coordenação de Aperfeiçoamento de Pessoal de Nível Superior—Brasil (CAPES)—Finance Code 001. NW thanks Coordenação de Aperfeiçoamento de Pessoal de Nível Superior (CAPES) for financial support. RAHD acknowledges FAPERJ (JCNE 2021–2024 and process number E-26/21.460/2019) and CNPq (productivity grant) for financial support. The implementation of the Projeto Pesquisa Marinha e Pesqueira is a compensatory measure established by the Conduct Adjustment Agreement under the responsibility of the PRIO company, conducted by the Federal Public Ministry – MPF/RJ.

REFERENCES

Agius, C. and Roberts, R.J. 2003. Melano-macrophage centres and their role in fish pathology. J. Fish Dis. 26(9): 499–509.

Akira, S. 2008. Innate immunity to pathogens: diversity in receptors for microbial recognition. Immunol. Rev. 227(1): 5–8.

Akira, S., Uematsu, S. and Takeuchi, O. 2006. Pathogen recognition and innate immunity. Cell. 124(4): 783–801.

Alexander, J.B. and Ingram, G.A. 1992. Noncellular nonspecific defense mechanisms of fish. Annu. Rev. Fish Dis. 249–279.

Almeida, J.A., Barreto, R.E., Novelli, E.L., Castro, F.J. and Moron, S.E., 2009. Oxidative stress biomarkers and aggressive behavior in fish exposed to aquatic cadmium contamination. Neotrop. Ichthyol. 7: 103–108.

Alves, L.M., Lemos, M.F.L., Correia, J.P.S., Costa, N.A.R. and Novais, S.C. 2015. The potential of cholinesterases as tools for biomonitoring studies with sharks: Biochemical characterization in brain and muscle tissues of *Prionace glauca*. J. Exp. Mar. Biol. Ecol. 465: 49–55.

Alves, L. M., Nunes, M., Marchand, P., Le Bizec, B., Mendes, S., Correia, J.P. et al. 2016. Blue sharks (*Prionace glauca*) as bioindicators of pollution and health in the Atlantic Ocean: Contamination levels and biochemical stress responses. Sci. Total Environ. 563: 282–292.

Atallah-Benson, L., Merly, L., Cray, C. and Hammerschlag, N. 2020. Serum protein analysis of nurse sharks. J. Aquat. Anim. Health 32(7): 77–82.

Ballantyne, J.S. and Robinson, J.W. 2010 Freshwater elasmobranchs: A review of their physiology and biochemistry. J. Comp. Physiol. B 180(4): 475–493.

Ballantyne, J.S. and Fraser, D.I. 2012. Euryhaline elasmobranchs. pp. 125–198. *In*: McCormick, S.D., Farrell, A.P. and Brauner, C.J. (eds). Fish Physiology: Euryhaline Fishes, Vol. 32. Academic Press.

Ballatori, N., Shi, C. and Boyer, J.L. 1988. Altered plasma membrane ion permeability in mercury-induced cell injury: Studies on hepatocytes of elasmobranch *Raja erinacea*. Toxicol. Appl. Pharmacol. 95(2): 279–291.

Ballatori, N. and Boyer, J.L. 1996. Disruption of cell volume regulation by mercuric chloride is mediated by an increase in sodium permeability and inhibition of an osmolyte channel in skate hepatocytes. Toxicol. Appl. Pharmacol. 140(2): 404–410.

Barone, G., Dambrosio, A., Storelli, A., Garofalo, R., Busco, V.P. and Storelli, M.M. 2018. Estimated dietary intake of trace metals from swordfish consumption: a human health problem. Toxics 6(2): 22.

Barrera-García, A., O'Hara, T., Galván-Magaña, F., Méndez-Rodríguez, L.C., Castellini, J.M. and Zenteno-Savín, T. 2012. Oxidative stress indicators and trace elements in the blue shark (*Prionace glauca*) off the east coast of the Mexican Pacific Ocean. Comp. Biochem. Physiol. C Pharmacol. Toxicol. Endocrinol. 156: 59–66.

Barrera-García, A., O'Hara, T., Galván-Magaña, F., Méndez-Rodríguez, L.C., Castellini, J.M. and Zenteno-Savín, T. 2013. Trace elements and oxidative stress indicators in the liver and kidney of the blue shark (*Prionace glauca*). Comp. Biochem. Physiol. A–Mol. Integr. Physiol. A 165: 483–490.

Begam, M. and Sengupta, M. 2015. Immunomodulation of intestinal macrophages by mercury involves oxidative damage and rise of pro-inflammatory cytokine release in the freshwater fish *Channa punctatus*. Fish Shellfish Immunol. 45: 378–385.

Bernal, D. and Lowe, C.G. 2015. Field studies of elasmobranch physiology. Fish Physiol. 34: 311–377.

Betka, M. and Callard, G.V. 1999. Stage-Dependent Accumulation of Cadmium and Induction of Metallothionein-Like Binding Activity in the Testis of the Dogfish Shark, *Squalus acanthias*. Biol. Reprod. 14–22.

Black, C., Merly, L. and Hammerschlag, N. 2021. Bacterial communities in multiple tissues across the body surface of three coastal shark species. Zool. Stud. 60: 69.

Bloom, N.S. 1992. On the chemical form of mercury in edible fish and marine invertebrate tissue. Can. J. Fish. Aquat.Sci. 49(5): 1010–1017.

Borucinska, J.D., Kotran, K., Shackett, M. and Barker, T. 2009. Melanomacrophages in three species of free-ranging sharks from the northwestern Atlantic, the blue shark *Prionacae glauca* (L.), the shortfin mako, *Isurusoxyrhinchus* Rafinesque, and the thresher, *Alopias vulpinus* (Bonnaterre). J. Fish Dis. 32(10): 883–891.

Caballero, S., Galeano, A.M., Lozano, J.D. and Vives, M. 2020. Description of the microbiota in epidermal mucus and skin of sharks (*Ginglymostomacirratum* and *Negaprionbrevirostris*) and one stingray (*Hypanus americanus*). PeerJ 8: e10240.

Cameron, J.N. and Iwama, G.K. 1989. Compromises between ionic regulation and acid–base regulation in aquatic animals. Can. J. Zool. 67(12): 3078–3084.

Carrier, J.C., Musick, J.A. and Heithaus, M.R. (eds). 2012. Biology of sharks and their relatives. CRC Press.

Cho, Y.S., Ha, E., Bang, I., Kim, D.S. and Nam, Y.K. 2005a. Expression of Cu/Zn Superoxide Dismutase (Cu/Zn-SOD) mRNA in Shark, *Schyliohinustorazame*, Liver during Acute Cadmium Exposure. J. Aquac. 18(2): 173–179.

Cho, Y.S., Chio, B.N., Ha, E.M., Kim, K.H., Kim, S.K., Kim, D.S., et al. 2005b. Shark (*Scyliorhinustorazame*) Metallothionein: cDNA Cloning, Genomic Sequence, and Expression Analysis. Mar. Biotechnol. 7: 350–362.

Choi, A.L. and Grandjean, P. 2008. Methylmercury exposure and health effects in humans. Environ. Chem. 5(2): 112–120.

Chunhabundit, R. 2016. Cadmium exposure and potential health risk from foods in contaminated area, Thailand. Toxicol. Res. 32(1): 65–72.

Coello, W.F. and Khan, M.A.Q. 1996. Protection against heavy metal toxicity by mucus and scales in fish. Arch. Environ. Contam. Toxicol. 30: 319–326.

Company, R., Felícia, H., Serafim, A., Almeida, A.J., Biscoito, M. and Bebianno, M.J., 2010. Metal concentrations and metallothionein-like protein levels in deep-sea fishes captured near hydrothermal vents in the Mid-Atlantic Ridge off Azores. Deep Sea Res. Pt. I. 57: 893–908.

Consales, G. and Marsili, L. 2021. Assessment of the conservation status of Chondrichthyans: underestimation of the pollution threat. Eur. Zool. J. 88(1): 165–180.

Cooke, S.J., Bergman, J.N., Madliger, C.L., Cramp, R.L., Beardall, J., Burness, G., et al. 2021. One hundred research questions in conservation physiology for generating actionable evidence to inform conservation policy and practice. Conserv. Physiol. 9(1): coab009.

Cortés, E. 1999. Standardised Diet Compositions and Trophic Levels of Sharks.

Coyle, P., Philcox, J.C., Carey, L.C. and Rofe, A.M. 2001. Metallothionein: The multipurpose protein. Cell. Mol. Life Sci. 59: 627–647.

Crowe, W., Allsopp, P.J., Watson, G.E., Magee, P.J., Strain, J.J., Armstrong, D.J., et al. 2016. Mercury as an environmental stimulus in the development of autoimmunity—A systematic review. Autoimmun. Rev. 16(1): 72–80.

De Boeck, G., Eyckmans, M., Lardon, I., Bobbaers, R., Sinha, A.K. and Blust, R. 2010. Metal accumulation and metallothionein induction in the spotted dogfish *Scyliorhinuscanicula*. Comp. Biochem. Physiol. Part A Mol. Integr. 155(4): 503–508.

Dent, F. and Clarke, S. 2015. State of the global market for shark products. FAO Fisheries and Aquaculture Technical Paper (590), I.

Desforges, J.W., Sonne, C., Levin, M., Siebert, U., De Guise, S., Dietz, R. 2016. Immunotoxic effects of environmental pollutants in marine mammals. Environ. Int. 86: 126–139.

Dhanapakiam, P. and Ramasamy, V.K. 2001. Toxic effects of copper and zinc mixtures on some hematological and biochemical parameters in common carp, *Cyprinus carpio* (Linn). J. Environ. Biol. 22(2): 105–111.

Dittmann, J., Fung, S.J., Vickers, J.C., Chuah, M.I., Chung, R.S. and West, A.K. 2005. Metallothionein biology in the ageing and neurodegenerative brain. Neurotox. Res. 7(1): 87–93.

Doane, M.P., Haggerty, J.M., Kacev, D., Papudeshi, B. and Dinsdale, E.A. 2017. The skin microbiome of the common thresher shark (*Alopias vulpinus*) has low taxonomic and gene function β-diversity. Environmental Microbiology Rep. 9: 357–373.

Dulvy, N.K., Pacoureau, N., Rigby, C.L., Pollom, R.A., Jabado, R.W., Ebert, D.A., et al. 2021. Overfishing drives over one-third of all sharks and rays toward a global extinction crisis. Curr. Biol. 31(21): 4773–4787.

Dunkelberger, J.R. and Song, W. 2010. Complement and its role in innate and adaptive immune responses. Cell Research 20: 34–50.

Edwards, S.L. and Marshall, W.S. 2012. Principles and Patterns of Osmoregulation and Euryhalinity in Fishes. Fish Physiology: Euryhaline Fishes 32: 1–44.

Eissa, B.L., Ossana, N.A., Ferrari, L. and Salibián, A., 2010. Quantitative behavioral parameters as toxicity biomarkers: fish responses to waterborne cadmium. Archives of Environmental Contamination and Toxicology, 58(4): 1032–1039.

Endo, T., Kimura, O., Ogasawara, H., Ohta, C., Koga, N., Kato, Y., et al. 2015. Mercury, cadmium, zinc and copper concentrations and stable isotope ratios of carbon and nitrogen in tiger sharks (Galeocerdo cuvier) culled off Ishigaki Island, Japan. Ecol. Indic. 55: 86–93.

Evans, D.H. 1984. Gill Na^+/H^+ and Cl^-/HCO_3^- exchange systems evolved before the vertebrates entered freshwater. J. Exp. Biol. 113: 465–469.

Evans, D.H. and Weingarten, K. 1990. The effect of cadmium and other metals on vascular smooth muscle of the dogfish shark, *Squalus acanthias*. Toxicology 61(3): 275–281.

Eyckmans, M., Lardon, I., Wood, C.M. and De Boeck, G. 2013. Physiological effects of waterborne lead exposure in spiny dogfish (*Squalus acanthias*). Aquat. Toxicol. 126: 373–381.

Flajnik, M.F. and Rumfelt, L.L. 2000. The immune system of cartilaginous fish. pp. 249–270. *In*: Pasquier, L. and Litman, G.W. (eds). Origin and Evolution of the Vertebrate Immune System, Vol. 248. Springer.

Förstner, U. and Wittmann, G.T. 2012. Metal Pollution in the Aquatic Environment. Springer Science & Business Media.

Fraser, M., Surette, C. and Vaillancourt, C. 2013. Fish and seafood availability in markets in the Baie des Chaleurs region, New Brunswick, Canada: a heavy metal contamination baseline study. Environ. Sci. Pollut. Res. 20(2): 761–770.

Genchi, G., Sinicropi, M.S., Lauria, G., Carocci, A. and Catalano, A. 2020. The effects of cadmium toxicity. Int. J. Environ. Res. Public Health 17(11): 3782.

Goyer, R.A. 1993. Lead toxicity: current concerns. Environ. Health Perspect. 100: 177–187.

Grant, K.R. 2015. Fish hematology and associated disorders. Vet. Clin. N. Am.: Exot. Anim. Pract. 18(1): 83–103.

Haines, A.N., Flajnik, M.F., Rumfelt, L.L. and Wourms, J.P. 2005. Immunoglobulins in the eggs of the nurse shark, *Ginglymostomacirratum*. Dev. Comp. Immunol. 29(5): 417–430.

Halliwell, B. and Gutteridge, J.M.C. 2007. Free Radicals in Biology and Medicine, 4th Ed. Oxford University Press, Oxford, UK.

Hammerschlag, N. and Sulikowski, J. 2011. Killing for conservation: The need for alternatives to lethal sampling of apex predatory sharks. Endanger. Species Res. 14(2): 135–140.

Hammerschlag, N., Davis, D.A., Mondo, K., Seely, M.S., Murch, S.J., Glover, W.B., et al. 2016. Cyanobacterial neurotoxin BMAA and mercury in sharks. Toxins 8(8): 238.

Hammerschlag, N., Schmitz, O.J., Flecker, A.S., Lafferty, K.D., Sih, A., Atwood, T.B., et al. 2019. Ecosystem function and services of aquatic predators in the Anthropocene. Trends Ecol. Evol. 34(4): 369–383.

Hammerschmidt, C.R., Sandheinrich, M.B., Wiener, J.G. and Rada, R.G. 2002. Effects of dietary methylmercury on reproduction of fathead minnows. Environ. Sci. Technol. 36: 877–883.

Hanada, T. and Yoshimura, A. 2002. Regulation of cytokine signaling and inflammation. Cytokine Growth Factor Rev. 13(4): 413–421.

Hara, Y., Yamaguchi, K., Onimaru, K., Kadota, M., Koyanagi, M., Keeley, S.D., et al. 2018. Shark genomes provide insights into elasmobranch evolution and the origin of vertebrates. Nat. Ecol. Evol. 2: 1761–1771.

Harayashiki, C.A.Y., Reichelt-Brushett, A., Cowden, K. and Benkendorff, K. 2018. Effects of oral exposure to inorganic mercury on the feeding behaviour and biochemical markers in yellowfin bream (*Acanthopagrus australis*). Mar. Environ. Res. 134: 1–15.

Harris, H.H., Pickering, I.J. and George, G.N. 2003. The chemical form of mercury in fish. Science 301(5637): 1203–1203.

Hauser-Davis, R.A., Rocha, R.C.C., Saint'Pierre, T.D. and Adams, D.H. 2021. Metal concentrations and metallothionein metal detoxification in blue sharks, *Prionace glauca* L. from the Western North Atlantic Ocean. J. Trace. Elem. Med. Biol. 68: 126813.

Hidalgo, J., Tort, L. and Flos, R. 1985. Cd-, Zn-, Cu-binding protein in the elasmobranch Scyliorhinuscanicula. Comp. Biochem. Physiol. C: Comparative Pharmacology 81(1): 159–165.

Hidalgo, J. and Flos, R. 1986. Dogfish metallothionein–I. Purification and characterization and comparison with rat metallothionein. Comp. Biochem. Physiol. C: Comparative Pharmacology 83(1): 99–103.

Hinds Vaughan, N. and Smith, S.L. 2013. Isolation and characterization of a c-type lysozyme from the nurse shark. Fish Shellfish Immunol. 35(6): 1824–1828.

Hossein-Khannazer, N., Azizi, G., Eslami, S., Mohammed, H.A., Fayyaz, F., Hosseinzadeh, R., et al. 2020. The effects of cadmium exposure in the induction of inflammation. Immunopharmacol. Immunotoxicol. 42(1): 1–8.

Hussey, N.E., MacNeil, M.A., McMeans, B.C., Olin, J.A., Dudley, S.F., Cliff, G., et al. 2014. Rescaling the trophic structure of marine food webs. Ecol. Lett. 17(2): 239–250.

IUCN, 2012. IUCN Red List Categories and Criteria: Version 3.1, 2nd ed. IUCN, Gland, Switzerland and Cambridge, UK. iv + 32 pp.

Jacoby, S.C., Gagnon, E., Caron, L., Chang, J. and Isenring, P. 1999. Inhibition of Na^+-K^+-$2Cl^-$ cotransport by mercury. Am. J. Physiol. Cell Physiol., 277(4): C684-C692.

Janeway Jr, C.A. and Medzhitov, R. 2002. Innate immune recognition. Annu. Rev. Immunol. 20(1): 197–216.

Jones, D.P. 2006. Redefining oxidative stress. Antioxid. Redox Signaling 8(9–10): 1865–1879.

Kägi, J.H.R. 1991. Overview on metallothionein. Meth. Enzymol. 205: 613–626.

Karimi, R., Fitzgerald, T.P. and Fisher, N.S. 2012. A quantitative synthesis of mercury in commercial seafood and implications for exposure in the United States. Environmental Health Perspectives 120(11): 1512–1519.

Keller, J.M., Meyer, J.N., Mattie, M., Augspurger, T., Rau, M., Dong, J. et al. 2000. Assessment of immunotoxicology in wild populations: Review and recommendations. Reviews in Toxicology 3(1): 167–212.

King, J.R., Agostini, V.N., Harvey, C.J., McFarlane, G.A., Foreman, M.G., Overland, J.E., et al. 2011. Climate forcing and the California current ecosystem. ICES Journal of Marine Science 68(6): 1199–1216.

Kinne-Saffran, E. and Kinne, R.K. 2001. Inhibition by mercuric chloride of Na-K-2Cl cotransport activity in rectal gland plasma membrane vesicles isolated from *Squalus acanthias*. Biochim. Biophys. Acta, Biomembr. 1510(1–2): 442–451.

Kong, X., Wang, S., Jiang, H., Nie, G. and Li, X. 2012. Responses of acid/alkaline phosphatase, lysozyme, and catalase activities and lipid peroxidation to mercury exposure during the embryonic development of goldfish Carassius auratus. Aquat. Toxicol. 120: 119–125.

Kumar, G. 2018. Mercury concentrations in fresh and canned tuna: a review. Reviews in Fisheries Science & Aquaculture 26(1): 111–120.

Kuraku, S. 2021. Shark and ray genomics for disentangling their morphological diversity and vertebrate evolution. Developmental Biology 477: 262–272.

Lee, C.S. and Fisher, N.S. 2016. Methylmercury uptake by diverse marine phytoplankton. Limnol. Oceanogr. 61(5): 1626–1639.

Lee, C.S. and Fisher, N.S. 2017. Bioaccumulation of methylmercury in a marine copepod. Environ. Toxicol. Chem. 36(5): 1287–1293.

Lehmann, I., Sack, U. and Lehmann, J. 2011. Metal ions affecting the immune system. Met. Ions Life Sci. 8: 157–185.

Lehnert, K., Desforges, J.P., Das, K. and Siebert, U. 2018. Ecotoxicological biomarkers and accumulation of contaminants in pinnipeds. pp. 261–289. *In*: Fossi, M.C. and Panti, C. (eds). Marine Mammal Ecotoxicology, Academic Press.

Li, R., Redmond, A.K., Wang, T., Bird, S., Dooley, H. and Secombes, C.J. 2015. Characterisation of the TNF superfamily members CD40L and BAFF in the small-spotted catshark (Scyliorhinuscanicula). Fish Shellfish Immunol. 47(1): 381–389.

Livingstone, D.R. 1993. Biotechnology and pollution monitoring: use of molecular biomarkers in the aquatic environment. J. Chem. Tech. Biotechnol. 57: 195–211.

Lochmiller, R.L. and Deerenberg, C. 2000. Trade-offs in evolutionary immunology: Just what is the cost of immunity? Oikos 88: 87–98.

Lohrer, A.M., Thrush, S.F. and Gibbs, M.M. 2004. Bioturbators enhance ecosystem function through complex biogeochemical interactions. Nature 431: 1092–1095.

López-Cruz, R.I., Dafre, A. and Filho, D.W. 2011. Oxidative stress in sharks and rays. pp. 157–164. *In*: Abele, D., Vázquez-Medina, J.P., and Zenteno-Savín, T. (eds.). Aquatic Ecosystems. Wiley.

Luster, M.I., Portier, C., Paît, D.G., White, K.L., Gennings, C., Munson, A.E., et al. 1992. Risk assessment in immunotoxicology: I. Sensitivity and predictability of immune tests. Fundam Appl Toxicol. 18(2): 200–210.

Lyons, K. and Wynne-Edwards, K.E. 2019. Steroid concentrations in maternal serum and uterine histotroph in round stingrays (Urobatishalleri). General and Comparative Endocrinology 274: 8–16.

Lyons, K., Kacev, D., Preti, A., Gillett, D., Dewar, H. and Kohin, S. 2019. Species-specific characteristics influence contaminant accumulation trajectories and signatures across ontogeny in three pelagic shark species. Environ. Sci. Technol. 53(12): 6997–7006.

Maceda-Veiga, A., Figuerola, J., Martínez-Silvestre, A., Viscor, G., Ferrari, N. and Pacheco, M. 2015. Inside the Redbox: applications of haematology in wildlife monitoring and ecosystem health assessment. Sci. Total Environ. 514: 322–332.

Mahbub, K.R., Krishnan, K., Naidu, R. and Megharaj, M. 2016. Mercury resistance and volatilization by *Pseudoxanthomonas* sp. SE1 isolated from soil. Environ. Technol. Innovation. 6: 94–104.

Mann, E.E., Manna, D., Mettetal, M.R., May, R.M., Dannemiller, E.M., Chung, K.K. et al. 2014. Surface micropattern limits bacterial contamination. Antimicrob. Resist. Infect. Control 3(1): 28.

Marchetti, C. 2013. Role of calcium channels in heavy metal toxicity. ISRN Toxicol. 2013: 184360.

Matulik, A.G., Kerstetter, D.W., Hammerschlag, N., Divoll, T., Hammerschmidt, C.R. and Evers, D.C. 2017. Bioaccumulation and biomagnification of mercury and methylmercury in four sympatric coastal sharks in a protected subtropical lagoon. Mar. Pollut. Bull. 116(1–2): 357–364.

McClusky, L.M. 2006. Stage-dependency of apoptosis and the blood-testis barrier in the dogfish shark (*Squalus acanthias*): cadmium-induced changes as assessed by vital fluorescence techniques. Cell Tissue Res. 325(3): 541–553.

Mearns, A.J., Reish, D.J., Oshida, P.S., Ginn, T., Rempel-Hester, M.A., Arthur, C., et al. 2015. Effects of pollution on marine organisms. Water Environ. Res. 87(10): 1718–1816.

Mergler, D., Anderson, H.A., Chan, L.H.M., Mahaffey, K.R., Murray, M., Sakamoto, M. et al. 2007. Methylmercury exposure and health effects in humans: a worldwide concern. AMBIO 36(1): 3–11.

Merly, L., Lange, L., Meÿer, M., Hewitt, A.M., Koen, P., Fischer, C., et al. 2019. Blood plasma levels of heavy metals and trace elements in white sharks (*Carcharodon carcharias*) and potential health consequences. Mar. Pollut. Bull. 142: 85–92.

Monserrat, J.M., Geracitano, L.A. and Bianchini, A. 2003. Current and future perspectives using biomarkers to assess pollution in aquatic ecosystems. Comments on Toxicology 9: 255–269.

Monserrat, J.M., Martínez, P.E., Geracitano, L.A., Amado, L.L., Martins, C.M.G., Pinho, G.L.L., et al. 2007. Pollution biomarkers in estuarine animals: Critical review and new perspectives. Comp. Biochem. Physiol. C Toxicol. Pharmacol. 146(1–2): 221–234.

Moore, M.N. 2004. Diet restriction induced autophagy: a lysosomal protective system against oxidative- and pollutant-stress and cell injury. Mar. Environ. Res. 58(2–5): 603–607.

Morales, K.A., Lasagna, M., Gribenko, A.V., Yoon, Y., Reinhart, G., Lee, J.C., et al. 2011. Pb^{2+} as Modulator of protein–membrane interactions. J. Am. Chem. Soc. 133(27): 10599–10611.

Morcillo, P., Esteban, M.A. and Cuesta, A. 2017. Mercury and its toxic effects on fish. AIMS Environ. Sci. 4(3): 386–402.

Müller-Eberhard, H.J. 1988. Molecular organization and function of the complement system. Annu. Rev. Biochem. 57(1): 321–347.

Nam, D., Adams, D.H., Flewelling, L.J. and Basu, N., 2010. Neurochemical alterations in lemon shark (*Negaprionbrevirostris*) brains in association with brevetoxin exposure. Aquatic Toxicol. 99: 351–359.

Nonaka, M. and Smith, S.L. 2000. Complement system of bony and cartilaginous fish. Fish Shellfish Immunol. 10(3): 215–228.

Norris, S.B., Reistad, N.A. and Rumbold, D.G. 2021. Mercury in neonatal and juvenile blacktip sharks (*Carcharhinus limbatus*). Part II: Effects assessment. Ecotoxicology 30(2): 311–322.

OECD. 2017. Marine Protected Areas: Economics, Management and Effective Policy Mixes. OECD. Available at: https://www.oecd.org/env/marine-protected-areas-9789264276208-en.htm.

O'Shea, O.R., Thums, M., Van Keulen, M. and Meekan, M. 2012. Bioturbation by stingrays at Ningaloo Reef, Western Australia. Mar. Freshw. Res. 63: 189–197.

Ososkov, I. and Weis, J.S. 1996. Development of social behavior in larval mummichogs after embryonic exposure to methylmercury. Trans. Am. Fish. Soc. 125: 983–987.

Pain, D.J., Mateo, R. and Green, R.E. 2019. Effects of lead from ammunition on birds and other wildlife: A review and update. Ambio 48(9): 935–953.

Pang, P.K., Griffith, R.W. and Atz, J.W. 1977. Osmoregulation in elasmobranchs. Am. Zool. 17(2): 365–377.

Papanikolaou, N.C., Hatzidaki, E.G., Belivanis, S., Tzanakakis, G.N. and Tsatsakis, A.M. 2005. Lead toxicity update. A brief review. Med. Sci. Monit. 11(10): RA329-RA336.

Patel, U.N., Patel, U.D., Khadayata, A.V., Vaja, R.K., Patel, H.B. and Modi, C.M. 2021. Assessment of neurotoxicity following single and co-exposure of cadmium and mercury in adult zebrafish: Behavior alterations, oxidative stress, gene expression, and histological impairment in brain. Water Air Soil Pollut. 232: 340.

Pereira Santos, C., Sampaio, E., Pereira, B.P., Pegado, M.R., Borges, F.O., Wheeler, C.R., et al. 2021. Elasmobranch responses to experimental warming, acidification, and oxygen loss—A meta-analysis. Front. Mar. Sci. 8: 735377. doi: 10.3389/fmars.2021.735377.

Pethybridge, H., Cossa, D. and Butler, E. 2010. Mercury in 16 demersal sharks from southeast Australia: Biotic and abiotic sources of variation and consumer health implications. Mar. Environ. Res. 69: 18–26.

Raschi, W. and Tabit, C. 1992. Functional aspects of placoid scales: A review and update. Aust. J. Mar. Freshw. Res. 43(1): 123–147.

Ratner, M.A., Decker, S.E., Aller, S.G., Weber, G. and Forrest Jr, J.N. 2006. Mercury toxicity in the shark (*Squalus acanthias*) rectal gland: apical CFTR chloride channels are inhibited by mercuric chloride. J. Exp. Zool. Part A: Comp. Exp. Biol. 305(3): 259–267.

Redmond, A.K., Zou, J., Secombes, C.J., Macqueen, D.J. and Dooley, H. 2019. Discovery of all three types in cartilaginous fishes enables phylogenetic resolution of the origins and evolution of interferons. Front. Immunol. 10: 1558.

Reverter, M., Tapissier-Bontemps, N., Lecchini, D., Banaigs, B. and Sasal, P. 2018. Biological and ecological roles of external fish mucus: A review. Fishes 3(4): 41.

Rodrigues, A.C.M., Gravato, C., Galvão, D., Silva, V.S., Soares, A.M.V.M., Gonçalves, J.M.S., et al. 2022. Ecophysiological effects of mercury bioaccumulation and biochemical stress in the deep-water mesopredator *Etmopterus spinax* (Elasmobranchii; Etmopteridae). J. Hazard. Mater. 423: 127245.

Rumbold, D., Wasno, R., Hammerschlag, N. and Volety, A. 2014. Mercury accumulation in sharks from the coastal waters of southwest Florida. Arch. Environ. Contam. Toxicol. 67: 402–412.

Rumfelt, L.L., McKinney, E.C., Taylor, E. and Flajnik, M.F. 2002. The development of primary and secondary lymphoid tissues in the nurse shark Ginglymostomacirratum: B-cell zones precede dendritic cell immigration and T-cell zone formation during ontogeny of the spleen. Scand. J. Immunol. 56(2): 130–148.

Sanchez-Dardon, J., Voccia, I., Hontela, A., Chilmonczyk, S., Dunier, M., Boermans, H., et al. 1999. Immunomodulation by heavy metals tested individually or in mixtures in rainbow trout (*Oncorhynchus mykiss*) exposed *in vivo*. Environ. Toxicol. Chem. 18: 1492–1497.

Sayın, D., Çakır, D.T., Gençer, N. and Arslan, O. 2012. Effects of some metals on paraoxonase activity from shark Scyliorhinuscanicula. J. Enzyme Inhib. Med. Chem. 27(4): 595–598.

Schmaier, A.H. 2019. Introduction to hematology. pp. 1–10. *In*: Lazarus, H. and Schmaier, A. (eds). Concise Guide to Hematology. Springer, Cham.

Scott, G.R., Sloman, K.A., Rouleau, C. and Wood, C.M. 2003. Cadmium disrupts behavioural and physiological responses to alarm substance in juvenile rainbow trout (*Oncorhynchus mykiss*). J. Exp. Biol. 206(11): 1779–1790.

Scott, G.R. and Sloman, K.A. 2004. The effects of environmental pollutants on complex fish behaviour: integrating behavioural and physiological indicators of toxicity. Aquat. Toxicol. 68(4): 369–392.

Seelinger, D. and Kalka, H. 1991. Nuclear Reaction Mechanisms - Proceedings of the XXth International Symposium on Nuclear Physics (Near Dresden, Germany, November 12-16, 1990). World Scientific Publishing Company.

Shipley, O.N., Lee, C.S., Fisher, N.S., Sternlicht, J.K., Kattan, S., Staaterman, E.R., et al. 2021. Metal concentrations in coastal sharks from the Bahamas with a focus on the Caribbean Reef shark. Sci. Rep. 11(1): 1–11.

Silva, P., Epstein, F.H. and Solomon, R.J. 1992. The effect of mercury on chloride secretion in the shark (*Squalus acanthias*) rectal gland. Comp. Biochem. Physiol. C Toxicol. Pharmacol. 103(3): 569–575.

Skomal, G.B. and Mandelman, J.W. 2012. The physiological response to anthropogenic stressors in marine elasmobranch fishes: a review with a focus on the secondary response. Comp. Biochem. Physiol. Part A Mol. Integr. 162(2): 146–155.

Sloman, K.A., Baker, D.W., Ho, C.G., McDonald, D.G. and Wood, C.M. 2003. The effects of trace metal exposure on agonistic encounters in juvenile rainbow trout, Oncorhynchus mykiss. Aquat. Toxicol. 63: 187–196.

Smith, S.L. 1998. Shark complement: an assessment. Immunol. Rev. 166 (1): 67–78.

Smith, S.L., Sim, R.B. and Flajnik, M.F. (eds) 2014. Immunobiology of the Shark. CRC Press.

Smith, N.C., Rise, M.L. and Christian, S.L. 2019. A comparison of the innate and adaptive immune systems in cartilaginous fish, ray-finned fish, and lobe-finned fish. Front. Immunol. 10: 2292. doi: 10.3389/fimmu.2019.02292.

Sokol, C.L. and Luster, A.D. 2015. The chemokine system in innate immunity. Cold Spring Harb. Perspect. Biol. 7(5): a016303.

Solé, M., Labera, G., Aljinovic, B., Ríos, J., García de la Parra, L.M., Maunoy, F., et al. 2008. Cholinesterases activities and lipid peroxidation levels in muscle from shelf and slope dwelling fish from the NW Mediterranean: Its potential use in pollution monitoring. Sci. Total Environ. 402: 306–317.

Somerville, R., Fisher, M., Persson, L., Ehnert-Russo, S., Gelsleichter, J. and Bielmyer-Fraser, G. 2020. Analysis of trace element concentrations and antioxidant enzyme activity in muscle tissue of the Atlantic Sharpnose Shark, Rhizoprionodon terraenovae. Arch. Environ. Contam. Toxicol. 79(4): 371–390.

Steinel, N.C. and Bolnick, D.I. 2017. Melanomacrophage centers as a histological indicator of immune function in fish and other poikilotherms. Front. Immunol. 8: 827.

Takeda, K., Kaisho, T. and Akira, S. 2003. Toll-like receptors. Annu. Rev. Immunol. 21: 335–376.

Tiktak, G.P., Butcher, D., Lawrence, P.J., Norrey, J., Bradley, L., Shaw, K., et al. 2020. Are concentrations of pollutants in sharks, rays and skates (Elasmobranchii) a cause for concern? A systematic review. Mar. Pollut. Bull. 160: 111701.

Tort, L. and Torres, P., 1988. The effects of sublethal concentrations of cadmium on haematological parameters in the dogfish, *Scyliorhinus canicula*. J. Fish. Biol. 32(2): 277–282.

UN, 2020. United Nations Decade of Ocean Science for Sustainable Development (2021–2030). Available at: https://en.unesco.org/ocean-decade.

UNDESA, 2014. How Oceans- and Seas-Related Measures Contribute to the Economic, Social and Environmental Dimensions of Sustainable Development: Local and Regional Experiences. Available at: https://sustainabledevelopment.un.org/index.php?page=view&type=400&nr=1339&menu=35.

Valko, M., Leibfritz, D., Moncol, J., Cronin, M.T.D., Mazur, M. and Telser, J. 2007. Free radicals and antioxidants in normal physiological functions and human disease. Int. J. Biochem. Cell Biol. 39: 44–84.

Van Beaumont, W. 1972. Evaluation of hemoconcentration from hematocrit measurements. J. Appl. Physiol. 32(5): 712–713.

Van der Oost, R., Beyer, J. and Vermeulen, N.P.E. 2003. Fish bioaccumulation and biomarkers in environmental risk assessment: A review. Environ. Toxicol. Pharmacol. 13(2): 57–149.

Van Kampen, E.J. and Zijlstra, W.G. 1966. Determination of hemoglobin and its derivatives. J. Appl. Physiol. 8: 141–187.

Van Rijn, J.A. and Reina, R.D. 2010. Distribution of leukocytes as indicators of stress in the Australian swellshark, *Cephaloscyllium laticeps*. Fish Shellfish Immunol. 29(3): 534–538.

Vas, P., Gordon, J.D., Fielden, P.R. and Overnell, J. 1993. The trace metal ecology of ichthyofauna in the Rockall Trough, north-eastern Atlantic. Mar. Pollut. Bull. 26(11): 607–612.

Vélez-Alavez, M., Labrada-Martagón, V., Méndez-Rodriguez, L.C., Galván-Magaña, F. and Zenteno-Savín, T. 2013. Oxidative stress indicators and trace element concentrations in tissues of mako shark (Isurus oxyrinchus). Comp. Biochem. Physiol. Part A Mol. Integr. 165(4): 508–514.

Vieira, L.R., Gravato, C., Soares, A.M.V.M., Morgado, F. and Guilhermino, L. 2009. Acute effects of copper and mercury on the estuarine fish Pomatoschistus microps: linking biomarkers to behaviour. Chemosphere 76(10): 1416–1427.

Vinodhini, M. and Narayanan, M. 2009. The impact of toxic heavy metals on the hematological parameters in common carp (*Cyprinus carpio* L.). Iran. J. Environ. Health Sci. Eng. 6: 23–28.

Virto, L.R. 2018. Conserve and sustainably use the oceans, seas and marine resources for sustainable development. Mar Policy 98: 47–57.

Walker, C.J., Gelsleichter, J., Adams, D.H. and Marine, C.A. 2014. Evaluation of the use of metallothionein as a biomarker for detecting physiological responses to mercury exposure in the bonnethead, *Sphyrna tiburo*. Fish Physiol. Biochem. 40(5): 1361–371.

Wang, R., Wang, X.T., Wu, L. and Mateescu, M.A. 1999. Toxic effects of cadmium and copper on the isolated heart of the dogfish shark, *Squalus acanthias*. J. Toxicol. Environ. Health Part A 57(7): 507–519.

Wang, D.D., Buerkel, D.M., Corbett, J.R. and Gurm, H.S. 2010. Fragmented QRS complex has poor sensitivity in detecting myocardial scar. Ann. Noninvasive Electrocardiol. 15(4): 308–314.

Wang, Z., Sun, Y., Yao, W., Ba, Q. and Wang, H. 2021. Effects of cadmium exposure on the immune system and immunoregulation. Front. Immunol. 12: a695484.

Webber, H.M. and Haines, T.A. 2003. Mercury effects on predator avoidance behavior of a forage fish, golden shiner (Notemigonus crysoleucas). Environ. Toxicol. Chem. 22(7): 1556–1561.

Weber, D.N. 1993. Exposure to sublethal levels of waterborne lead alters reproductive behavior patterns in fathead minnows (Pimephales promelas). Neurotoxicology Summer-Fall 14(2–3): 347–358.

Wikelski, M. and Cooke, S.J. 2006. Conservation physiology. Trends Ecol. Evol. 21(1): 38–46.

Wosnick, N. and Freire, C.A. 2013. Some euryhalinity may be more common than expected in marine elasmobranchs: The example of the South American skate *Zapteryx brevirostris* (Elasmobranchii, Rajiformes, Rhinobatidae). Comp. Biochem. Physiol., Part A, Mol. Integr. Physiol. 166(1): 36–43.

Wosnick, N., Niella, Y., Hammerschlag, N., Chaves, A.P., Hauser-Davis, R.A., da Rocha, R.C.C., et al. 2021a. Negative metal bioaccumulation impacts on systemic shark health and homeostatic balance. Mar. Pollut. Bull. 168: 112398.

Wosnick, N., Chaves, A.P., leite, R.D., Nunes, J.L.S., Saint'Pierre, T.D., Willmer, I.Q., et al. 2021b. Nurse sharks, space rockets and cargo ships: Metals and oxidative stress in a benthic, resident and large-sized mesopredator, *Gymglimostoma cirratum*. Env. Poll. 288: 117784.

Wu, P., Zakem, E.J., Dutkiewicz, S. and Zhang, Y. 2020. Biomagnification of methylmercury in a marine plankton ecosystem. Environ. Sci. Technol. 54(9): 5446–5455.

Yancey, P.H. and Somero, G.N. 1979. Counteraction of urea destabilization of protein structure by methylamine osmoregulatory compounds of elasmobranch fishes. Bioche. J. 183(2): 317–323.

Zelikoff, J.T. 1993. Metal pollution-induced immunomodulation in fish. Annu. Rev. Fish Dis. 3: 305–325.

Zheng, N., Wang, S., Dong, W., Hua, X., Li, Y., Song, X., et al. 2019. The toxicological effects of mercury exposure in marine fish. Bull. Environ. Contam. Toxicol. 102: 714–720.

Zou, J. and Secombes, C.J. 2016. The function of fish cytokines. Biology 5(2): 23.

Chapter 4

Cadmium, Lead and Mercury in Crustacea: Environmental and Risk Assessments

Isabella C. Bordon*[1], Mariana V. Capparelli[2], Joseane A. Marques[3], Anieli C. Maraschi[4] and Rogério O. Faleiros[5]

[1]Instituto de Ciências Biomédicas,Universidade de São Paulo, São Paulo-SP, Brazil.

[2]Estación el Carmen, Instituto de Ciencias del Mar y Limnología, Universidad Nacional Autónoma de México, Carretera Carmen-Puerto Real, Ciudad del Carmen, Campeche, México.

[3]Zuckerberg Institute for Water Research (ZIWR) Ben-Gurion University of the Negev, Israel.

[4]Centro de Biologia Marinha, Universidade de São Paulo, São Sebastião-SP, Brazil.

[5]Departamento de Ciências Agrárias e Biológicas, Universidade Federal do Espírito Santo, São Mateus-ES, Brazil.

INTRODUCTION

Crustaceans are arthropods found along the five continents, representing one of the most species-rich taxonomic groups from both oceans and freshwater bodies, displaying asignificant diversity of sizes, morphologies, behaviors, and life cycles. Crustacean representatives have colonized many ecosystems, from the deepest ocean trenches and hydrothermal vents, across the vast water bodies of the world's oceans, to intertidal and supratidal coastal habitats. They also inhabit anchialine

*Corresponding author: isabella.bordon@gmail.com

caves, inland freshwater ecosystems including endorheic lakes, and terrestrial habitats such as desert saltpans, epiphytic bromeliads in mountain forests, and rocky plateaus of coastal and oceanic island (Brusca et al., 2018; Melo, 1996).

Although some Crustacea species can be endemic, most of them are cosmopolitan and many families exhibit economic importance. Their ecological relevance is well-known, since many families occupy important trophic niches, from decomposers to predators. Crustaceans play central roles in various habitats, taking part in ecosystem functions by controlling primary production and transferring energy and nutrients from primary producers to higher trophic levels. Many decapod crustaceans, for example, represent key species in coastal, estuarine, and intertidal habitats, where they may be so abundant that changes in their population structure may directly influence the structure of the entire ecosystem (Burggren and McMahon, 1988). Despite their broad distribution and crucial ecosystem maintenance contribution, crustaceans are influenced by changes in physical-chemical environmental parameters, such as temperature, salinity, and acidification (Dewiyanti et al., 2018). Therefore, crustaceans are interesting models to evaluate anthropogenic stressor effects, such as metal releases into aquatic environments (Capparelli et al., 2016).

Crustacea undergo at least one development phase as aquatic organisms, with some species fully aquatic and others, semi-terrestrial lifestyle. Regarding feeding habits, several species are detritivores, essential for the maintenance of the structure and quality of aquatic ecosystems. Detritivorous crustaceans feed on dead plants and animals (or on their detritus), and are, thus, exposed to many contaminants, such as metals, through the dietary route. Uptake from the dissolved water phase and from ingested food may both be important routes of metal accumulation (Vitale et al., 1999). Dissolved metals can accumulate by direct adsorption to body surfaces and by absorption across the respiratory epithelia while particulate metals can accumulate following the ingestion and ingestion of food (Wang and Fisher, 1999). In crustaceans, most studies have assessed metals dissolved in the water phase (Viarengo and Nott, 1993), and only a very few investigations have examined the effects of metal contaminated diets (Bordon et al., 2018; Capparelli et al., 2020; Sá et al., 2008; Sabatini et al., 2009). The dietary metal toxicity needs further efforts for better understanding, and consequently, the dietary contamination route is not usually considered in existing regulations regarding environmental contamination or in risk assessments (Borgmann et al., 2005; De Schamphelaere et al., 2007).

The tolerance of environmental variation makes this group an important model for understanding how species can be affected by the physiological challenges of either natural or anthropogenic disturbances. Some species are resilient components even in degraded ecosystems (Capparelli et al., 2016) and crustaceans occupying contaminated areas do make physiological adjustments (Klerks and Weis, 1987). They can be used both in a model of resistance to contaminants, as environmental biomonitors and bioindicators. They are easily collected (Figure 4.1a,b) and maintained in laboratory conditions (Figure 4.2a, b), and by the wide geographic distribution, can be strong candidates for comparative investigations.

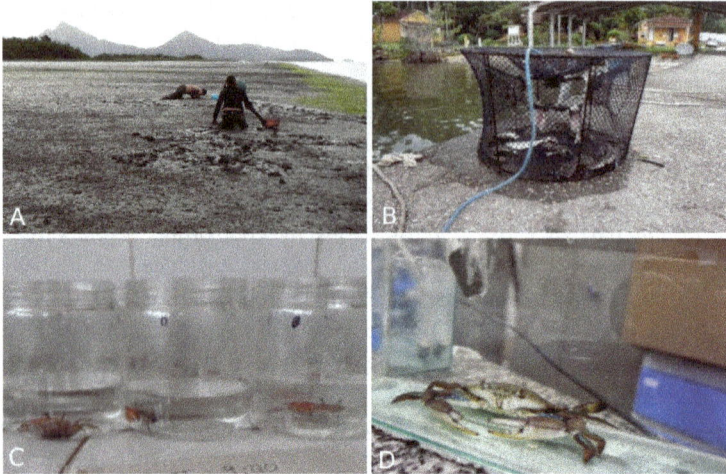

Figure 4.1 Sampling of *Uca maracoani*. (catching) (A) and *Callinectes* sp. (trapping), (B) indifferent estuarine site, (C) *Minuca rapax* (D) and *Callinectes danae* (b) maintenance in laboratory conditions.

AQUATIC BIOTA METAL EXPOSURE PATHWAYS

Metals from both natural and anthropogenic sources are continuously and increasingly released in the aquatic environment, becoming a serious threat due to their toxicity, long persistency, bioaccumulation, and potential biomagnification (Wang, 2002). Thus, metals in contaminated habitats may accumulate throughout the aquatic ecosystem, reaching humans through the consumption of contaminated seafood and resulting in environmental and public health concerns (Gupta et al., 2009).

Metals are involved in many distinct chemical processes. Several metals, however, although essential, become toxic to organisms when their total bioavailability has reached specific thresholds, which vary among organisms and metals (Casado-Martinez et al., 2010; Luoma and Rainbow, 2008; Rainbow, 2002, 2007; Van Straalen et al., 2005). Accumulated metals can be categorized into two forms, detoxified fraction and bioavailable fraction (Luoma and Rainbow, 2008; Rainbow, 2002, 2007; Rainbow and Luoma, 2011; Vijver et al., 2004). Metals are taken up and accumulate in aquatic Crustacea directly from their environment across the gills or other body surfaces, and from their diet via the intestine (Henry et al., 2012). Someelements are initially metabolically available, but most can be excreted or detoxified. Toxic effects in aquatic invertebrates occur when metal uptake rates exceed combined excretion and detoxification rates, leading to unacceptable metal buildup in metabolically available forms (Adams et al., 2010; Casado-Martinez et al., 2010; Luoma and Rainbow, 2008; Rainbow, 1998, Rainbow, 2002; 2007; Vijver et al., 2004; Wang and Rainbow, 2006). Some metals are essential micronutrient required by all living organisms for a variety of physiological and biochemical processes. Those metals are a co-factor in multiple enzymatic processes but is potentially toxic to aquatic organisms above certain levels (Martins et al., 2011). However, some metals are not required for any physiological activity and are toxic at very low concentrations, for example cadmium (Cd), lead (Pb) and mercury (Hg).

CADMIUM (Cd), LEAD (Pb) AND MERCURY (Hg) BIOACCUMULATION IN AQUATIC ORGANISMS

Bioaccumulation comprises the permanence of certain contaminants in living tissues, which may become bioavailable due to biochemical transformation processes, thus causing toxicity and/or lethality to exposed biota. Regarding metal persistence in aquatic organisms, contaminated producers may be preyed upon by primary consumers, subsequently up the trophic chain, and some metals may increase concentrations in this manner, resulting in biomagnification (Costa et al., 2022). Both processes can directly affect human health due to the consumption of contaminated seafood (Marques et al., 2010).

In ecological terms, the accumulation and persistence of metals can cause cellular and tissue damage, as well as neurological and reproductive issues, among other effects, compromising community survival, and affecting important niches responsible for ecosystem maintenance and conservation (O'Mara et al., 2019). Metals can be categorized as essential and nonessential (Ali et al., 2019). For example, Cu is a well-known essential metallic element, participating in several enzymatic and non-enzymatic processes in different taxonomic groups. In Crustacea, for example, Cu participates in the hemocyanin synthesis, responsible for oxygen transport in these animals. (Engel, 1987). Cd, Pb and Hg are, however, non-essential metals. They are released in the environment mainly due to anthropogenic activities such as metallurgic industries, agriculture, urbanization, and mining. Cadmium effects include oxidative stress, compromising important enzymatic processes and DNA integrity, while neurological Pb and Hg effects are well-known in aquatic organisms. The Hg biomagnification in aquatic organisms occurs mainly in the organic and most toxic form of Hg, methylmercury (Ali et al., 2019).

Cd, Pb AND Hg BIOACCUMULATION IN CRUSTACEANS

Metal uptake and accumulation in aquatic Crustacea takes place through water, across the gills or other body surfaces, and via the dietary route (Henry et al., 2012). Aquatic crustaceans, as other organisms, inhabit estuarine and coastal lagoon ecosystems that are recognized as areas in which their physicochemical characteristics contribute to metal retention these organisms' uptake from sediment, water column, and preys; the main metal input is related to their feeding habits and ecological lifestyles (Livingstone, 2001).

Crustaceans can accumulate metals to high body concentrations, depending on the relationship between metal uptake and excretion and the rate of diluting body growth (Frías-Espericueta et al., 2022; Rainbow, 2018). Tissuesof different invertebrates accumulate trace metals at different concentrations (Eisler, 1981), even from closely related taxa to species in the same genus (Moore and Rainbow, 1987; Rainbow, 1993, 1998). Because of this, they are routinely employed as sentinel species, reflecting metal contamination levels in surface sediment in many biomonitoring assessments (Ololade et al., 2011). For example, crabs are scavengers that tends to feed on detritus (Ip et al., 2005), plus exhibiting higher metal-bioaccumulation in their hepatopancreas, one of the most important detoxifying organs (Wang, 2002). Gills provide an interface between external environment and internal milieu, constituting a multifunctional organ serving in gas exchange, osmolyte transport, nitrogenous waste excretion, volume regulation, and acid-base regulation (Freire et al., 2008; Henry et al., 2012).

The proportion of each intake route is related to the crustacean physiological needs/endogenous factors (e.g., sex, age, condition and tissue) and to the element's bioavailability in seawater and diet, which is influenced by external environmental factors (e.g., season, location, substrate, depth, salinity, temperature and anthropogenic pressure) (Rainbow, 2002).

CADMIUM (Cd)

Cd is a non-essential and highly toxic metal that does not break down naturally in the environment, making ecosystems vulnerable to its persistence. Many studies have reported that Cd is one of the most frequently detected metals found in crustacean tissues, accumulating in soft tissues, such as the hepatopancreas, as well as the exoskeleton (Rainbow, 1997). Physicochemical and physiological effects interact to determine the uptake rates of Cd by crustaceans. Cd uptake rates of marine crustaceans are predominantly determined almost exclusively by physicochemical equilibria releasing the free metal ion. Most euryhaline crustaceans may vary their apparent water permeability to differing degrees, and thereby affect both absolute trace metal uptake rates and the pattern of their change with salinity variation (Rainbow, 1995). Previous reports concerning Cd contamination and bioaccumulation effects in crustaceans have focused on decapods, such as Portunidae (Annabi et al., 2018; Bordon et al., 2012a, b, 2016; Fakhri et al., 2021; Karar et al., 2019) and Ocypodidae crabs (Capparelli et al., 2016; Silva et al., 2018) as well as in Palaemonidae (Ezemonye et al., 2019; Tu et al., 2008) and Penaeidae shrimps (Baki et al., 2018; Di Beneditto et al., 2019). Reported values range from 0.001 to 0.9 μg g^{-1} Cd in Portunidae crabs, 0.004 to 0.7 μg g^{-1} in ocypodid crabs, 0.006 to 0.93 μg g^{-1} in palaemonid shrimps, 0.01 to 0.71 μg g^{-1} in penaeid shrimps, and 0.1 to 1.07 μg g^{-1} in representatives of other crustacean families (see Cabrini et al., 2018; Shaiek et al., 2018). As with the most metals, Cd contents differ among crustacean tissues. In the blue crab *Callinectes danae,* for example, Cd concentrations were 0.021 μg g^{-1} in gills, 0.041 μg g^{-1} in hepatopancreas and 0.004 μg g^{-1} in muscle (Bordon et al., 2016). The hepatopancreas has been described as the determinant organ in metal detoxification process, with a high capacity to accumulate metals (Amiard et al., 2006). In freshwater and marine shrimps, Cd concentrations, however, were higher in the cephalotorax than in the abdomen as consequence of the higher bioaccumulation bias in gill and hepatopancreas compared to muscle (Bertrand et al., 2016).

Cd exposure causes several deleterious effects in crustaceans, such as respiratory disruption, molt inhibition and growth decreases. This element can also cause cytological damage, mainly gill necrosis, as this organ is directly exposed to metals, consequently affecting oxygen consumption (Felten et al., 2008). Moreover, Cd exposure has also been reported as inhibiting the regulation of hormones responsible for molting process (Moreno et al., 2003). Survival, life expectancy, fecundity, age at first reproduction and growth rates may be reduced in the presence of Cd (Chandini, 1989; Kadiene et al., 2019). Cadmium also induces metallothionein (MT) and metal-rich granule (MRG) synthesis in crustaceans, suggesting that MT and MRG are the main responsible Cd detoxification pathways in this taxonomic group (Boudet et al., 2019).

LEAD (Pb)

Lead is atoxic metal with the ability to bioaccumulate in many crustaceans. The main Pb sources in the environment comprise of anthropogenic activities, such as mining activities and the result ingore tailings and contaminated effluents. Although Pb was banned from oil fuels many years ago, the presence of this element from combustion processes in the environment is still being reported, indicating significant concerns.

Regarding Pb bioaccumulation in crustaceans, several studies have reported Pb bioaccumulation in the gills, hepatopancreas and muscles of *Callinectes danae* sampled from the Southeastern coast of Brazil (Bordon et al., 2012a, 2012b; 2016). Some studies have suggested that this element is aggregated as non-soluble deposits (or granules) in *Penaeus monodon* shrimp (Vogt and Quinitio, 1994) and in *Penaeus vannamei* shrimp hepatopancreas (Núñez-Nogueira et al., 2010, 2012). Rainbow (1998) pointed out that isopods also accumulate metals in the form of granules. Laboratory assessments have also been conducted regarding Pb exposure in crustaceans. Bordon et al. (2018), for example, reported indicated Pb hepatopancreas bioaccumulation in *Callinectes danae* exposed to contaminated water at 2.0 μg g^{-1} after 7 and 14 days, also reporting that Pb bioaccumulation in gills occurred effectively after both exposure times, statistically higher after 14 days of exposure.

Previous biomonitoring approaches have reported Pb bioaccumulation, however, levels may vary depending on the crab and species, also depending on type of tissue and geographic distribution. For example, Bordon et al. (2016), reported that the mean Pb concentrations in muscle of the Portunidae crab *Callinectes danae* (0.16 μg g^{-1}) was equal to the hepatopancreas Pb concentration (0.16 μg g^{-1}), but lower then gills Pb concentration (0.54 μg g^{-1}), while Cabrini et al. (2018) reported higher concentrations of 0.63 μg g^{-1} Pb in the soft tissues of the same species. Considering the *Callinectes* genus, Çogun et al. (2017) reported 77.8 μg g^{-1} Pb in hepatopancreas and 3.7 μg g^{-1} Pb in the muscle tissue of

Callinectes sapidus, while Fakhri et al. (2021) reported Pb at 0.238 µg g^{-1} in the muscle tissue of the same species. Regarding the Grapsidae crab, both Carneiro et al. (2018) and Costa et al. (2018) evaluated Pb concentrations in *Goniopsis cruentata* gills, reporting very different mean values (3.5 and 0.17 µg g^{-1}, respectively). For the Palaemonidae crab, Tu et al. (2008) reported the maximum Pb concentrations in *Macrobrachium rosenbergii* muscle (0.03 µg g^{-1}) lower than in the hepatopancreas (0.109 µg g^{-1}), in turn lower than in the exoskeleton (0.134 µg g^{-1}). Regarding Penaeidae shrimps, Baki et al. (2018) reported Pb concentrations of 0.69 µg g^{-1} in *Parapenaeopsis sculptilis* muscle tissue, whereas Di Beneditto et al. (2019) reported concentrations ranging from 0.87 to 1.17 µg g^{-1} Pb in *Xiphopenaeus kroyeri* muscle tissue, and from 0.46 to 0.94 µg g^{-1} of Pb in *Artemisia longinaris* muscle tissue.

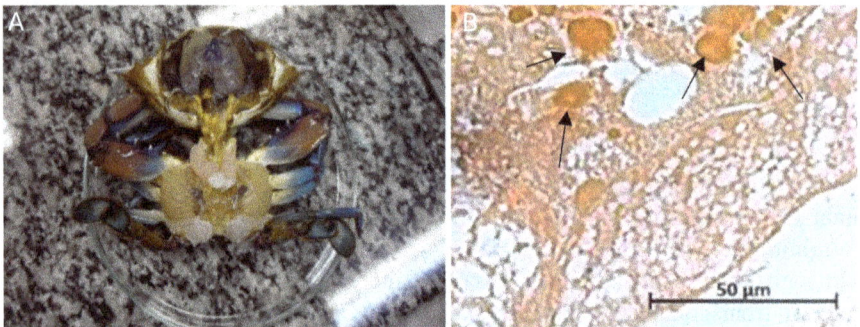

Figure 4.2   Dissection of *Callinectes danae* tissues (A) and Pb granules (arrows) in the hepatopancreas (B) of *C. danae* exposed to Pb.

MERCURY (Hg)

Mercury contamination is of significant concern for environmental and public health, due to its toxicity and accumulation in aquatic organisms. This element is listed as a high priority environmental pollutant within the Convention for the Protection of the Marine Environment of the North-East Atlantic (OSPAR Convention) and the United States Environmental Protection Agency (U.S. EPA) (Lillebø et al., 2011). It is a biologically, chemically, and geologically active element, released by both natural and anthropogenic activities, such as mining, gold mining and coal burning, final disposal of solid waste in landfills, and deforestation (Capparelli et al., 2020, 2021; Cruz et al., 2019).

Hg in aquatic ecosystems is transformed into organic and inorganic mercury forms due to different environmental conditions and reaction mechanism (Burgess and Meyer, 2008; De Almeida Rodrigues et al., 2019). Hg in sediments can be more readily transferred and bioaccumulated in aquatic biota, biomagnified in aquatic ecosystems, causing health risks to humans (Peña-Fernandes et al., 2014). The organic form, methylmercury (MeHg) is a biomagnifying neurotoxin of significant concern, due to its toxicity and biomagnification potential (Castoldi et al., 2001).

Previous studies have reported differences in Hg concentrations in different species depending on their habitats and ecological niches. Shrimps have been reported as transfer vectors of this element to main marine predators in lower trophic food chains, for example, female crabs (Hosseini et al., 2014).

In general, previous Hg bioaccumulation studies in crustaceans indicate that different tissues display different Hg accumulation, more likely to accumulate this metal in the hepatopancreas, followed by the gills and muscle tissue, respectively (Gerç and Yilma, 2015; Hosseini et al., 2014; Parsa et al., 2014). Most studies have focused on decapod representatives, such as the Portunidae (Baki et al., 2018; Bordon et al., 2012a, 2012b, 2016; Gerç and Yilma, 2015; Parsa et al., 2014), Cancridae and Matutidae crabs, and Nephropidae crayfish and Palinuridael obster (Baki et al., 2018), as well as in Palaemonidae (Tu et al., 2008) and Penaeidae (Baki et al., 2018; Hosseini et al., 2014;) shrimps. These findings revealed a set of means ranging from 0.02 to 1.3 μg g^{-1} Cd in portunid crabs, 0.02 to 0.22 μg g^{-1} in cancri dand matutid crabs, 0.19 to 0.22 μg g^{-1} in nephropid crayfish and palinurid lobster, and 0.07 to 0.23 μg g^{-1} in palemonid shrimps and 0.03 to 0.92 μg g^{-1} in penaeid shrimp.

HUMAN HEALTH RISKS REGARDING Cd–, Hg– AND Pb-CONTAMINATED CRUSTACEA CONSUMPTION

The trace metal contamination of food is gradually becoming a global crisis given that water is vulnerable to the growing discharges of pollutants at the bay on almost every continental and marine waters around the world (Ahmed et al., 2015; Lai et al., 2020). Crustaceans comprise an important fishery and aquaculture protein source, increasingly consumed by humans (Baki et al.,, 2018). The human health risk associated with the consumption of food contaminated by toxic metals have been known for a long time. Populations that often ingest crustaceans as their only protein source in estuarine and marine zones have been noted as particularly vulnerable.

Crustaceans taken from areas close to contamination sources, such as mining, industrial and agricultural areas, are likely to be exposed and accumulate metal. Food, water and sediment-traced metal can accumulatein marine organisms, such as crustaceans (crabs, lobster and shrimp) (Mansour and Sidky, 2002). The aquatic environment may be deteriorated with the increasingly release of metals from natural and anthropogenic sources, and aquatic organisms may consume metal residues from the contaminated habitats, compromising food chain and increasing health risk to humans (Gupta et al., 2009).

For example, Hg biomagnification and trophic transfer must be evaluated in areas where seafood is an important source of protein in human population's diet, as MeHg (the organic mercury) may induce neurological diseases in adults and neurocognitive dysfunctions in developing fetuses (Budnik and Casteleyn, 2019). This was observed in Japan in the 1950s due Minamata Disease, awaking humans to the danger of Hg in the environment (Eriksen and Perrez, 2014).

Previous studies have warned about the potential risks of MeHg contamination in sensitive populations, such as pregnant women, the elderly, and children, even when exposed to low doses (Dai et al., 2021). The values of metal accumulation in muscle and carapace from crustaceans were used by many researchers to calculate the estimated daily intake of metals, target hazard quotients, hazard index and target cancer risk separately for locals and tourists.

Specific legislation on the maximum concentration of Cd, Hg and Pb, as well as other metals, have been set for human consumption, to mitigate deleterious effects due to toxic metal effects. Regarding international regulatory agencies, the recommended or allowed metal concentrations for crustacea consumption by humans differ significantly, as follows: [Cd] = 0.0025 μg g^{-1} FAO/WHO, 0.001 μg g^{-1} FDA, 0.5 μg g^{-1} Brazil and EC; [Pb] = 0.005 μg g^{-1} FAO/WHO, 0.015 μg g^{-1} FDA, 0.5 μg g^{-1} Brazil and EC; [Hg] = 0,005 μg g^{-1} FAO/WHO, 0.01 μg g^{-1} FDA, 0.5 μg g^{-1} Brazil and EC. In general, several reports indicate that Cd, Pb and Hg in crustacean tissues worldwide routinely exceed these limits established, indicating significant consumer health risks. In review, aquatic crustaceans can take up trace metals from the water and from food. According to Brazilian Health Regulatory Agency (ANVISA), Cd and Hg concentration were above the maximum limits allowed by the RDC 42/2013 in fish and crustaceans collected from areas impacted by the Fundão dam failure, MG, Brazil (2–6% to Cd; 0.5–2.5% to Hg) (Brazil, 2019).

Thus, human health risk concerns regarding Crustacea consumption by humans isstill a concern, and Cd, Pb and Hg concentrations above the limits established by different regulatory agencies have been reported in the literature. Despite several studies reporting the presence of contaminants in the tissues of crustaceans were published, only few evaluated their risks for human consumption (Barrento et al., 2009; Sivaperumal et al., 2007; Turoczy et al., 2001).

Most of the risk assessments on human health have been carried out with a focus on marine and estuarine crustaceans. For freshwater, there are still few works found, information on the commercial value and consumption data of species are insufficient. *Macrobrachium amazonicum,* for example, is a crustacean distributed by a wide range of environments of fresh and brackish water in tropical and subtropical regions. In Amazon, this species has great economic relevance, and its consumption is greater in states, located in the North and Atlantic Coast region of Brazil (Costa et al., 2020), areas highly impacted by mining activity (Giusti, 2009; Gouveia and Prado, 2010). Thus, consumption of *M. amazonicum* by the local population also could be mercury exposure route.

In the right concentration, many metals are essential to life. However, in excessive or high concentration, the same metals can be considered toxic. Residues of toxic metals in the environment, if not well controlled and managed, can be hazardous to the public. Food chains are affected when a mining disaster occurs; therefore, other countries require screening of chemicals in fishery products, especially if fish and crustaceans are part of the population's diet. However, the regular screening of metals in fish and other agricultural products has not been a regular practice and is not part of a regulation in Latin American countries. Metals become toxic if not metabolized by the body and accumulate in soft tissues.

Chronic low-level exposure to unmetabolized metals and discharge from the body becomes a public health problem (Agarin et al., 2021). Routine biomonitoring efforts of impacted areas, adequate wastewater treatment, and revised maximum permissible concentration limits for Cd, Hg and Pb are required. Furthermore, laws to guarantee crustacean diversity and the protection of breeding and spawning areas should also be implemented.

REFERENCES

Adams, W.J., Blust, R., Borgmann, U., Brix, K.V., DeForest, D.K., Green, A.S. et al. 2010. Utility of tissue residuesfor predicting effects of metals on aquatic organisms. Integr. Environ. Assess. Manag. 7: 75–98.

Agarin, C.J.M., Mascareñas, D.R., Nolos, R., Chan, E. and Senoro, D.B. 2021. Transition Metals in Freshwater Crustaceans, Tilapia, and Inland Water: Hazardous to the Population of the Small Island Province. Toxics 9(4): 71.

Ahmed, M., Baki, M.A., Islam, M., Kundu, G.K., Habibullah-Al-Mamun, M., Sarkar, S.K., et al. 2015. Human health risk assessment of heavy metals in tropical fish and shellfish collected from the river Buriganga, Bangladesh. Environ. Sci. Pollut. Res. 22(20): 15880–15890.

Ali, H., Khan, E. and Ilahi, I. 2019. Environmental chemistry and ecotoxicology of hazardous heavy metals: environmental persistence, toxicity, and bioaccumulation. J. Chem. 2019: 14. Article ID 6730305. https://doi.org/10.1155/2019/6730305.

Amiard, J.C., Amiard-Triquet, C., Barka, S., Pellerin, J. and Rainbow, P.S. 2006. Metallothioneins in aquatic invertebrates: their role in metal detoxification and their use as biomarkers. Aquat. Toxicol. 76: 160–202.

Annabi, A., Bardelli, R., Vizzini, S. and Mancinelli, G. 2018. Baseline assessment of heavy metals content and trophic position of the invasive blue swimming crab Portunussegnis (Forskål, 1775) in the Gulf of Gabès (Tunisia). Mar. Pollut. Bull. 136: 454–463.

Baki, M.A., Hossain, M.M., Akter, J., Quraishi, S.B., Haque-Shojib, M.F., Atique-Ullah, A.K.M. et al. 2018. Concentration of heavy metals in seafood (fishes, shrimp, lobster and crabs) and human health assessment in Saint Martin Island, Bangladesh. Ecotoxicol. Environ. Saf. 159: 153–163.

Barrento, S., Marques, A., Teixeira, B., Carvalho, M.L., Vaz-Pires, P. and Nunes, M.L. 2009. Accumulation of elements (S, As, Br, Sr, Cd, Hg, Pb) in two populations of Cancer pagurus: ecological implications to human consumption. Food Chem. Toxicol. 47(1): 150–156.

Bertrand, L., Asis, R., Monferrán, M.V. and Amé, M.V. 2016. Bioaccumulation and biochemical response in South American native species exposed to zinc: Boosted regression trees as novel tool for biomarkers selection. Ecol. Indic. 67: 769–778.

Bordon, I.C., Sarkis, J.E., Tomás, A.R., Souza, M.R., Scalco, A., Lima, M. et al. 2012a. A preliminary assessment of metal bioaccumulation in the blue crab, *Callinectes danae* S., from the Sao Vicente Channel, Sao Paulo State, Brazil. Bull. Environ. Contam. Toxicol. 88(4): 577–581.

Bordon, I.C.A.C., Sarkis, J.E.S., Tomás, A.R.G., Scalco, A., Lima, M., Hortellani, M.A., et al. 2012b. Assessment of metal concentrations in muscles of the blue crab, *Callinectes danae* S., from the santos estuarine system. Bull. Environ. Contam. Toxicol. 89(3): 484–488.

Bordon, I.C.A.C., Sarkis, J.E.S., Andrade, N.P., Hortellani, M.A., Favaro, D.I.T., Kakazu, M.H. et al. 2016. An environmental forensic approach for tropical estuaries based on metal bioaccumulation in tissues of *Callinectes danae*. Ecotoxicology 25: 91–104.

Bordon, I.C., Emerenciano, A.K., Melo, J.R.C., da Silva, J.R.M.C., Favaro, D.I.T., Gusso-Choueri, P.K. et al. 2018. Implications on the Pb bioaccumulation and metallothionein levels due to dietary and waterborne exposures: the *Callinectes danae* case. Ecotoxicol. Environ. Saf. 162: 415–422.

Borgmann, U., Couillard, Y., Doyle, P. and Dixon, D.G. 2005. Toxicity of sixty-three metals and metalloids to *Hyalellaazteca* at two levels of water hardness. Environ. Toxicol. Chem. 24(3): 641–652.

Boudet, L.C., Mendieta, J., Romero, M.B., Carricavur, A.D., Polizzi, P., Marcovecchio, J.E., et al. 2019. Strategies for cadmium detoxification in the white shrimp Palaemonargentinus from clean and polluted field locations. Chemosphere 236: 124224.

Brazil, 2013. RESOLUÇÃO-RDC No.-42, DE 29 DE AGOSTO DE 2013- dispõe sobre o Regulamento Técnico MERCOSUL sobre Limites Máximos de Contaminantes Inorgânicos em Alimentos, Ministério da Saúde. Available in https://bvsms.saude.gov. br/bvs/saudelegis/anvisa/2013/rdc0042_29_08_2013.html#:~:text=Disp%C3%B5e%20 sobre%20o%20Regulamento%20T%C3%A9cnico,III%20e%20IV%2C%20do%20art.

Brazil, 2019. NOTA TÉCNICA Nº 8/2019/SEI/GEARE/GGALI/DIRE2/ANVISA- Avaliação de Risco: Consumo de pescado proveniente de regiões afetadaspelo rompimento da Barragem do Fundão/MG. Available in https://www.gov.br/anvisa/pt-br/arquivos-noticias-anvisa/1850json-file-1/view

Brusca, R.C., Moore, W. and Shuster, S.M. 2018. Invertebrados. 3a edição. Editora Guanabara-Koogan, Rio de Janeiro. 1010.

Budnik, L.T. and Casteleyn, L. 2019. Mercury pollution in modern times and its socio-medical consequences. Sci. Total Environ. 654: 720–734.

Burgess, N.M. and Meyer, M.W. 2008. Methylmercury exposure associated with reduced productivity in common loons. Ecotoxicology 17(2): 83–91.

Burggren, W.W. and McMahon. 1988. The Biology of Land Crabs. Cambridge University Press.

Cabrini, T.M.B., Barboza, C.A.M., Skinner, V.B., Hauser-Davis, R.A., Rocha, R.C., Saint'Pierre, T.D., et al. 2018. Investigating heavy metal bioaccumulation by macrofauna species from different feeding guilds from sandy beaches in Rio de Janeiro, Brazil. Ecotoxicol. Environ. Saf. 162: 655–662.

Capparelli, M.V., Abessa, D.M. and McNamara, J.C. 2016. Effects of metal contamination in situ on osmoregulation and oxygen consumption in the mudflat fiddler crab *Uca rapax* (Ocypodidae, Brachyura). Comp. Biochem. Physiol. Part C: Toxicol. Pharmacol. 185: 102–111.

Capparelli, M.V., Moulatlet, G.M., de Souza Abessa, D.M., Lucas-Solis, O., Rosero, B., Galarza, E. et al. 2020. An integrative approach to identify the impacts of multiple metal contamination sources on the Eastern Andean foothills of the Ecuadorian Amazonia. Sci. Total Environ. 709: 136088.

Capparelli, M.V., Cabrera, M., Rico, A., Lucas-Solis, O., Alvear-S, D., Vasco, S. et al. 2021. An Integrative Approach to Assess the Environmental Impacts of Gold Mining Contamination in the Amazon. Toxics 9: 149. https://doi.org/10.3390/toxics9070149

Carneiro, L.M., da Silva, D.J., dos Reis, L.C., de Oliveira, D.A., Maciel, L.D.C., Garcia, K.S. et al. 2018. Distribuição de elementos traço em tecidos de *Goniopsis cruentata* (latreille, 1803) capturados nos manguezais do Sul da Bahia-Brasil e avaliação do potencial de risco no consumo. Química Nova 41: 959–968.

Casado-Martinez, M.C., Smith, B.D., Luoma, S.N. and Rainbow, P.S. 2010. Metal toxicity in a sediment-dwelling polychaete: threshold body concentrations or overwhelming accumulation rates? Environ. Pollut. 158(10): 3071–3076.

Castoldi, A.F., Coccini, T., Ceccatelli, S. and Manzo, L. 2001. Neurotoxicity and molecular effects of methylmercury. Brain Res. Bull. 55(2): 197–203.

Chandini, T. 1989. Survival, growth and reproduction of *Daphnia carinata* (Crustacea: Cladocera) exposed to chronic cadmium stress at different food (Chlorella) levels. Environ. Pollut. 60(1–2): 29–45.

Çogun, H.Y., Firat, O., Aytekin, T., Firidin, G., Firat, O., Varkal, H. et al. 2017. Heavy metals in the blue crab (*Callinectes sapidus*) in Mersin Bay, Turkey. Bull. Environ. Contam. Toxicol. 98: 824–829.

Costa, B.N.S., Almeida, H.P., da Silva, B.C.P., de Figueiredo, L.G., de Oliveira, A.M. and de Oliveira Lima, M. 2020. *Macrobrachium amazonicum* (Crustacea, Decapoda) used to biomonitor mercury contamination in rivers. Arch. Environ. Contam. Toxicol. 78(2): 245–253.

Costa, R.G., Araújo, C.F.D.S., Ferreol Bah, A.H., Junior, E.A.G., Rodrigues, Y.J.D.M. and Menezes-Filho, J.A. 2018. Lead in mangrove root crab (*Goniopsis cruentata*) and risk assessment due to exposure for estuarine villagers. Food Addit. Contam.: B Surveill. 11(4): 293–301.

Costa, P.G., Marube, L.C., Artifon, V., Escarrone, A.L., Hernandes, J.C., Zebral, Y.D. et al. 2022. Temporal and spatial variations in metals and arsenic contamination in water, sediment and biota of freshwater, marine and coastal environments after the Fundão dam failure. Sci. Total Environ. 806: 151340.

Cruz, A.C.F., Gusso-Choueri, P., de Araujo, G.S., Campos, B.G. and de Sousa Abessa, D.M. 2019. Levels of metals and toxicity in sediments from a Ramsar site influenced by former mining activities. Ecotoxicol. Environ. Saf. 171: 162–172.

Dai, S.S., Yang, Z., Tong, Y., Chen, L., Liu, S.Y., Pan, R. et al. 2021. Global distribution and environmental drivers of methylmercury production in sediments. J. Hazard. Mater. 407: 124700.

De Almeida Rodrigues, P., Ferrari, R.G., Dos Santos, L.N. and Junior, C.A.C. 2019. Mercury in aquatic fauna contamination: a systematic review on its dynamics and potential health risks. J. Environ. Sci. 84: 205–218.

De Schamphelaere, K.A.C., Forrez, I., Dierckens, K., Sorgeloos, P. and Janssen, C.R. 2007. Chronic toxicity of dietary copper to *Daphnia magna*. Aquat. Toxicol. 81(4): 409–418.

Dewiyanti, I., Suryani, D. and Nurfadillah, N. 2018. Community structure of crustacean in mangrove ecosystem rehabilitation in Banda Aceh and Aceh Besar district, Indonesia. *In*: IOP Conf. Ser.: Earth Environ. Sci. 216(1): 012001. IOP Publishing.

Devi, M., Thomas, D.A., Barber, J.T. and Fingerman, M. 1996. Accumulation and physiological and biochemical effects of cadmium in a simple aquatic food chain. Ecotoxicology and Environmental Safety 33(1): 38–43.

Di Beneditto, A.P.M., Semensato, X.E.G., Carvalho, C.E.V. and de Rezende, C.E. 2019. Trace metals in two commercial shrimps from southeast Brazil: Baseline records before large port activities in coastal waters. Marine Pollution Bulletin 146: 667–670.

EFSA Scientific Committee. 2012. Guidance on selected default values to be used by the EFSA scientific committee, scientific panels and units in the absence of actual measured data. EFSA Journal 10(3): 2579.

Eisler, R. 1981. Trace Metal Concentrations in Marine Organisms. Pergamon Press. Inc., Elmsford, New York.

Ellison, J.C. 2015. Vulnerability assessment of mangroves to climate change and sea-level rise impacts. Wetl. Ecol. Manag. 23(2): 115–137.

Engel, D.W. 1987. Metal regulation and molting in the blue crab, *Callinectes sapidus*: copper, zinc, and metallothionein. Biol. Bull. 172(1): 69–82.

Eriksen, H.H. and Perrez, F.X. 2014. The minamata convention: A comprehensive response to a global problem. Rev. Eur. Comp. Int. Environ. Law 23(2): 195–210.

European Comission (E.C.) 2006. Regulamento (CE) n° 1881/2006 da comissão de 19 de Dezembro de 2006 que fixa os teores máximos de certos contaminantes presentes nos gêneros alimentícios. Jornal Oficial da UniãoEuropeia L 364: 5–24.

Ezemonye, L.I., Adebayo, P.O., Enuneku, A.A., Tongo, I. and Ogbomida, E. 2019. Potential health risk consequences of heavy metal concentrations in surface water, shrimp (*Macrobrachium macrobrachion*) and fish (*Brycinus longipinnis*) from Benin River, Nigeria. Toxicol Rep. 6: 1–9.

Fakhri, Y., Hoseinvandtabar, S., Heidarinejad, Z., Borzoei, M., Bagheri, M., Dehbandi, R. et al. 2021. The concentration of potentially hazardous elements (PHEs) in the muscle of blue crabs (*Callinectes sapidus*) and associated health risk. Chemosphere 279: 130431.

FAO, W. 2009. Principles and methods for the risk assessment of chemicals in food. Environ. Health Criteria 240.

Felten, V., Charmantier, G., Mons, R., Geffard, A., Rousselle, P., Coquery, M. et al. 2008. Physiological and behavioural responses of *Gammarus pulex* (Crustacea: Amphipoda) exposed to cadmium. Aquat. Toxicol. 86(3): 413–425.

Food and Agriculture Organization/Word Oraganization Health (FAO/WHO). 1982. Evaluation of certain food additives and contaminants, Twenty-Sixth Report of the Joint FAO/WHO Expert Committee on Food Additives; World Health Organization: Geneva, Switzerland.

Food and Agriculture Organization/Word Oraganization Health (FAO/WHO). 2011. Evaluation of certain food additives and contaminants, Seventy-Third Report of the Joint FAO/WHO Expert Committee on Food Additives; World Health Organization: Geneva, Switzerland, 201.

Freire, C.A., Onken, H. and McNamara, J.C. 2008. A structure-function analysis of ion transport in crustacean gills and excretory organs. Comp. Biochem. Physiol. 151A: 272–304.

Frías-Espericueta, M.G., Bautista-Covarrubias, J.C., Osuna-Martínez, C.C., Delgado-Alvarez, C., Bojórquez, C., Aguilar-Juárez, M. et al. 2022. Metals and oxidative stress in aquatic decapod crustaceans: A review with special reference to shrimp and crabs. Aquat. Toxicol. 242: 106024.

Genç, T.O. and Yilma, F. 2015. Bioaccumulation indexes of metals in blue crab inhabiting specially protected area Koycegiz Lagoon (Turkey). Indian J. Anim. Sci. 85(1): 994–999.

Giusti, L. 2009. A review of waste management practices and their impact on human health. Waste Manag. 29(8): 2227–2239.

Gouveia, N. and Prado, R.R.D. 2010. Health risks in areas close to urban solid waste landfill sites. Revista de saude publica. 44: 859–866.

Gupta, A., Rai, D.K., Pandey, R.S. and Sharma, B. 2009. Analysis of some heavy metals in the riverine water, sediments and fish from river Ganges at Allahabad. Environ. Monit. Assess., 157(1): 449-458.

Henry, R.P., Lucu, C., Onken, H. and Weihrauch, D. 2012. Multiple functions of the crustacean gill: osmotic/ionic regulation, acid-base balance, ammonia excretion, and bioaccumulation of toxic metals. Front. Physio. 3: 431. doi: 10.3389/fphys.2012.00431.

Hosseini, M., Nabavi, S.M.B., Parsa, Y. and Ardashir, R.A. 2014. Mercury accumulation in selected tissues of shrimp *Penaeus merguiensis* from Musa estuary, Persian Gulf: variations related to sex, size, and season. Environ. Monit. Assess 186: 5439–5446.

Ip, C.C.M., Li, X.D., Zhang, G., Wong, C.S.C. and Zhang, W.L., 2005. Heavy metal and Pb isotopic compositions of aquatic organisms in the Pearl River Estuary, South China. Environ. Pollut. 138: 494–504.

Kadiene, E.U., Meng, P.J., Hwang, J.S. and Souissi, S. 2019. Acute and chronic toxicity of cadmium on the copepod Pseudodiaptomus annandalei: A life history traits approach. Chemosphere 233: 396–404.

Karar, S., Hazra, S. and Das, S. 2019. Assessment of the heavy metal accumulation in the Blue Swimmer Crab (*Portunus pelagicus*), northern Bay of Bengal: Role of salinity. Mar. Pollut. Bull. 143: 101–108.

Klerks, P.L. and Weis, J.S. 1987. Genetic adaptation to heavy metals in aquatic organisms: A review. Environ. Pollut. 45(3): 173–205.

Lai, Q.T., Irwin, E.R. and Zhang, Y. 2020. Quantifying harvestable fish and crustacean production and associated economic values provided by oyster reefs. Ocean Coast. Manag. 187: 105104.

Lillebo, A.I., Coelho, P.J., Pato, P., Válega, M., Margalho, R., Reis, M. et al. 2011. Assessment of mercury in water, sediments and biota of a Southern European Estuary (Sado Estuary, Portugal). Wat. Air and Soil Poll. 214(1): 667–680.

Livingstone, D.R. 2001. Contaminant-stimulated reactive oxygen species production and oxidative damage in aquatic organisms. Mar. Pollut. Bull. 42(8): 656–666.

Luoma, S.N. and Rainbow, P.S. 2008. Metal Contamination in Aquatic Environments: Science and Lateral Management. Cambridge, New York.

Mansour, S.A. and Sidky, M.M. 2002. Ecotoxicological studies. 3. Heavy metals contaminating water and fish from Fayoum Governorate, Egypt. Food Chem. 78(1): 15–22.

Marques, A., Nunes, M.L., Moore, S.K. and Strom, M.S. 2010. Climate change and seafood safety: Human health implications. Food Res. Int. 43(7): 1766–1779.

Martins, C.M., Menezes, E.J., Giacomin, M.M., Wood, C.M. and Bianchini, A. 2011. Toxicity and tissue distribution and accumulation of copper in the blue crab *Callinectes sapidus* acclimated to different salinities: *in vivo* and in *vitro studies*. Aquat. Toxicol. 101: 88–99.

Maunoury-Danger, F., Felten, V., Bojic, C., Fraysse, F., Ponce, M.C., Dedourge-Geffard, O., et al. 2018. Metal release from contaminated leaf litter and leachate toxicity for the freshwater crustacean *Gammarus fossarum*. Environ. Sci. Pollut. Res. 25(12): 11281–11294.

Melo, G.A.S. 1996. Manual de identificação dos Brachyura (caranguejos e siris) do litoral brasileiro. Plêiade/FAPESP Ed., São Paulo.

Moore, P.G. and Rainbow, P.S. 1987. Copper and zinc in an ecological series of talitroidean Amphipoda (Crustacea). Oecologia 73: 120–126.

Moreno, P.A.R., Medesani, D.A. and Rodrıguez, E.M. 2003. Inhibition of molting by cadmium in the crab *Chasmagnathus granulata* (Decapoda Brachyura). Aquat. Toxicol. 64(2): 155–164.

Núñez-Nogueira, G., Mouneyrac, C., Muntz, A. and Fernandezbringas, L. 2010. Metallothionein-like proteins and energy reserve levels after Ni and Pb exposure in the Pacific White Prawn *Penaeus vannamei*. J. Toxicol. 2010: Article ID 407360.

Núñez-Nogueira, G., Fernández-Bringas, L., Ordiano-Flores, A., Gómez-Ponce, A., De León-Hill, C.P. and González-Farías, F. 2012. Accumulation and regulation effects from the metal mixture of Zn, Pb, and Cd in the tropical shrimp Penaeus vannamei. Biol.Trace. Elem. Res. 150: 208–213.

Ololade, I.A., Lajide, L., Olumekun, V.O., Ololade, O.O. and Ejelonu, B.C. 2011. Influence of diffuse and chronic metal pollution in water and sediments on edible seafoods within Ondo oil-polluted coastal region, Nigeria. J. Environ. Sci. Health, Part A 46(8): 898–908.

O'Mara, K., Adams, M., Burford, M.A., Fry, B. and Cresswell, T. 2019. Uptake and accumulation of cadmium, manganese and zinc by fisheries species: Trophic differences in sensitivity to environmental metal accumulation. Sci. Total Environ. 690: 867–877.

Parsa, Y., Nabavi, S.S.M.B., Nabavi, S.N. and Hosseini, M. 2014. Mercury accumulation in food chain of fish, crab and sea bird from Arvand River. J. Mar. Sci., Res. Dev. 4: 1000148.

Peña-Fernández, A., González-Muñoz, M.J. and Lobo-Bedmar, M.C. 2014. Establishing the importance of human health risk assessment for metals and metalloids in urban environments. Environ. Int. 72: 176–185.

Rainbow, P.S. 1993. The significance of trace metal concentration in marine invertebrates. pp. 3–23. *In*: Dallinger, R. and Rainbow, P.S. (eds). Ecotoxicology of Metals in Invertebrates. SETAC Special Publication, Lewis Publishers, Boca Raton, Florida.

Rainbow, P.S. 1995. Physiology, physicochemistry and metal uptake—A crustacean perspective. Mar. Pollut. Bull. 31(1–3): 55–59.

Rainbow, P.S. 1997. Ecophysiology of trace metal uptake in crustaceans. Estuar. Coast. Shelf Sci. 44(2): 169–176.

Rainbow, P.S. 1998. Phylogeny of trace metal accumulation in crustaceans. pp. 285–319. *In*: Langston, W.J. and Bebianno, M.J. (eds). Metal Metabolism in Aquatic Environments. Springer, Boston, MA.

Rainbow, P.S. 2002. Trace metal concentrations in aquatic invertebrates: why and so what? Environ. Pollut. 120: 497–507.

Rainbow, P.S. 2007. Trace metal bioaccumulation: models, metabolic availability and toxicity. Environ. Int. 33(4): 576–582.

Rainbow, P.S. 2018. Heavy metal levels in marine invertebrates. pp. 67–79. *In*: Furness, R.W. and Rainbow, P.S. (eds). Heavy Metals in the Marine Environment. CRC Press, Boca Raton, FL.

Rainbow, P.S. and Luoma, S.N. 2011. Metal toxicity, uptake and bioaccumulation in aquatic invertebrates—modelling zinc in crustaceans. Aquat. Toxicol. 105(3-4): 455–465.

Ramos, R.J., Tadokoro, CE., de Carvalho Gomes, L. and Leite, G.R. 2021. Efficiency in heavy metal purge in crustaceans during the ecdysis. Environ. Dev. Sustain. 1–3S0.

Sá, M.G., Valenti, W.C. and Zanotto, F.P. 2008. Dietary copper absorption and excretion in three semi-terrestrial grapsoid crabs with different levels of terrestrial adaptation. Comp. Biochem. Physiol. C: Toxicol. Pharmacol. 148(2): 112–116.

Sabatini, S.E., Chaufan, G., Juárez, Á.B., Coalova, I., Bianchi, L., Eppis, M.R., et al. 2009. Dietary copper effects in the estuarine crab, *Neohelice (Chasmagnathus) granulata*, maintained at two different salinities. Comp. Biochem. Physiol. C: Toxicol. Pharmacol. 150(4): 521–527.

Shaiek, M., Zaaboub, N., Ayas, D., Martins, M.V.A. and Romdhane, M.S. 2018. Crabs as bioindicators of trace element accumulation in Mediterranean Lagoon (Bizerte Lagoon, Tunisia). J. Sediment. Environ. 3(1): 1–11.

Silva, B.M.D.S., Morales, G.P., Gutjahr, A.L.N., Faial, K.D.C.F. and Carneiro, B.S. 2018. Bioacumulation of trace elements in the crab *Ucidescordatus* (Linnaeus, 1763) from the macrotidal mangrove coast region of the Brazilian Amazon. Environ. Monit. Assess. 190(4): 1–15.

Sivaperumal, P., Sankar, T.V. and Nair, P.V. 2007. Heavy metal concentrations in fish, shellfish and fish products from internal markets of India vis-a-vis international standards. Food Chem. 102(3): 612–620.

Truchet, D.M., Buzzi, N.S., Simonetti, P. and Marcovecchio, J.E. 2020. Uptake and detoxification of trace metals in estuarine crabs: Insights into the role of metallothioneins. Environ. Sci. Pollut. Res. 27: 31905–31917.

Tu, N.P.C., Ha, N.N., Ikemoto, T., Tuken, B.C., Tanabe, S. and Takeuchi, I. 2008. Bioaccumulation and distribution of trace elements in tissues of giant river prawn *Macrobrachium rosenbergii* (Decapoda: Palaemonidae) from South Vietnam. Fish. Res. 74: 109–119.

Turoczy, N.J., Mitchell, B.D., Levings, A.H. and Rajendram, V.S. 2001. Cadmium, copper, mercury, and zinc concentrations in tissues of the King Crab (*Pseudocarcinus gigas*) from southeast Australian waters. Environ. Int. 27(4): 327–334.

United States Environmental Protection Ambietal (USEPA). 2007. Tolerable daily intake by metals. Available in: http://www.popstoolkit.com/tools/HHRA/TDI_USEPA.aspx. Switzerland. 2007

Van Strallen, N.M., Donker, M.H., Vijver, M.G. and Van Gestel, C.A.M. 2005. Bioavailability of contaminants estimated from uptake rates into soil invertebrates. Environ. Pollut. 136(3): 409–417.

Viarengo, A. and Nott, J.A., 1993. Mechanisms of heavy metal cation homeostasis in marine invertebrates. Comp. Biochem. Physiol. 104C: 355–372.

Vijver, M.G., van Gestel, C.A.M., Lanno, R.P., van Straalen, N.M. and Peijnenburg, W.J.G.M. 2004. Internal metal sequestration and its ecotoxicological relevance: An overview. Environ. Sci. Technol. 38: 4705–4712.

Vitale, A.M., Monserrat, J.M., Castilho, P. and Rodriguez, E.M. 1999. Inhibitory effects of cadmium on carbonic anhydrase activity and ionic regulation of the estuarine crab *Chasmagnathus granulata* (Decapoda, Grapsidae). Comp. Biochem. Physiol. C: Pharmacol. Toxicol. Endocrinol. 122(1): 121–129.

Vogt, G. and Quinitio, E.T. 1994. Accumulation and excretion of metal granules in the prawn, *Penaeus monodon*, exposed to water-borne copper, lead, iron and calcium. Aquat. Toxicol. 28(3–4): 223–241.

Wang, W.X. 2002. Interactions of trace metals and different marine food chains. Mar. Ecol. Prog. 243: 295–309.

Wang, W.X. and Fisher, N.S. 1999. Assimilation efficiencies of chemical contaminants in aquatic invertebrates: a synthesis. Environ. Toxicol. Chem. 18: 2034–45.

Wang, W.X. and Rainbow, P.S. 2006. Subcellular partitioning and the prediction of cadmium toxicity to aquatic organisms. Environ. Chem. 3: 395–399.

Zhao, S., Feng, C., Quan, W., Chen, X., Niu, J. and Shen, Z. 2012. Role of living environments in the accumulation characteristics of heavy metals in fishes and crabs in the Yangtze River Estuary, China. Mar. Pollut. Bull. 64(6): 1163–1171.

Chapter 5

Accumulation of Lead, Mercury and Cadmium in Coastal Sediments in the Eastern Mediterranean Sea

Debra Ramon[1,2], Malka Britzi[3], Nadav Davidovich[1,2,4], Dan Tchernov[1,2] and Danny Morick*[1,2,5]

[1]Department of Marine Biology, Leon H. Charney School of Marine Sciences, University of Haifa, Haifa, 3498838 Israel.

[2]Morris Kahn Marine Research Station, University of Haifa, Haifa, 3498838 Israel.

[3]National Residue Control Laboratory, Kimron Veterinary Institute, Bet Dagan, 5025001, Israel.

[4]Israeli Veterinary Services, Bet Dagan, 5025001, Israel.

[5]Hong Kong Branch of Southern Marine Science and Engineering Guangdong Laboratory (Guangzhou), Hong Kong, China.

MEDITERRANEAN SEA: A BRIEF BACKGROUND

The Mediterranean Sea is a semi-enclosed oligotrophic basin fed by the Atlantic Ocean through the narrow Straights of Gibraltar and is characterized by its high salinity and temperatures. It is bound between three different continents with Europe to the north, Africa to the south, and Asia to the east. Compared to the world's ocean, the continental shelf of the Mediterranean Sea is more dominant, making up 20% of the sea floor (Coll et al., 2010; Pinardi and Zavatarelli, 2006)

---

*Corresponding author: dmorick@univ.haifa.ac.il

While the Mediterranean Sea is considered small compared to other areas of the ocean, contributing to only 0.82% of its surface area and 0.32% of the volume (El-Geziry and Bryden, 2010), it hosts 4–18% of the world's marine biodiversity that includes many endemic species (Durrieu de Madron et al., 2011). It is also home to numerous emblematic species and habitats (Coll et al., 2010) including the Mediterranean monk seal, cetaceans, sea turtles, bluefin tuna, elasmobranchs, seagrass beds, and unique deep sea habitats.

The sea is considered an evaporate basin due to its low precipitation and high evaporation, which influences its unique circulation and biogeochemical processes. Upper Atlantic Ocean waters, already depleted in nutrients, enter the Mediterranean Sea through the narrow Straights of Gibraltar, which is 13 km wide and does not exceed 1,000 m in depth. A strong west-east gradient exits, with surface waters becoming increasingly warmer, saltier, and depleted of nutrients. This results in a corresponding decrease in the productivity as well. External sources of water input into the Mediterranean include the Black Sea and river basins. While the Nile River is one of the largest rivers in the world, the construction of the Aswan Dam has limited its flow into the Mediterranean. Other river inputs are mainly from the European boundary as very few perennial rivers flow from the African Coastline. However, the geology of the coast combined with its short shelf means that these European drainage systems contribute very little to the overall input (El-Geziry and Bryden, 2010).

The basin can be divided into two sub-basins, a western and an eastern basin (El-Geziry and Bryden, 2010) with the Straits of Sicily acting as a physical divider between the two. The geographic structure of the Sicilian-Tunisian sill in the Straits of Sicily, found at 400 m in depth, limits the hydrological transfer between the two (El-Geziry and Bryden, 2010). Thus, essentially, the Mediterranean Sea acts as a lagoon for the Atlantic Ocean, and the eastern basin a lagoon for the western basin. The two differ in their characteristics, with the eastern basin being larger by two folds and is more complex than its western companion. While the western basin is characterized by five different seas, the connectivity between them is non-confining, unlike in the eastern basin. The eastern basin comprises of four sub-seas: the Ionian Sea, the Levantine Sea, the Adriatic Sea, and the Aegean Sea (Figure 5.1). Directly eastward to the Straits of Sicily, the basin opens into the Ionian Sea, located between Italy and Greece. The deepest point of eastern basin is found here with a depth of 5 km just south of Greece (El-Geziry and Bryden, 2010). Eastward to the Ionian Sea lies the Levantine Sea, with its deepest point near Rhodes at 4.5 km. North of the Ionian Sea is the Adriatic Sea, which is separated by the 75 m wide Strait of Otranto with a sill at 800 m depth. The Adriatic Sea elongates into the European continent and is confined on either side by two major mountain ranges. To the north of the Levantine Sea lies the Aegean Sea, bound by Greece to its west and Turkey to its east. The Aegean Sea is quite shallow compared to the rest of the basin, with a maximum depth of 1.5 km. It is not an open basin, and unlike the other sub-basins in the eastern basin, it consists of many islands scattered within. The Black Sea highly influences the Aegean Sea, which flows directly into it, bringing a significant load of fresh water and nutrients.

There is immense human pressure around the entire Mediterranean basin. Coastal urbanization is particularly intense as the coast hosts a large population with 132 million people (Durrieu de Madron et al., 2011). Agricultural runoff via rivers brings heavy loads of pollutants (i.e., hydrocarbons, pesticides, herbicides, fertilizers, litter) from both coastal and inland areas, with rivers acting as transboundary transports. Industrial activities are scattered around the basin, with 161 identified marine pollutant hotspots often associated with large industrial areas and harbors (Antonio et al., 2014; EEA, 2006). While not up-to-date, the European Environmental Agency (EEA) has summarized such hotspots around the entire Mediterranean Sea, many of which are still relevant today (EEA, 2006). These industries include sectors such as petro-chemical infrastructure, chemical production, plastic production, metal treatment, and more (EEA, 2006). There is also particular concern with obsolete chemicals which are stockpiled near coastal areas and can leach into the marine environment (EEA, 2006). With urbanization and industrialization, treatment of both domestic and industrial wastewaters are problematic (EEA, 2006). The Mediterranean Sea is utilized by all coastal countries as a final release point for different levels of treated waste, with wastewater treatment regulations unique to each individual continent and country. Often, large harbors are coupled with heavily industrialized areas in order to logistically transfer goods between countries. Maritime traffic in the Mediterranean is particularly dense as the Suez Canal acts as a maritime bridge between the east and the west. Approximately 30% of the global merchant shipping fleet and 20% of the oil shipping fleet passes through the Mediterranean Sea each year (Antonio, et al., 2014). With intense maritime traffic, marine-based pollution is a prominent threat, resulting in chronic shipping discharge as well as acute incidents such as oil spills. In addition to the added anthropogenic input of metals, it has been suggested that natural background levels of metals in the Mediterranean Sea may be higher compared to other areas of the ocean due to tectonic activity around the central and western basin (Copat et al., 2014). Furthermore, due to the oligotrophic nature of the of the Mediterranean, the system is more sensitive to bio-accumulation processes as the bio-dilution of contaminants by organic carbon is reduced (Durrieu de Madron et al., 2011).

METAL RESEARCH IN THE EASTERN MEDITERRANEAN SEA

The Eastern Mediterranean Sea (EMS) is an interesting study area in many aspects of marine sciences, yet research in this entire area is highly lacking. Sixteen countries border the EMS including Italy, Malta, Slovenia, Croatia, Bosnia and Herzegovina, Montenegro, Albania, Greece, Turkey, Cyprus, Syria, Lebanon, Israel, Egypt, Libya, and Tunisia. Some countries have had more focus on metal research in their local waters (i.e., Italy, Turkey, and Greece), others with moderate research (such as Israel, Libya, and Tunisia), while some have almost no studies to come by (i.e., Syria, Albania, and Montenegro). As basin wide studies are extremely limited, there is little understanding of basin wide

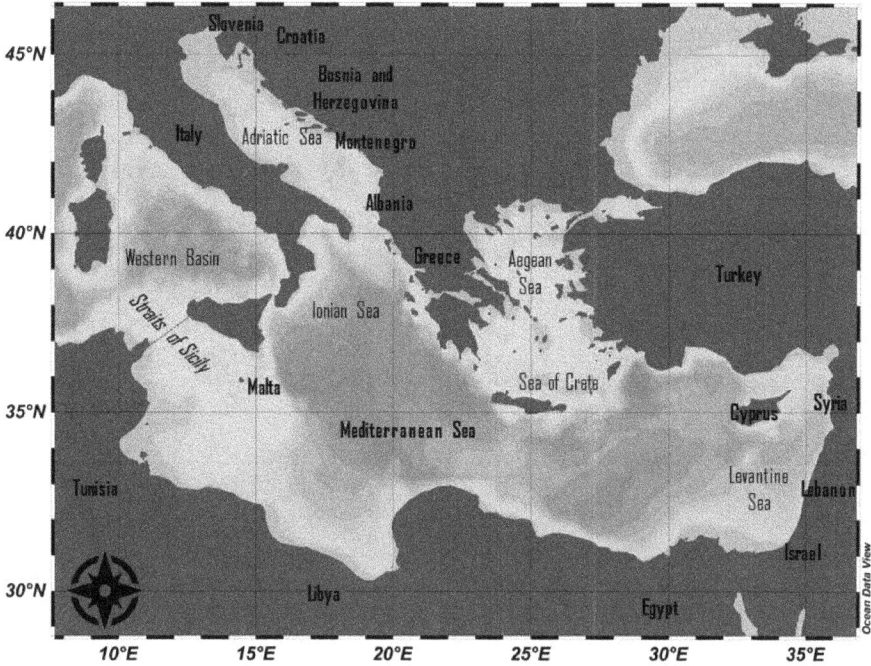

Figure 5.1 Map of the Eastern Mediterranean Sea.

metal dynamics. Regional focused studies (such as those assessing the Aegean) are much more common, with a wealth of information coming from these areas. Yet, these regions are just a minor makeup of the entire EMS, still limiting our understanding of accumulation patterns. Therefore, understanding metal accumulation in the EMS comes from locally produced research originating from individual countries. Unfortunately, these studies, especially more recent ones, have never been collected and assessed on a more regional scale. Thus, through this chapter, we hope to contribute to a more regional understanding as we provide a brief summary of metal assessment studies in sediments from the different sub-regions of the EMS.

This summary focuses on the accumulation of mercury, cadmium and lead in sediments as an indicator of ecotoxicological potential due to both the availability of studies in the field as well as the comparison ability due to a relatively normalized set of standards. Currently, sediment analysis can be advantageous over biota studies as precise sediment pollution indexes exist that allow for the determination of enrichment, contamination, and potential transfer to biota. In comparison, metal levels in marine biota are open to ecotoxicological interpretation, with existing standards focusing on the human health aspect. Sediments play a major role in metal accumulation and are highly relevant to ecotoxicology studies. While the major route in which metals reach the marine environment is through either aerosols or land-based effluents, and the initial interface interaction is within the aquatic phase, only a small fraction of these contaminants remains suspended in the water column (Bonsignore et al., 2018). Instead, it is the sediments which act as the major sink for metals through processes such as adsorption and absorption onto sediment, precipitation, and interaction with the organic fraction (Gargouri et al., 2011). More than 90% of marine inorganic pollutants bind to sediments (Damak et al., 2019), thus acting as major reservoirs for pollutants. The level of absorption depends on its sediment qualities such as grain size, geochemical composition, and organic fracture, as well as environmental factors like salinity, temperature, and pH (summarized in El Baz and Khalil, 2018). With sediments acting as a concentrator for metals, as well as acting as sites for chemical speciation which can improve its bioavailability, they are an important source of metal transfer to biological systems. Therefore, by assessing sediment concentrations, the level of contamination can be assessed for potential ecological risk.

Coastal areas in particular are adequately studied as they are logistically more accessible and point sources of pollution are more easily identifiable. Seawater enrichment of metals often occurs intensively throughout coastal areas due to the high urbanization of coastlines, which contribute to industrial, urban, and agricultural inputs into the marine environment (EEA, 2006; UNEP/MAP, 2012). Regardless of the origin of the land-based metal source, contaminated coastal sediments constitute an important secondary non-point pollution source as they release metals into the overlying water. As metals tend to precipitate after their introduction into the coastal marine environment they accumulate in sediments and biota. This occurs especially in sheltered areas that enhance accumulation such as harbors and semi-enclosed bays (Merhaby et al., 2018) in the vicinity of land-based metal sources. Increased metal concentrations have been identified in many coastal areas in the Mediterranean Sea, such as the coast of Tuscany (Tyrrhenian Sea), Kastella Bay (Adriatic Sea), Haifa Bay and the coast of Alexandria (eastern Mediterranean), and Izmir Bay and Elefsina Bay (Aegean Sea) (EEA, 2006). With the EMS being used by numerous countries with varying social economic backgrounds and research advancements, coastal waters remain the primary focus for contaminant studies.

In this chapter we provide a brief overview of the different regions of the EMS including the; (1) north African coastline, (2) Levantine Sea, (3) Aegean Sea, (4) Adriatic Sea, and (5) Ionian Sea. Though the north African coastline is not considered an official sub-region of the EMS, it is included as a sub-division in this chapter due to the similarities between the countries it borders. Within each region, a brief description is provided for each individual country on the state of the marine environment as well as the anthropogenic activity that may affect local pollution levels. Additionally, a brief overview of the metal research in sediments is provided.

NORTH AFRICAN COASTLINE

The North African coastline of the EMS is bordered by Tunisia, Libya, and Egypt (Figure 5.2). Despite the emergence of increased research of the marine system from this region over recent years, the scientific literature is still quite limited with baseline studies on biodiversity often lacking. Sampling efforts for this region center mostly in the vicinity of industrial areas, especially surrounding the harbors and ports of sizable cities. Therefore, while some studies can identify specific industries as sources of metal enrichment of marine coastal sediments, others are influenced by a cocktail of polluters from both the industrial and domestic sector. Most studies focus on areas identified as hotspots according to the EEA (EEA, 2006). Sediment enrichment along the North African coastline is shown in Figure 5.2 and concentrations are shown in Table 5.1.

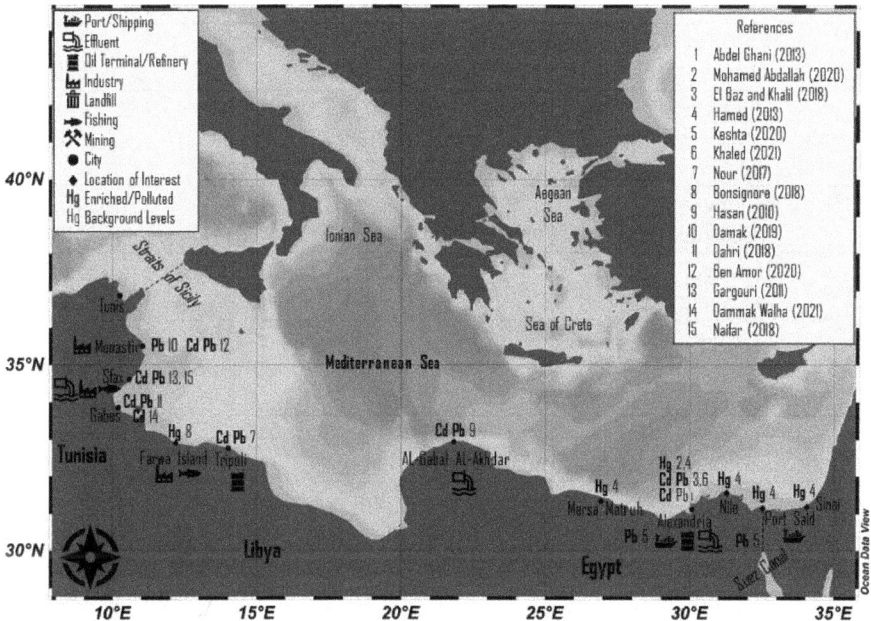

Figure 5.2   Metal enrichment in sediments along the North African Coastline including Tunisia, Libya, and Egypt.

Tunisia

Tunisia divides its 1,300 km coastline between both the western and eastern basin of the Mediterranean Sea. It is currently undergoing rapid urban growth and industrialization along the coastline, and the marine system suffers negative impacts from wastewater discharge, industrial activities (i.e., textiles, cement, and phosphate), agriculture, and urban development. Additionally, Tunisia hosts Mediterranean tapeweed (*Posidonia oceanica*) meadows, which compose a highly valuable marine ecosystem for many species (Telesca et al., 2015).

Over the recent years, Tunisia has shown a boost in research focusing on marine contaminants. In the eastern basin, the waters in front of the cities Gabes and Sfax have been identified as pollution hot spots (Figure 5.2) (EEA, 2006). Both cities are situated on the Gulf of Gabes, a highly productive area responsible

Table 5.1 Studies assessing metal accumulation in sediments along the North African coastline. Metal concentrations ($\mu g.g^{-1}$, dry weight – dw) are reported depending on data provided from within the studies as either mean value ±SD, range within parentheses, or both. Values in bold indicate that sediments were considered either contaminated or enriched by the study

Location	Hg	Cd	Pb	References
Egypt (Abu-Qir Bay – Alexandria)	– (–)	**2.93** **(0.31–4.89)**	8.2 (1.9–16.79)	Abdel Ghani et al. (2013)
Egypt (Eastern Harbor – Alexandria)	– (–)	**1.11** **(0.3–1.83)**	**40.57** **(1.3–112.09)**	Abdel Ghani et al. (2013)
Egypt (Western Harbor –Alexandria)	**4.07** **(1.01–6.6)**	– (–)	– (–)	Abdallah (2020)
Egypt (Central Zone – El Mex to Port Said)	– (–)	0.16 (0.06–0.42)	14.75 (5.3–57)	El Baz et al. (2018)
Egypt (Coastline)	**0.01–0.02**	– (–)	– (–)	Hamed et al. (2013)
Egypt (Alexandria)	– (–)	0.29±0.07 (–)	32.55±6.49 (–)	Khaled et al. (2021)
Libya (Sabratha)	– (–)	0.83 (0.19–2.19)	**11.69** **(2.16–38.22)**	Nour and El-Sorogy (2017)
Libya (Farwa)	– **(0.01–0.16)**	– (–)	– (–)	Bonsignore et al. (2018)
Libya (Al-Gabal Al-Akhda)	– (–)	– **(0.79–1.4)**	– **(1.4–7.5)**	Hasan et al. (2010)
Tunisia (Monastir Bay)	– (–)	– (–)	**5.29±3.54** **(1.48–13.04)**	Damak et al. (2019)
Tunisia (Gabes Catchment)	– (–)	**0.68** **(0–5.62)**	28.12 (0.13–162.7)	Dahri et al. (2018)
Tunisia (Monastir Bay)	– (–)	– **(0.01–3.45)**	– **(0–47)**	Ben Amor et al. (2020)
Tunisia (Sfax Coast)	– (–)	5.9±0.5 (5.5–7)	32±17 (18–88)	Gargouri et al. (2011)
Tunisia (Gulf of Gabes)	– (–)	**0.4±0.14** (–)	– (–)	Dammak Walha et al. (2021)
Tunisia (Gulf of Gabes)	– (–)	– **(0.55–24.52)**	– (3.68–39.7)	Naifar et al. (2018)

for 40% of the country's fish landings and important habitats for many marine organisms (Béjaoui et al., 2019). Due to its economic and ecological importance, the gulf has been well researched. This area in particular is negatively impacted by industrial activity like phosphate factories (EEA, 2006), which has been attributed to elevated Cd contamination in sediments (Dammak Walha et al., 2021; El Zrelli et al., 2018). The metals Pb and Cd were also found elevated near channel outlets leading into the gulf due to effluents released throughout the catchment by both industrial and domestic activity (Ben Amor et al., 2020; Naifar et al., 2018), though industrial sources appear to be the major source (El Zrelli et al., 2018). Concentrations were higher especially in muddy sediments (Ben Amor et al., 2020; Gargouri et al., 2011), with decreasing levels of contamination further from shore with increasing grain size and organic matter levels. Additionally, a study investigating the impacts of fish farming showed that Pb was enriched directly below the fish cages (Damak et al., 2019).

Libya

Libya is particularly limited in research despite having the most dominant coastline in the southern border of the Mediterranean Sea, comprising of approximately 1,970 kilometers (IUCN, 2011). Many pristine areas exist along the coast with the gray literature describing the Libyan marine system as a wealthy one with three key Mediterranean hotspots (IUCN, 2011) home to a range of marine species such as fish, larger pelagic fish, elasmobranchs, benthic organisms, sea turtles, and sea birds (Regional Activity Centre for Specially Protected Areas, 2016). Anthropogenic activity including coastal urbanization, discharge of sewage water, construction debris, petrochemical industry, oil terminals, ports, and other industries (Bonsignore et al., 2018; Regional Activity Centre for Specially Protected Areas, 2016; IUCN, 2011) have been noted as considerable threats to the health of the marine ecosystem and the fisheries that depend on it. The country has no perennial rivers that provide a constant supply of water throughout the year (EEA, 2006).

Libyan research is highly limited with only a handful of articles published on sediment contamination (Figure 5.2, Table 5.1). Pollution hotspots have been identified around the ports of Tripoli, Benghazi, and Tobruk (EEA, 2006) with Tripoli and Benghazi being the two largest cities in the country with populations exceeding one million inhabitants. On the western border of Libya, an area with active chemical industries as well as known fishing area (Farwa Island) observed that sediments closer to shore were found to be enriched and diluted with depth (Bonsignore et al., 2018). Additionally, commercially important fish species showed Hg accumulation, with sediment sifting *Mullus* species (a known bioindicator for metals in the Mediterranean Sea), particularly susceptible to bioaccumulation (Bonsignore et al., 2018). Less than 100 km east of Farwa Island is the city of Sabratha, which hosts a fishing port as well as the Mellitah Complex Oil and Gas just beyond the city, which is the main point in western Libya for gas and oil export to Italy. Sediments here were shown to be enriched with both Pb and Cd, with particularly elevated Cd levels near the fishing port and highest levels of Pb were observed near the oil and gas complex (Nour and El-Sorogy, 2017).

Egypt

Egypt's 950 km coastline is mostly dominated by sandy beaches with occasional rocky shores (Fouda, 2017). It is the endpoint of the Nile River, which drains into the Mediterranean Sea. It also hosts the Suez Canal, which creates a water bridge between the Red and Mediterranean Sea, thus making the Mediterranean one of the busiest waterways in the world and susceptible to an elevated risk of oil spills. While it's marine biodiversity has been described as relatively poor in comparison to other regions of the Mediterranean Sea, it still hosts a diversity of marine organisms many of which are endangered (Fouda, 2017). The Sinai Mediterranean coast has remained quite pristine, while other sections of Egypt's coastline are influenced by anthropogenic activity. Due to urbanization, 20% of the population have settled in coastal areas and over 40% of the local industry centered around coastal zones includes ports, petroleum, and mining (Fouda, 2017). River drainage systems ultimately reaching the sea faces pollution threats from wastewater discharge (domestic and industrial) as well as from agricultural runoff (Fouda, 2017). Additionally, large lagoons/urban lakes situated along the coast act as accumulation points of metals before flowing into the sea (Keshta, et al., 2020).

In Egypt, major research focus has been placed on the central zone, defined between Alexandria and Port Said (Figure 5.2). This area is considered some of the most industrialized areas of the Mediterranean, particularly in port cities, due to the amount of shipping that passes through Egypt's Suez Canal. As a result, this area is considered a pollution hotspot by the EEA (EEA, 2006). The sediment of Alexandria has been particularly well sampled, showing enrichment of Hg (Abdallah, 2020; Hamed et al., 2013), Cd (Abdel Ghani et al., 2013; Khaled et al., 2021), and Pb (Abdel Ghani et al., 2013; Khaled et al., 2021) (Figure 5.2, Table 5.1). While Alexandria's harbor receives a cocktail of inputs including wastewater, petrochemical, and cement factory discharge, the elevated contamination levels of Hg within the harbor has been shown to come from a local chlor-alkali plant (Abdallah, 2020). As a result fish feeding on the organic material in the bottom sediments were shown to have corresponding elevated Hg levels (Abdallah, 2020). Cadmium is believed to originate from the industrial sludge from wastewater treatment while Pb is a result of the manufacturing of batteries and petrol fuel (El Baz and Khalil, 2018). These metals also accumulate in the Egypt's northern lakes located in urban areas adjacent to the coastline (Keshta et al., 2020), showing the contamination potential in such urban environments from runoff.

THE LEVANTINE SEA

The Levantine Sea is the eastern most section of the Mediterranean Sea. It is bordered by Egypt, Gaza, Israel, Lebanon, Syria, the island of Cyprus, and eastern Turkey (Turkey's Mediterranean region). This area is considered extremely oligotrophic with saline and warm waters. Due to the long shore current traveling north along the eastern boundary, the Nile River plays a major influence in this area. Additionally, the opening of the Suez Canal, which allows ship passage between the Red Sea and Mediterranean Sea, has brought with it a number of invasive species (i.e., *Pterois miles*—common lionfish, *Surida lessepsianus*—lizardfish, *Siganus rivulatus*—marbled spinefoot rabbitfish) in the Lessepsian migration (EEA, 2006) resulting in changes in the marine ecology of the entire region. The region has become a major source of economic interest with the discovery of gas fields, leading to both exploration of this resource as well as the establishment of infrastructure used for its extraction. Importantly, this region is highly impacted by socio-economic issues, ultimately influencing the quality of marine management and the research of the marine system. For this section, we focus on exploring Gaza Strip Palestine, Israel, Lebanon, Syria, Cyprus, and the Mediterranean region of Turkey. The accumulation of metals in this region is summarized in Figure 5.3 and Table 5.2.

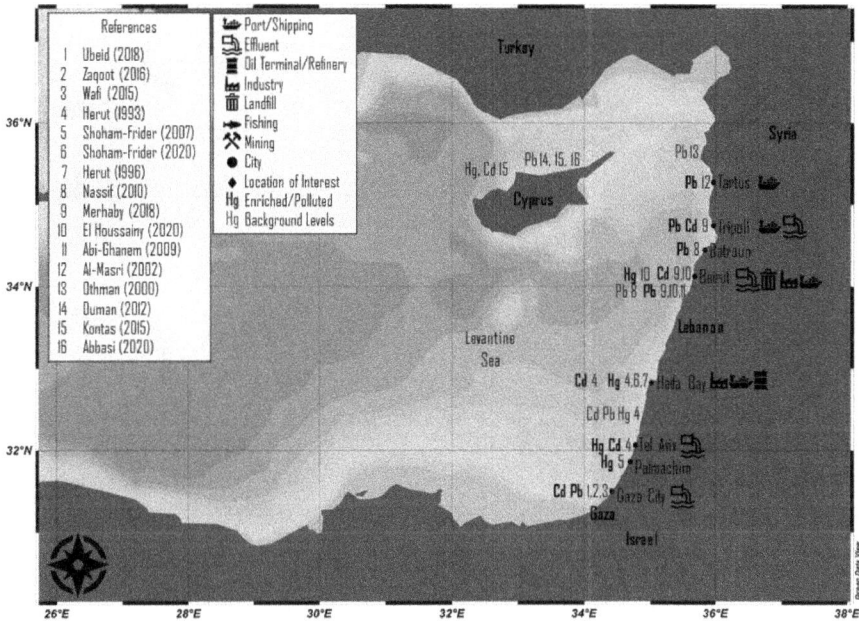

Figure 5.3   Metal enrichment in sediments along within the Levantine Sea including Gaza, Israel, Lebanon, Syria, Cyprus, and Turkey.

Gaza

The coastline of Gaza extends only 42 km and is characterized by sandy habitats (Ali, 2002) similar to that of southern Israel. Unfortunately, there is little information available on the status of the marine environment for Gaza. However, it has been reported that the coastal marine ecosystem has been greatly impacted, leaving little to no pristine areas (Adel Zaqoot et al., 2012). Despite this, artisanal fisheries fleet do exist in this area and there has been documentation in the gray literature (Ali, 2002) as well as local reports of elasmobranch species

Table 5.2 Studies assessing metal accumulation in sediments along the Levantine Sea. Metal concentrations ($\mu g\ g^{-1}$ dw) are reported depending on data provided from within the studies as either mean value ±SD, range (), or both. Values in bold indicate that sediments were considered either contaminated or enriched by the study.

Location	Hg	Cd	Pb	Reference
Gaza (Gaza Coastline)	– (–)	**1.1 ± 0.63** **(0–3.72)**	15 ± 9.7 (3.4–63.7)	Ubeid et al. (2018)
Gaza (Gaza Fishing Harbor)	– (–)	1.07 (–)	7.27 (–)	Wafi et al. (2015)
Israel (Coast)	– (0.004–0.07)	– (0.04–0.31)	– (3.59–10.3)	Herut et al. (1993)
Israel (River Mouths)	– (0.002–0.46)	– (0.04–1.8)	– (3.19–48.3)	Herut et al. (1993)
Israel (Palmachim)	**0.26 ± 0.24** **0-1**	– (–)	– (–)	Shoham-Frider et al. (2007)
Israel (Haifa Bay)	– **(0.09–2.2)**	– (–)	– (–)	Shoham-Frider et al. (2020)
Lebanon (Coast)	– (–)	– (3.29–13.47)	– (1.2–443.3)	Merhaby et al. (2018)
Lebanon (Beirut Harbor)	– (0.1–0.7)	– (0.4–1.1)	– (25.1–376.1)	El Houssainy et al. (2020)
Lebanon (Akaar)	– (–)	– (–)	– (6.2–15.7)	Abi-Ghanem et al. (2009)
Lebanon (Dora)	– (–)	– (–)	– (70.7–101.4)	Abi-Ghanem et al. (2009)
Lebanon Selaata	– (–)	– (–)	– (4.8–34.5)	Abi-Ghanem et al. (2009)
Syria	– (–)	– (–)	– (2.6–24)	Othman et al. (2000)
Northern Cyprus	– (–)	– (–)	– (1.6-9.2)	Duman et al. (2012)
Northern Cyprus	– (0.02–0.07)	– (0.03–0.32)	– (1.1–2.4)	Kontaş et al. (2015)
Northern Cyprus	– (–)	– (–)	– (11–22)	Abbasi and Mirekhtiary (2020)

caught in Gaza that have been less frequently observed in Israeli waters. The human population of the entire area is one of the densest in the world resulting in numerous environmental problems that impact the marine environment. The release of untreated wastewater is one of the greatest problems to the area (Adel Zaqoot et al., 2012) with 80% of it untreated and released directly into the sea (IUCN, 2015). This includes local waste as well as trans-boundary pollutants that originate from beyond Gaza's borders (Israel and West Bank) that flow to the coast via seasonal rivers. The central area of Gaza is particularly polluted owing to the discharge of wastewaters from Gaza City (Adel Zaqoot et al., 2012). Industrial pollution also contributes to the release of metals and other pollutants into the marine system (EEA, 2006) via wastewater treatment which also remains untreated similar to domestic wastewaters. Industries shown to release Pb, Cd, and As include the pharmaceutical industry, cosmetic industry, textile industry, electroplating factories, galvanic factories, detergent factories, textile washing factories, soft drink factories, and car washing workshops (Adel Zaqoot et al., 2012). Additionally, solid hazardous waste disposed in open landfills also act as a source for pollutants entering the marine environment.

Research on metal accumulation in both sediments and biota are limited to a few papers. Focus has been emphasized around Gaza Fishing Harbor, which is a major outlet for wastewater disposal. Overall, wastewaters have been blamed as the major culprit contaminating the waters, sediments, and fish for this area. Sediments along the Gaza coastline show enriched levels of Pb and Cd, with concentrations increasing deeper off shore where grain size decreases (Ubeid et al., 2018). Work assessing the all three mediums (water, sediment, and biota) indicate that the accumulation within the indicator species flathead grey mullet (*Mugil cephalus)* is higher than the local seawater yet lower than that of sediments (Wafi 2015; Zaqoot et al., 2017). Sediment concentrations were particularly elevated following rainy periods (Wafi, 2015) possibly as rivers are flushed from the entire inland watershed (Israel and Palestine/West Bank) out into Gaza Harbor.

Israel

Israel's coastline extends approximately 200 km and is characterized by mostly sandy beaches with occasional rocky habitats (Scheinin et al., 2013). While the marine richness is less compared to the western basin, and despite the harsh conditions of this area, research on the marine environment shows that it sustains a wealth of life and is an important habitat for larger marine species including elasmobranchs, cetaceans, sea turtles, and large pelagic fish (Scheinin et al., 2013). The coastline is highly urbanized with over half of the population living along the coastal plain (Israeli Ministry of Foreign Affairs). There are two major ports along its waters with intense industrial activity centered in these areas. Numerous rivers flow through Israel into the Mediterranean Sea with some of the larger ones including the Yarkon, Alexander, and Kishon rivers. While the release of untreated sewage has dramatically decreased over the years, the release of raw sewage either into rivers or directly into the sea does still occur, especially during the winter months. Additionally, as numerous rivers originate in the Palestinian West Bank, where wastewater management is less efficient compared to Israel, untreated urban and industrial effluents are also transferred via river systems directly into the sea. Pollution hotspots are concentrated around areas like harbors and river outlets (EEA, 2006).

Long term monitoring of metals in Israel's marine environment has been conducted continuously for almost 40 years by Israel's National Monitoring Program (Herut et al., 1993). Coastal sandy sediments were found to be relatively

Lead, Mercury and Cadmium in the Aquatic Environment

unpolluted, except for hotspot areas like river outlets along the coastline (Herut et al., 1993). A major research focus has been placed on the pollution hotspot of Haifa Bay (Bareket et al., 2016; Herut et al., 1993, 1996; Shoham-Frider et al., 2020) (Figure 5.3), which is a major harbor, fishing port, river outlet, and industrial area in Israel. Due to the intense industrial activities in the area, a considerable Hg signature was observed in the sediments and biota of the harbor (Herut et al., 1996). However, the level of metals entering the marine environment in Israel has dramatically decreased over recent years with the reduction of over 99% of land-based sources, specifically the disposal of raw sewage, due to improved legislation on source water treatment (Malster, 2019). This reduction can be directly observed in both the sediment and biota of Haifa Bay (Herut et al., 1996). Additionally, locally consumed seafood show low presence of Cd and Pb, though Hg is still present (Ramon et al., 2021). Despite this, point problems still remain which can be seen, for example, with the passive release of Hg from an abandoned electrochemical plant which has led to elevated levels in the Haifa Bay sediments (2,200 ng.g^{-1}) and local biota (Shoham-Frider et al., 2020). While the land-based sources may have decreased, sediments continue to act as a secondary source as winter storms result in the resuspension of buried contaminated sediments (Bareket et al., 2016).

Lebanon

The Lebanese coastline extends 235 km and is characterized by mostly rocky formations (Abboud Abi-Saab, 2012), differing from the sandy habitats dominating Gaza and Israel to its south. With these unique habitats, Lebanon's marine environment is a hotspot for marine biodiversity (Badreddine, 2018; Bitar et al., 2018). Over 1,500 species have been recorded in Lebanese waters, from macrophytes, cetaceans, elasmobranchs, bony fish, sea birds, and sea turtles (Bitar et al., 2018). However, these natural resources also compete against rapid urbanization, with over 55% of the population living on the coast (Abboud Abi-Saab, 2012). The shoreline hosts four commercial ports, 15 fishing harbors, dozens of oil terminals, three power plants, and intense agriculture (Abboud Abi-Saab 2012; Badreddine, 2018). Effluents from wastewaters from both urban and industrial sources are released untreated into the coastal waters (Badreddine, 2018; Abboud Abi-Saab, 2012) while rivers carry effluents from inland areas (Abboud Abi-Saab, 2012). Coastal dumping of both solid wastes are a major issue (EEA 2006), with landfills situated directly on the shoreline (Ghosn et al., 2020). Such dumpsites in Lebanon, many of them illegal, have been reported to be major sources of trace elements being released into the marine environment (Ghosn et al., 2020; Merhaby et al., 2018) Political conflict, like the Syrian Civil War, has brought in a large influx of refugees, further pushing population stress and urbanization while national conflict has resulted in industrial development with little overseeing management (Badreddine, 2018).

Research on metals in Lebanon has focused on major cities like Beirut and Tripoli, both of which are characterized as major industrial areas with commercial harbors (Figure 5.3). Beirut, in particular, is faced with a combination of anthropogenic inputs that has led to metal enrichment of Hg, Cd, and Pb in the marine sediments and is considered a highly polluted area (El Houssainy et al. 2020; Merhaby et al., 2018). With solid wastes in coastal areas still a major national issue, a local dumpsite in close proximity to Beirut has been identified as a source of leached Pb with extremely high coastal marine sediment concentrations of 70.7–101.4 µg.g^{-1} (Abi-Ghanem et al., 2009). These elevated concentrations have been shown to have similar isotopic signatures to that of gasoline (Abi-Ghanem et al., 2009). Despite these high levels, the biota-sediment transfer factor to commonly consumed seafood species like spiny oysters (*Spondylus spinosus*) and kuruma shrimp (*Marsupenaeus japonicus*) collected in Beirut show a low transfer of Hg, Cd, and Pb, indicating that highly contaminated sediment does not always equate to certainty of bioaccumulation (Ghosn et al., 2020). In comparison, Tripoli's biota-sediment transfer factor does indicate that Hg does indeed transfer from sediments to kuruma shrimp, perhaps due to the shrimp's lifestyle which is associated with the benthos (Ghosn et al., 2020). While some marine species have been investigated in Lebanon, studies are still quite limited, and focus on species relevant to human consumption. Additionally, no recent comparisons have been conducted on highly polluted areas of Lebanon compared to more pristine areas. In order to gain a more comprehensive view of the state of marine environment of the country, more detailed local research is required.

Syria

The Syrian coastline stretches 182 km with sandy beaches making up 20% of the coastline (Ibrahim, 2009) and the rest characterized as rocky. As little information exists in the literature on the marine biodiversity of Syria, its similarities to that of Lebanon may indicate a comparable potential as a marine biodiversity hotspot. In contrast, there is less urbanization compared to Lebanon with around 10% of the population living on the coastline (EEA, 2006; Ibrahim, 2009), and a quarter of coastal population is centered in four cities (Ibrahim, 2009). Only three major cities are situated along the coastline; Latakia, Baniyas and Tartus (Othman, et al., 2000). Along the coastline there are four commercial ports and 14 fishing harbors, as well as a variety of industrial activities such as half of the country's oil processing industry (Ibrahim, 2009). Direct disposal of both domestic and industrial untreated wastewater mas well as the slicks from the oil industry are direct inputs into the marine environment (EEA, 2006). More recently, in August 2021, a major oil spill was reported in the international media (including CNN, BBC) along the Syrian coastline, spreading to deeper waters and nearing neighboring countries like Cyprus and Turkey.

Unfortunately, published research on the topic of metals in Syria's marine environment is limited to a few studies from 20 years ago. This is probably attributed to the nation's political situation leading to a lower priority in environmental research. The little scientific information available shows radionuclide Pb enrichment around Tartus, which hosts the second largest port in Syria and is particularly known for its phosphate loading (Al-Masri et al., 2002).

Beyond this, very little scientific information on metals along Syria's coastline is provided in the literature.

Cyprus

Cyprus is the third largest island in the Mediterranean Sea. Its coastline extends 735 km and is characterized by rocky shores and fringed with sandy beaches. Similar to other countries of this region, there is limited published literature on the marine habitat of Cyprus. Literature describes the soft sandy substrates hosting seagrass beds of *Posidonia oceanica*, holding important ecological significance to species like turtles, monk seals, and dolphins (Kletou et al. 2020). While Cyprus is characterized by a relatively small industrial sector, a mining industry does exist on the island (EEA, 2006). Additionally, anthropogenic activities like fish farming, cement factories, and oil terminals have been reported adjacent to sensitive habitats adjacent of seagrass meadows (Kletou et al. 2020). While urban wastewater has been noted as a coastal environmental problem in the past (EEA, 2006), as the country is water limited, much effort is placed by waste water treatment facilities to treat waters for both agricultural use and replenishing aquifers (Nicos Neocleous, 2018), thus wastewater discharge directly to the sea has been practically eliminated.

Cyprus also lacks published scientific literature addressing the accumulation of metals in its marine system, though some sediment and biota studies do exist. Sediment studies focus on northern Cyprus, which observed low Pb pollution (1.6–9.2 mg kg^{-1} Duman et al., 2012) (1.12–23.8 µg.g^{-1}; Kontaş et al., 2015) (1.6–9.2 mg.kg^{-1}; Abbasi and Mirekhtiary, 2020), Hg (0.02–0.07 µg.g^{-1}; Kontaş et al., 2015), and Cd (0.03–0.32 µg.g^{-1}; Kontaş et al., 2015). It would appear that Cyprus' almost non-existant industrial complexes, as seen in other countries around the Mediterranean Sea, plays a major role on the local contamination levels. In regards to biota, one of the only published studies is on the local nesting sea turtle population which showed that concentrations of metals appear to impact turtle health (Godley et al., 1999). While turtles may be using Cyprus as a nesting site, their accumulation of contaminants is not local, and thus does not provide direct indication of the local pollution. However, as they are a sentinel species that travel throughout the Mediterranean Sea, they can be important indicators to the overall marine health.

Turkey

Turkey's coastline extends for 8,333 km and is divided between the Black Sea, Sea of Marmara, Aegean Sea, and the Mediterranean Sea. Only 1,707 km of coastline is situated along the Mediterranean Sea (PAP/REP, 2005). Its marine environment has been described as a rich one due to its strategic location (Sustainable Development Turkey, 2012). It is also an important exporter in fisheries and aquaculture (Sustainable Development Turkey, 2012), though only 5% of its catch comes from the Mediterranean Sea (PAP/REP, 2005). The coastal dunes of the Mediterranean Sea act as key nesting grounds for multiple sea turtle species while limestone structures provide a habitat for the endangered Mediterranean monk seal (Sustainable Development Turkey, 2012). The seas that Turkey borders are influenced differently by pollutants, yet ultimately all these water bodies are in contact with one another and flow into the Mediterranean Sea. Coastal pollution is mainly due to land-based sources from both direct inputs in coastal areas as well as transfer by local and trans-boundary rivers (Sustainable Development Turkey, 2012). Similar to many other countries around the Mediterranean, wastewater from major cities were once directly released into the sea, though untreated sewage outflow from the major cities of Istanbul and Izmir has been reportedly stopped since 2002 (Sustainable Development Turkey, 2012). Though industrial wastewater output is relatively smaller compared to domestic, it has been noted that they contain high loads of metals, thus creating major hotspots around coastal industrial areas (Sustainable Development Turkey, 2012). Agricultural runoff also continues to be a major contributor of pollutants that eventually reach the sea (Sustainable Development Turkey, 2012). There is a strong presence of marine traffic, with the Turkish Mediterranean region hosting numerous seaports including the major contributors Port of Mersin, İskenderun Harbor, and Port of Isdemir, and the other minor contributors Port of Antalya, Taşucu Seka Harbor, Port of Assan Iskenderun, and Port of Yesilovacik. Additionally, the Port of Ceyhan hosts three oil terminals that transfer crude oil from landlocked areas.

The coastal regions of Turkey relevant to this chapter include the Mediterranean region, the Aegean region, the Marmara region, and the Black Sea region. For this sub-section, the relevant region for the Levant Basin is the Mediterranean region alone. Despite Turkey hosting a wealth of information with regard to metal research, it appears that a majority of the scientific information focuses on other regions (particularly the Aegean and Black Sea regions) and emphasis is placed on biota-based studies rather than sediments. This emphasis may be attributed to the economic importance of fisheries to the local public as well as exports. However, these studies often focus on the human health effects rather than the ecotoxicological significance. From the few investigations on the bioaccumulation of metals from the Mediterranean region of Turkey, Cd and Pb levels have decreased in edible fish species compared to prior studies (Korkmaz et al., 2019; Türkmen et al., 2005) perhaps due to a decrease of industry in this area (Türkmen et al., 2005). In contrast, shrimp and fish species were found to exceed permissible human health standards (Aytekin et al., 2019). Overall, this area remains relatively uninvestigated with regard to the metal accumulation and the potential impacts on the environment.

AEGEAN SEA

The Aegean Sea is a 215,000 km^2 semi-closed extension of the Mediterranean Sea located between Turkey and Greece, with its southern limits being the islands of Crete and Rhodes. It is one of the few locations in the Eastern Mediterranean where deep water is formed, specifically near Rhodes (Simboura et al., 2018). The coastline is characteristically irregular with thousands of islands distributed throughout its waters (Tanhua et al., 2013). It is considered a biodiversity hotspot, hosting important habitats like seagrass meadows, coralligene reefs, and marine caves (Panayiotis et al., 2020). This area is considered highly rich in species, with species richness comparing to that of the richer western basin (Simboura et al., 2018). Additionally, it hosts marine mammal species including numerous cetacean species and the Mediterranean monk seal (Simboura et al., 2018). Besides the Mediterranean, the Aegean Sea is fed by Black Sea and the Sea of Marmara. Due to the introduction of nutrient rich waters from the Black Sea, the Northern Aegean in considered particularly productive (Ozsoy et al., 2016). Similar to other areas around the Mediterranean Sea, the Aegean Sea is threatened by anthropogenic activity including coastal development, pollution, shipping, and more (Katagan et al., 2015). It is a major maritime route that links the land-based industries to the Black Sea and out to the Mediterranean, and major ports in the Aegean Sea exist both in Turkey and Greece. Additionally, the Aegean Sea plays an important role connecting the route for oil between the Black and Mediterranean Seas. One fifth of the cargo ships passing through the Turkish straights carry transporting dangerous and hazardous cargo (Oral, 2015). The Black Sea is highly polluted as different countries use it as a resource, the intense industrialization of the area, and the number of large rivers that act as a transport for inland pollutants. The Sea of Marmara is polluted from maritime traffic as well as inputs of untreated domestic and industrial wastewaters (Ozsoy et al., 2016). Research conducted on sediment accumulation mostly targets major cities and harbors, and particular interest around the Maramara and Black Sea (Figure 5.4). The accumulation of metals in this region is summarized in Figure 5.4 and Table 5.3.

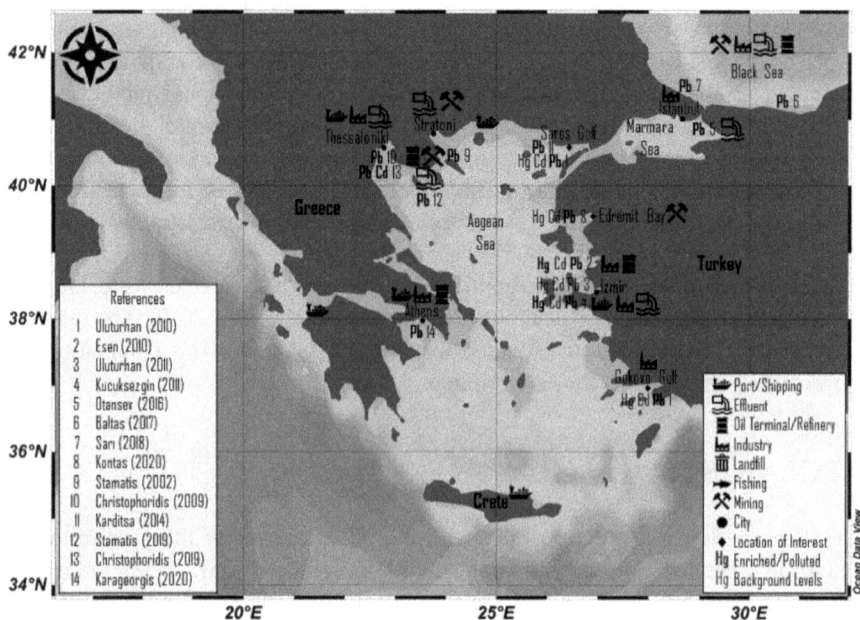

Figure 5.4 Metal enrichment in sediments along the Aegean Sea including Turkey and Greece.

Table 5.3 Studies assessing metal accumulation in sediments along the Aegean Sea. Metal concentrations ($\mu g\ g^{-1}$ dw) are reported depending on data provided from within the studies as either mean value ±SD, range (), or both. Values in bold indicate that sediments were considered either contaminated or enriched by the study

Location	Hg	Cd	Pb	References
Turkey (Black Sea)	– (–)	– (0.02–0.9)	– (<0.05–84.2)	Balkıs et al. (2007)
Turkey (Saros Gulf)	– (0.07–0.19)	– (0.01–0.04)	– **(3.9–48.2)**	Uluturhan (2010)
Turkey (Gokova Gulf)	– (0.07–0.17)	– (0.01–0.05)	– **(10.0–21.8)**	Uluturhan (2010)
Turkey (Nemru Bay)	**(1.70–9.6)**	(0.005–0.25)	**(22.3–89.4)**	Esen et al. (2010)
Turkey (Homa Lagoon-Izmir)	0.33 ± 0.08 (0.22–0.48)	0.107 ± 0.03 (0.06–0.19)	10.49 ± 5.05 (2.43–17.2)	Uluturhan et al. (2011)
Turkey (Izmir Bay-Outer)	– (0.05–0.99)	– (0.005–0.25)	– (9.8–119)	Kucuksezgin et al. (2011)
Turkey (Izmir Bay-Inner)	– (0.12–1.3)	– (0.01–0.82)	– (3.1–94)	Kucuksezgin et al. (2011)
Turkey (Sea of Marmara)	– (–)	– (–)	32.9 (9.1–73.1)	Otansev et al. (2016)
Turkey (Eastern Black Sea)	– (–)	– (–)	97.33 ± 3.2 (3.7–177.75)	Baltas et al. (2017)
Turkey (Black Sea)	– (–)	– (0.5–14.3)	– (17–129)	Sarı et al. (2018)
Turkey (Edremit Bay)	– (0.03–017)	– (0.10–0.27)	– (5.32–47)	Kontas et al. (2020)
Greece (Athens Sewage Outfall)	– (0.38–3.1)	– (–)	– (–)	Papakostidis (1975)
Greece (Thermaikos Gulf)	(2.9–8.88)	(0.87–1.08)	(20.9–27.8)	Fytianos and Vasilikiotis (1983)
Greece (Strymonikos and Ierissos Gulf)	– (–)	– (–)	61.9 (23.4–92.5)	Stamatis et al. (2002)
Greece (Thermaikos Gulf)	– (–)	– (–)	72 (10–218)	Christophoridis et al. (2009)
Greece (Alexandroupolis Gulf)	– (–)	– (–)	50.5 ± 30.1 **(9–113)**	Karditsa et al. (2014)
Greece	– (–)	– (–)	166.23 (–)	Stamatis et al. (2019)
Greece (Kavala Gulf)	– (–)	– (–)	52.79 ± 36.21 (18.12–203.28)	Stamatis et al. (2019)
Greece (Strymonikos Gulf)	– (–)	– (–)	91.1 ± 35.06 (23.41–130.46)	Stamatis et al. (2019)
Greece (Ierissos Gulf)	– (–)	– (–)	637.66 ± 713.82 (52.91–2233.09)	Stamatis et al. (2019)
Greece (Thessaloniki Bay)	– (–)	**2.51** **(0.2–13)**	**84.19** **(29.4–195.4)**	Christophoridis et al. (2019)
Greece (Saronikos Gulf)	– (–)	– (–)	**69 ± 69** **(5–374)**	Karageorgis et al. (2020)

Turkey (Aegean Sea)

Expanding on the background provided in Turkey in the Levantine Basin section, the Turkish regions relevant to the Aegean Sea include the Aegean, Marmara, and Black Sea regions. These regions are collectively highly influenced by anthropogenic activity, with major urbanization and industrialization. Many rural areas have rapidly developed over the years due to tourism, often without adequate infrastructure to accommodate its expansion (Ozsoy et al., 2016). Maritime traffic is particularly prominent in this region, with Turkey bordering a large segment of the Black Sea as well as the entire Sea of Marmara. The major Turkish port sitting directly on the Aegean Sea is the Izmir Port, which also hosts a major oil terminal. However, the bulk of maritime movement is seen in the Marmara Sea, with major ports including Port of Haydarpaşa (Istanbul) and numerous oil terminals. With the major industrial activities taking place in both the Black and Marmara Sea area (Ozsoy et al., 2016), these basins can act as pollution accumulation points and as a secondary pollutant source into the Mediterranean Sea. Additionally, the behavior of the currents in the Aegean Sea bring pollutants from the rest of the Aegean Sea as well as the Mediterranean towards Turkey's Aegean region (Izdar et al., 2015). These regions also encompass large catchment areas with high urbanization, resulting in fresh water drainages to the sea acting as transport of land-based pollutants (Izdar et al., 2015). Istanbul, in particular, has been noted as a major polluter to this area with over 60% of its industry directly located on the coast of the Sea of Marmara.

Investigations in the Aegean region of Turkey have focused on metals in industrialized areas and harbors rather than pristine regions along the coast. Particular focus has also been placed on the Black and Marmara Sea and have been shown to accumulate metals in their surface sediments (Balkıs et al., 2007; Otansev et al., 2016; Sarı et al., 2018). Through sediment cores, a major increase in metals was shown to have taken place in the 1970s/1980s as anthropogenic activities increased in this area (Sarı et al., 2018). In addition to the direct industrial and domestic waste acting as a major contributor (Otansev et al., 2016) pollution sources are not only local, and European rivers have been shown to act as a means of transport to the western and north-western Black Sea (Sarı et al., 2018). Such studies emphasize how the Black and Marmara Seas act as accumulation points and are a source of metals into the Aegean Sea. Looking directly at the Aegean Sea, the coastal areas are less industrialized compared to the Black Sea. One of the largest Turkish cities on the Aegean coastline, Izmir, is home to Izmir Bay which is a major commercial harbor and a release point of domestic wastewater discharge (Katagan et al., 2015). The bay was found to have enriched levels of Hg from mining and enriched levels of Pb due to previous use of leaded fuels (Kucuksezgin et al., 2011). In comparison, the less industrialized Edremit Gulf showed a decrease in Pb levels since the early 1990s, which has been reported to be potentially related to the removal of leaded gasoline in Turkey since 2005 (Kontas et al., 2020). The Homa Lagoon, located near to Izmir Bay, is one of the most productive lagoons in the Eastern Aegean Sea, and has important ecological and economic value for the area. Despite no point sources of Hg, Cd, and Pb pollution observed in the area, sediments contamination was reported to originate from landlocked cities via the Gediz River.

Greece (Aegean Sea)

Greece has an extensive coastline estimated to exceed 18,000 km (Simboura et al., 2018). Due to the extent of its thousands of islands, a majority of the Aegean Sea is filled with Greek islands. Greek waters are considered rich in biodiversity and are an important resource for the country (Simboura et al., 2018). A majority of its southern islands, categorized by a high quality ecological status, with much of the Greek Aegean coastline categorized with good to high ecological status (Simboura et al., 2018). However, the Greek coastline is highly urbanized with over a third of the population living just a few kilometers from the coast. Similar to other regions, such intense coastal urbanization developed without accommodating adequate infrastructure (Simboura et al., 2018). The major Aegean coastal cities include Athens, Thessaloniki, and Volos also act as important ports for the region. These city ports are important industrial areas, with over 80% of the industry taking place in these urban centers (Simboura et al., 2018). Land-based pollution sources include electric power plants, sewage treatment plants, industrial waste, and agricultural activity (Simboura et al., 2018). With many small communities around the Aegean Sea, especially on islands, domestic wastewater is required to undergo at least secondary treatment for population sizes exceeding 2,000, yet this requirement is not always complied with (Simboura et al., 2018). Many of the major river catchments that reach the Aegean Sea along the Greek coastline are trans-boundary, originating in other countries (Simboura et al., 2018). The use of the marine realm in Greek culture is highly important, with intense ship traffic and the marine based pollution that comes with it. There are over 900 ports (22 international) throughout Greece that are important in both international trade and connecting more than 100 inhabited islands to important resources (Simboura et al., 2018). Harbors and ports are a major source of pollution from the mechanical activity that takes place there. Additionally, the marine transport of liquid goods like oil and energy products are a dominant trade (Simboura et al., 2018).

Similar to Turkey, pristine areas from Greece are also data deficient. With Greek ports acting as major centers of urbanization, industry, and maritime traffic, they are a major focus in sediment research. The Saronikos Gulf, home to the busiest Greek port Piraeus (Athens), presents enriched Pb levels that have been decreasing over the past 20 years due to tougher pollution management policies (Karageorgis et al., 2020). The second largest Greek Port in Thessaloniki, has been frequently assessed and shows that sediments are heavily polluted with both Cd and Pb (Christophoridis et al., 2019). Unlike in the Levantine Basin and North African coast, an active mining industry still exists in Greece. Mining activity often requires coastal access to load and transfer raw materials. With mining comes effluent discharge directly from the mining operations, and marine sediments around unloading terminals have been shown to be enriched with Pb (Stamatis et al., 2019). Recently, fish framing has become more pronounced throughout the basin, with research indicating that metals such as Cd and Pb are enriched directly below the cages (Kalantzi et al., 2013; 2021). In a similar trend within harbors, concentrations of metals below the cages decrease with distance from the point source.

ADRIATIC-IONIAN SEA

The Adriatic Sea is a shallow, semi-closed sea situated between the Apennine and Balkan Peninsulas. It comprises 5.6% of the Mediterranean Sea, and is considered its most extensive gulf, connected by the Straits of Otranto in its southern limits. Six countries border the Adriatic Sea including Italy (1,272 km coastline) to its west, and to its east Slovenia (47 km), Croatia (5,835 km), Bosnia and Herzegovina (21 km), Montenegro (294 km), and Albania (406 km) (Joksimović et al., 2021). Similar to the Aegean Sea, the Adriatic is full of islands (over 1,200) that are mostly located on the eastern coastline. Over 20 large rivers flow into the Adriatic, most significant being Italy's Po River, which is responsible for a third of the freshwater and a fourth of the sediment input into the Adriatic Sea (Lopes-Rocha et al., 2017). Due to the environmental conditions, the Adriatic Sea is considered rich with over 7,000 marine species (Zonn et al., 2021) and is a habitat for numerous endemic species in addition to endangered species. The richness of this habitat is expressed through the importance of fishing in the region, especially in the northern Adriatic (Zonn et al., 2021). A large population dwells along the Adriatic coastline, approximately 3.5 million people, with major cities including Italy's Bari, Venice, Trieste, Ravenna, and Rimini; Croatia's Split, Rijeka, and Zadar; Albania's Durrës and Vlorë; Slovenia's Koper; and Montenegro's Budva and Bar (Zonn et al., 2021). The Northern Adriatic is considered particularly urbanized with pollution hotspots in Italy, Slovenia, and Croatia (EEA, 2006). Intense urbanization in this area, especially due to the encouragement of tourism, has been connected to inadequate infrastructure management of coastal areas. Ports also play an important role in this region, 19 major ports, with Trieste in Northern Italy being the largest (Zonn et al., 2021). Large bays in the region include the Gulf of Venice, Gulf of Trieste, Gulf of Manfredonia, and Bay of Kotor. A unique problem in this area are obsolete chemical stockpiles (EEA, 2006) as well as marine buried chemical weapons from WWII (Zonn et al., 2021). Directly south of the Adriatic Sea, divided by the Straits of Otranto, lies the Ionian Sea, which is highly influenced by the Adriatic Sea. It is bounded by southern Italy to its west, and southern Albania and Greece to its east. From all the sub-basins withing the Mediterranean Sea, it is the deepest and reaches a depth of 5,000 meters (UNEP/MAP, 2012). The shoreline is characterized by mostly rocky shores with occasional coarse beaches (Rivaro et al., 2004). Unfortunately, the literature describing the overall marine ecology of this area is highly lacking.

Metal research in the Adriatic Sea is dominant on its western side (Italy) compared to east and is summarized in Figure 5.5 and Table 5.4. Additionally, much research is focused on the northern Adriatic where major pressures are located including the Po River, in addition to a more developed coastal industry.

With rivers in the Adriatic holding such importance in the area, understanding the potential transfer of metals, especially with fine sediments, has been prioritized. The Po River has been shown to have an impact on not only the coastal sediments of the areas surrounding the Po River Delta, but of the entire Adriatic Sea, with a contamination signal taking 10 years to reach the southern Adriatic Sea (450 km south) (Lopes-Rocha et al., 2017). By the Otranto Strait, where the Aegean meets the Ionian Sea, the signal reduces to background levels. Anthropogenic contaminants are shown to reach the Po River from land-based sources, with sediments from tributaries of the Po River (like the Lambro River) showing elevated Pb, Hg, and Cd (Marziali et al., 2021). However, with Italy establishing regulations on wastewater treatment in the mid 80s, the signal intensity of the region has gradually decreased over time (Lopes-Rocha et al., 2017). In Albania, rivers have been shown to be a source of metal pollution from mining activities up-stream, though the coastal Pb in sediments did not appear to be enriched by this output (Rivaro et al., 2004). A long-term study conducted in Italy shows the enrichment in the coastal sediments began around 1910s, potentially associated with the industry surrounding World War I (Lopes-Rocha et al., 2017). Enrichment of Pb was shown to be related to shipping activity, particularly around shipyards and marinas (Obhodaš and Valković, 2010). Lead in the sediments collected from the Sicilian Channel were also attributed to the ship traffic in the area (Tranchida et al., 2010). In Croatia, enriched Pb levels were attributed to an oil terminal, though levels have noticeably decreased with the elimination of leaded fuels (Cukrov et al., 2011). In regard to Hg, numerous studies conducted in the area focus on the historical release of Hg from industry and the residual levels and enrichment that it continues to have today. Point sources identified include oil terminals (Cukrov et al., 2011), sewage outflow (Cukrov et al., 2011), mining enrichment (Acquavita et al., 2012; Covelli et al., 2001) and chlor-alkali plants (Acquavita et al., 2012). Such combination of sources makes areas like the Marano-Grado Lagoon (Italy) hotspots for metal accumulation and will continue to act as a secondary source for many years to come.

Studies in the Ionian Sea are highly lacking, with research coming mostly from Italy summarized in Figure 5.6 and Table 5.5. Hotspots within the Ionian Sea include Taranto Bay and Sicily's Augusta Harbor (EEA, 2006), both considered highly industrialized areas. Taranto Bay, in particular, is considered by the Italian government as an area of high environmental risk (Di Leo et al., 2013) with the presence of steel factories, petroleum refineries, marine traffic, and urban waste. Mercury pollution in the bay has been considered problematic (Di Leo et al., 2013; Spada et al., 2012) with military areas, harbors, and shipbuilding work considered major contributors. Augusta Harbor became a major industrial center following WWII, considered an European petro-chemical hub (Romano et al., 2021), with an active chlor-alkali plant active between 1958–2003. Additionally, breakwater construction enclosing the bay has led to the entrapment of fine terrestrial sediments that also allow for the accumulation of metals within the bay (Romano et al., 2021) and can act as a secondary source of metal pollution. Despite the high bioaccumulation factor of Cd and Hg within these sediments, sediment contamination alone was shown to not be sufficient enough to lead to biological transfer, and the sediment's physicochemical properties was reported to be a major driving force (Signa et al., 2017). Though not within the Ionian Sea itself, the Straights of Sicily have shown overall low Hg levels, yet are still considered enriched (Di Leonardo et al., 2006; Tranchida et al., 2010).

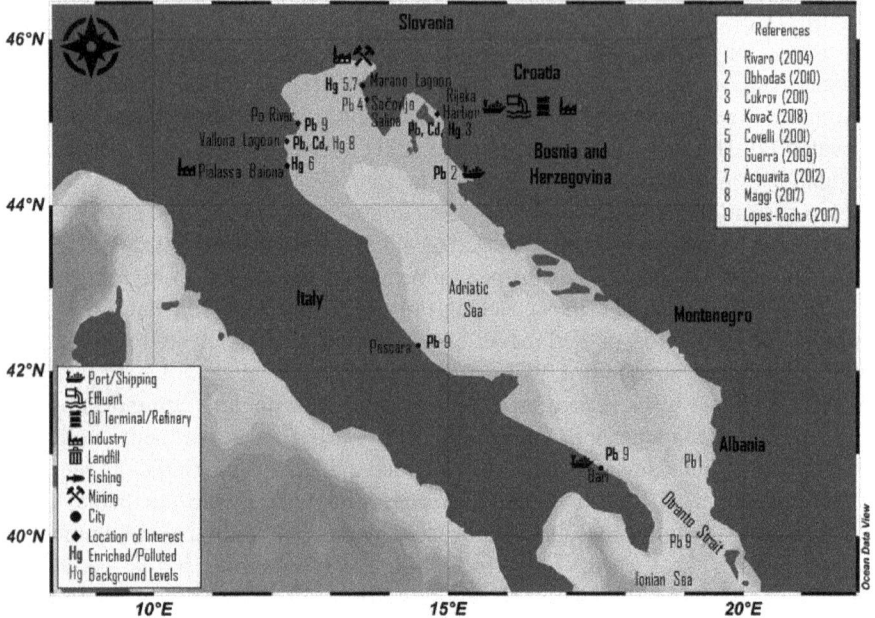

Figure 5.5   Metal enrichment in sediments along the Adriatic Sea including Italy, Slovenia, Croatia, Bosnia and Herzegovina, Montenegro, and Albania.

Compared to the other countries in the area, studies from Italy have given great attention to investigating metal accumulation in sediments from both the Adriatic and Ionian Sea. Italy recognizes the intense and uncontrolled industrialization that took place following World War II as a major influence on the accumulation of marine pollutants of all kinds in their local waters. While most sediment-oriented studies focus on the surface sediments around particular hotspots to provide insight on specific point sources, Italy has provided numerous long term monitoring studies that evaluate the change of metal levels over longer time periods, some even dating back to 1920s (following World War I). These studies provide immense input, and have allowed for the proper legislative measures to be established. Distinctively, from all the papers reviewed throughout the entire Eastern Mediterranean region, studies from Italy describe the Environmental Quality Standard for the metals Pb, Hg, and Cd in marine sediments and coastal zones as established by the Italian Parliament. With defined measures and strict legislation in place, Italy has actively reduced metal discharge into the Adriatic-Ionian Sea (Lopes-Rocha et al., 2017).

Table 5.4 Studies assessing metal accumulation in sediments along the Adriatic Sea. Metal concentrations ($\mu g\ g^{-1}$ dw) are reported depending on data provided from within the studies as either mean value ±SD, range (), or both. Values in bold indicate that sediments were considered either contaminated or enriched by the study

Location	Hg	Cd	Pb	References
Albania (Coastline)	– (–)	– **(0.07–0.753)**	– (22–69)	Rivaro et al. (2004)
Croatia (Coastline)	– (–)	– (–)	9.8 (2.1–65.6)	Obhodaš et al. (2010)
Croatia (Rijeka harbor)	**2.2 ± 1.99** **(0.1–8.06)**	1.07 ± 0.87 (0.14–4.66)	**227 ± 152** **(23.6–637)**	Cukrov et al. (2011)
Slovania (Sečovlje Salin)	– (–)	– (–)	24.88 ± 2.97 (20.32–28.29)	Kovač et al. (2018)
Italy (Isonzo River Mouth – Gulf of Trieste)	**5.04** **(0.06–30.38)**	– (–)	– (–)	Covelli et al. (2001)
Italy (Pialassa Baiona lagoon)	– (0.4–5.5)	– (–)	– (–)	Guerra et al. (2009)
Italy (Marano & Grado Lagoon)	– (0.68–9.95)	– (–)	– (–)	Acquavita et al. (2012)
Italy (Vallona Lagoon)	– (0.01–0.27)	– **(0.09–0.58)**	– **(4.7–31.05)**	Maggi (2017)
Italy (Po River Delta)	– (–)	– (–)	**28 ± 6** **(13–39)**	Lopes-Rocha et al. (2017)
Italy (Pescara)	– (–)	– (–)	**16 ± 3.6** **(11–26)**	Lopes-Rocha et al. (2017)
Italy (Bari)	– (–)	– (–)	24 ± 4 (19–36)	Lopes-Rocha et al. (2017)
Otranto Strait	– (–)	– (–)	11 ± 2 (9–18)	Lopes-Rocha et al. (2017)

Table 5.5 Studies assessing metal accumulation in sediments along the Ionian Sea. Metal concentrations ($\mu g\ g^{-1}$ dw) are reported depending on data provided from within the studies as either mean value ±SD, range (), or both. Values in bold indicate that sediments were considered either contaminated or enriched by the study.

Location	Hg	Cd	Pb	References
Italy (Taranto Gulf)	0.12 (0.04–0.41)	– (–)	57.8 (44.7–74.8)	Buccolieri et al. (2006)
Straits of Sicily	**48** **(15–70)**	– (–)	– (–)	Di Leonardo et al. (2006)
Italy (Sicily-Augusta Industrial Area)	– **(0.02–1.67)**	– (–)	– (–)	Di Leonardo et al. (2008)
Italy (Taranto Gulf)	– **(0.10–1.79)**	– (0.12–0.17)	– (14.28–29.19)	Di Leo et al. (2013)
Italy (Taranto Gulf)	**2.7** **(0.04–7.73)**	– (–)	– (–)	Signa et al. (2017)

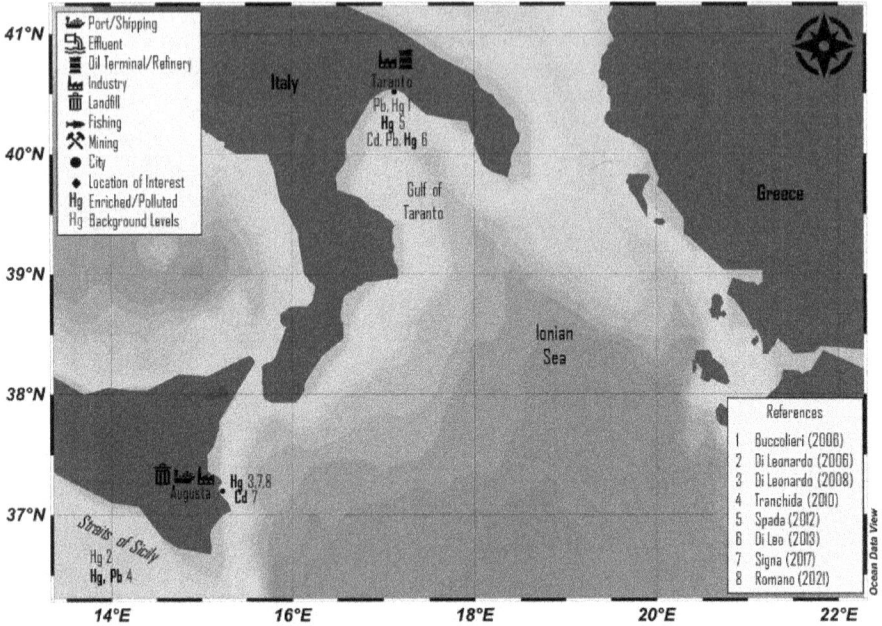

Figure 5.6 Metal enrichment in sediments along the Ionian including Italy and Greece.

CONCLUSION

This chapter provides a brief summary on the accumulation of metals in the coastal sediments of the EMS. While pollution inputs are influenced by the unique socio-economic conditions of each country, there are evident similarities between studies of different sub-regions. Areas of high industrial and economic importance are a major research focus in all countries, and indeed show enrichment of metals as a result. Unfortunately, many pristine areas that could provide a baseline comparison are lacking in almost all countries of the EMS. Even basic information describing the marine environment for this region is highly lacking, emphasizing the importance of further research efforts in the future. Major reviews per individual country are also lacking, and summaries of basin wide pollutants are mainly conducted by bodies like the United Nation Environmental Program. Another major issue, which is just briefly touched upon throughout this chapter, is legislation. With many different countries belonging to multiple continents using the Mediterranean Sea, uniform legislation is challenging. Despite this obstacle, the Barcelona Convention (The Convention for the Protection of the Mediterranean Sea Against Pollution) aims to reduce eliminate pollution in the Mediterranean Sea and protect the marine environment. From the EMS, all countries are contracting parties of the Barcelona Convention including Albania, Bosnia and Herzegovina, Croatia, Cyprus, Egypt, Greece, Israel, Italy, Lebanon,

Libya, Malta, Montenegro, Slovenia, Syria, Tunisia, Turkey, and the European Union. On a final note, it is important to emphasize that sediment studies can provide highly important information regarding the accumulation of metals in local sediments in a way that is relatively feasible logistically. There are also uniform standards that have been accepted internationally regarding both protocols and measurement indexes that indicate sediment pollution or enrichment. However, sediment contamination does not always equate to biological transfer. Therefore, in regard to ecotoxicology, while acceptable concentrations in food stuffs exists for marine organisms, ecotoxicological context is indeed lacking.

REFERENCES

Abbasi, A. and Mirekhtiary, F. 2020. Heavy metals and natural radioactivity concentration in sediments of the Mediterranean Sea coast. Mar. Pollut. Bull. 154: 111041. https://doi.org/10.1016/j.marpolbul.2020.111041

Abboud Abi-Saab, M. 2012. Marine biodiversity in coastal waters. Review and Perspectives of Environmental Studies in Lebanon. INCAM-EU/CNRS Lebanon, pp. 1–29.

Abdallah, M.A.M. 2020. Mercury speciation in aquatic environment south eastern coast of the Mediterranean Sea, Egypt. Emerging Contam. 6: 194–203. https://doi.org/10.1016/j.emcon.2020.04.003

Abdel Ghani, S., El Zokm, G., Shobier, A., Othman, T. and Shreadah, M. 2013. Metal pollution in surface sediments of Abu-Qir Bay and Eastern Harbour of Alexandria, Egypt. Egypt. J. Aquat. Res. 39(1): 1–12. https://doi.org/10.1016/j.ejar.2013.03.001

Abi-Ghanem, C., Chiffoleau, J.F., Bermond, A., Nakhlé, K., Khalaf, G., Borschneck, D., et al. 2009. Lead and its isotopes in the sediment of three sites on the Lebanese coast: Identification of contamination sources and mobility. Appl. Geochem. 24(10): 1990–1999. https://doi.org/10.1016/j.apgeochem.2009.07.012

Acquavita, A., Covelli, S., Emili, A., Berto, D., Faganeli, J., Giani, M., et al. 2012. Mercury in the sediments of the Marano and Grado Lagoon (northern Adriatic Sea): Sources, distribution and speciation. Estuarine, Coastal and Shelf Science 113: 20–31. https://doi.org/10.1016/j.ecss.2012.02.012

Adel Zaqoot, H., Saleem Hujair, T., Khalique Ansari, A. and Hayat Khan, S. 2012. Assessment of land-based pollution sources in the Mediterranean Sea along Gaza Coast-Palestine. pp. 175–189. *In*: Md. Uqaili, A. and Harijan, K. (eds). Energy, Environment and Sustainable Development, Springer Vienna. https://doi.org/10.1007/978-3-7091-0109-4

Al-Masri, M.S., Mamish, S. and Budeir, Y. 2002. The impact of phosphate loading activities on near marine environment: The Syrian coast. J. Environ. Radioact. 58(1): 35–44. https://doi.org/10.1016/S0265-931X(01)00074-1

Ali, M. 2002. The Coastal Zone of Gaza strip-Palestine Management and Problems. In Presentation for MAMA First Kick-off Meeting. Retrieved from https://www.humanitarianresponse.info/sites/www.humanitarianresponse.info/files/documents/files/Gaza Urban Profile.pdf

Antonio, M., Rosa, R. and Leonor Nunes, M. 2014. Seafood safety and human health implications. pp. 589–603. *In*: Goffredo, S. and Dubinsky, Z. (eds). The Mediterranean Sea. Springer, Dordrecht. https://doi.org/10.1007/978-94-007-6704-1

Aytekin, T., Kargın, D., Çoğun, H.Y., Temiz, Ö., Varkal, H.S. and Kargın, F. 2019. Accumulation and health risk assessment of heavy metals in tissues of the shrimp and fish species from the Yumurtalik coast of Iskenderun Gulf, Turkey. Heliyon 5(8): e02131. doi: 10.1016/j.heliyon.2019.e02131.

Badreddine, A. 2018. Coastal ecosystems of the Lebanese coast: ecological status, conservation, evolution. Retrieved from https://tel.archives-ouvertes.fr/tel-01887770/

Balkıs, N., Topcuoğlu, S., Güven, K.C. and Öztürk, B. 2007. Heavy metals in shallow sediments from the Black Sea, Marmara Sea and Aegean Sea regions of Turkey. Environment 13: 147–153.

Baltas, H., Sirin, M., Dalgic, G., Bayrak, E.Y. and Akdeniz, A. 2017. Assessment of metal concentrations (Cu, Zn, and Pb) in seawater, sediment and biota samples in the coastal area of Eastern Black Sea, Turkey. Mar. Pollut. Bull. 122(1–2): 475–482. https://doi.org/10.1016/j.marpolbul.2017.06.059

Bareket, M.M., Bookman, R., Katsman, R., de Stigter, H. and Herut, B. 2016. The role of transport processes of particulate mercury in modifying marine anthropogenic secondary sources, the case of Haifa bay, Israel. Mar. Pollut. Bull. 105(1): 286–291. https://doi.org/10.1016/j.marpolbul.2016.02.014

Béjaoui, B., Ben Ismail, S., Othmani, A., Ben Abdallah-Ben Hadj Hamida, O., Chevalier, C., Feki-Sahnoun, W. et al. 2019. Synthesis review of the Gulf of Gabes (eastern Mediterranean Sea, Tunisia): Morphological, climatic, physical oceanographic, biogeochemical and fisheries features. Estuarine Coastal Shelf Sci. 219: 395–408. https://doi.org/10.1016/j.ecss.2019.01.006

Ben Amor, R., Jerbi, H., Abidi, M. and Gueddari, M. 2020. Assessment of trace metal contamination, total organic carbon and nutrient accumulation in surface sediments of Monastir Bay (Eastern Tunisia, Mediterranean Sea). Reg. Stud. Mar. Sci. 34: 101089. https://doi.org/10.1016/j.rsma.2020.101089

Bitar, G., Ramadan Jaradi, G., Hraoui-Bloquet, S. and Lteif, M. 2018. National monitoring programme for marine Biodiversity in Lebanon. Tunis. Retrieved from http://www.rac-spa.org/sites/default/files/ecap/imap_lebanon/imap_liban_eng_2019.pdf

Bonsignore, M., Salvagio Manta, D., Al-Tayeb Sharif, E.A., D'Agostino, F., Traina, A., Quinci, E.M., et al. 2018. Marine pollution in the Libyan coastal area: Environmental and risk assessment. Mar. Pollut. Bull. 128: 340–352. https://doi.org/10.1016/j.marpolbul.2018.01.043

Buccolieri, A., Buccolieri, G., Cardellicchio, N., Dell'Atti, A., Di Leo, A. and Maci, A. 2006. Heavy metals in marine sediments of Taranto Gulf (Ionian Sea, Southern Italy). Mar. Chem. 99(1–4): 227–235. https://doi.org/10.1016/j.marchem.2005.09.009

Canu, D. and Rosati, G. 2017. Long-term scenarios of mercury budgeting and exports for a Mediterranean hot spot (Marano-Grado Lagoon, Adriatic Sea). Estuarine Coastal Shelf Sci. 198: 518–528. https://doi.org/10.1016/j.ecss.2016.12.005

Christophoridis, C., Bourliva, A., Evgenakis, E., Papadopoulou, L. and Fytianos, K. 2019. Effects of anthropogenic activities on the levels of heavy metals in marine surface sediments of the Thessaloniki Bay, Northern Greece: Spatial distribution, sources and contamination assessment. Microchem. J. 149: 104001. https://doi.org/10.1016/j.microc.2019.104001

Christophoridis, C., Dedepsidis, D. and Fytianos, K. 2009. Occurrence and distribution of selected heavy metals in the surface sediments of Thermaikos Gulf, N. Greece. Assessment using pollution indicators. J. Hazard. Mater. 168(2–3): 1082–1091. https://doi.org/10.1016/j.jhazmat.2009.02.154

Coll, M., Piroddi, C., Steenbeek, J., Kaschner, K., Lasram, F.B.R., Aguzzi, J. et al. 2010. The biodiversity of the Mediterranean Sea: Estimates, patterns, and threats. PLoS ONE 5(8): e11842. https://doi.org/10.1371/journal.pone.0011842

Covelli, S., Faganeli, J., Horvat, M. and Brambati, A. 2001. Mercury contamination of coastal sediments as the result of long-term cinnabar mining activity (Gulf of Trieste, northern Adriatic sea). Appl. Geochem. 16(5): 541–558. https://doi.org/10.1016/S0883-2927(00)00042-1

Cukrov, N., Frančišković-Bilinski, S., Hlača, B. and Barišić, D. 2011. A recent history of metal accumulation in the sediments of Rijeka harbor, Adriatic Sea, Croatia. Mar. Pollut. Bull. 62(1): 154–167. https://doi.org/10.1016/j.marpolbul.2010.08.020

Dahri, N., Atoui, A., Ellouze, M. and Abida, H. 2018. Assessment of streambed sediment contamination by heavy metals: The case of the Gabes Catchment, South-eastern Tunisia. J. Afr. Earth Sci. 140: 29–41. https://doi.org/10.1016/j.jafrearsci.2017.12.033

Damak, M., Fourati, R., Ellech, B. and Kallel, M. 2019. Assessment of organic and metallic contamination in the surface sediment of Monastir Bay (Eastern Tunisia): Spatial distribution, potential sources, and ecological risk assessment. Mar. Pollut. Bull. 149: 110500. https://doi.org/10.1016/j.marpolbul.2019.110500

Dammak Walha, L., Hamza, A., Abdmouleh Keskes, F., Cibic, T., Mechi, A., Mahfoudi, M., et al. 2021. Heavy metals accumulation in environmental matrices and their influence on potentially harmful dinoflagellates development in the Gulf of Gabes (Tunisia). Estuarine Coastal Shelf Sci. 254: 107317. https://doi.org/10.1016/j.ecss.2021.107317

Di Leo, A., Annicchiarico, C., Cardellicchio, N., Spada, L. and Giandomenico, S. 2013. Trace metal distributions in Posidonia oceanica and sediments from Taranto Gulf (Ionian Sea, Southern Italy). Mediterr. Mar. Sci. 14(1): 204–213. https://doi.org/10.12681/mms.v0i0.316

Di Leonardo, R., Bellanca, A., Angelone, M., Leonardi, M. and Neri, R. 2008. Impact of human activities on the central Mediterranean offshore: Evidence from Hg distribution in box-core sediments from the Ionian Sea. Appl. Geochem. 23(12): 3756–3766. https://doi.org/10.1016/j.apgeochem.2008.09.010

Di Leonardo, R., Tranchida, G., Bellanca, A., Neri, R., Angelone, M. and Mazzola, S. 2006. Mercury levels in sediments of central Mediterranean Sea: A 150+ year record from box-cores recovered in the Strait of Sicily. Chemosphere 65(11): 2366–2376. https://doi.org/10.1016/j.chemosphere.2006.04.076

Duman, M., Kucuksezgin, F., Atalar, M. and Akcali, B. 2012. Geochemistry of the northern Cyprus (NE Mediterranean) shelf sediments: Implications for anthropogenic and lithogenic impact. Mar. Pollut. Bull. 64(10): 2245–2250. https://doi.org/10.1016/j.marpolbul.2012.06.025

Durrieu de Madron, X., Guieu, C., Sempéré, R., Conan, P., Cossa, D., D'Ortenzio, F., et al. 2011. Marine ecosystems' responses to climatic and anthropogenic forcings in the Mediterranean. Prog. Oceanogr. 91(2): 97–166. https://doi.org/10.1016/j.pocean.2011.02.003

EEA. 2006. Priority issues in the Mediterranean environment. EEA Report No. 4/2006. Copenhagen. Retrieved from https://www.eea.europa.eu/publications/eea_report_2006_4

El-Geziry, T.M. and Bryden, I.G. 2010. The circulation pattern in the Mediterranean Sea: Issues for modeller consideration. J. Oper. Oceanogr. 3(2): 39–46. https://doi.org/10.1080/1755876X.2010.11020116

El Baz, S.M. and Khalil, M.M. 2018. Assessment of trace metals contamination in the coastal sediments of the Egyptian Mediterranean coast. J. Afr. Earth Sci. 143: 195–200. https:// doi.org/10.1016/j.jafrearsci.2018.03.029

El Houssainy, A., Abi-Ghanem, C., Dang, D.H., Mahfouz, C., Omanović, D., Khalaf, G. et al. 2020. Distribution and diagenesis of trace metals in marine sediments of a coastal Mediterranean area: St Georges Bay (Lebanon). Mar. Pollut. Bull. 155: 111066. https:// doi.org/10.1016/j.marpolbul.2020.111066

El Zrelli, R., Rabaoui, L., Ben Alaya, M., Daghbouj, N., Castet, S., Besson, P. et al. 2018. Seawater quality assessment and identification of pollution sources along the central coastal area of Gabes Gulf (SE Tunisia): Evidence of industrial impact and implications for marine environment protection. Mar. Pollut. Bull. 127: 445–452. https://doi.org/ 10.1016/j.marpolbul.2017.12.012

Esen, E., Kucuksezgin, F. and Uluturhan, E. 2010. Assessment of trace metal pollution in surface sediments of Nemrut Bay, Aegean Sea. Environ. Monit. Assess. 160(1): 257– 266. https://doi.org/10.1007/s10661-008-0692-9

Fouda, M.M. 2017. National monitoring program for biodiversity and non-indigenous species in Egypt. Retrieved from www.iucn.org

Fytianos, K. and Vasilikiotis, G.S. 1983. Concentration of heavy metals in seawater and sediments from the Northern Aegean Sea, Greece. Chemosphere 12(l): 83–91. https:// doi.org/https://doi.org/10.1016/0045-6535(83)90183-2

Gargouri, D., Azri, C., Serbaji, M.M., Jedoui, Y. and Montacer, M. 2011. Heavy metal concentrations in the surface marine sediments of Sfax Coast, Tunisia. Environ. Monit. Assess. 175(1): 519–530. https://doi.org/10.1007/s10661-010-1548-7

Ghosn, M., Mahfouz, C., Chekri, R., Ouddane, B., Khalaf, G., Guérin, T., et al. 2020. Assessment of trace element contamination and bioaccumulation in algae (*Ulva lactuca*), bivalves (*Spondylus spinosus*) and shrimps (*Marsupenaeus japonicus*) from the Lebanese coast. Reg. Stud. Mar. Sci. 39: 101478. https://doi.org/ 10.1016/j. rsma.2020.101478

Godley, B.J., Thompson, D.R. and Furness, R.W. 1999. Do heavy metal concentrations pose a threat to marine turtles from the Mediterranean Sea? Mar. Pollut. Bull. 38(6): 497–502. https://doi.org/10.1016/S0025-326X(98)00184-2

Guerra, R., Pasteris, A. and Ponti, M. 2009. Impacts of maintenance channel dredging in a northern Adriatic coastal lagoon. I: Effects on sediment properties, contamination and toxicity. Estuar. Coast. Shelf Sci. 85(1): 134–142. https://doi.org/10.1016/j.ecss. 2009.05.021

Hamed, M.A., Mohamedein, L.I., El-Sawy, M.A. and El-Moselhy, K.M. 2013. Mercury and tin contents in water and sediments along the Mediterranean shoreline of Egypt. Egypt. J. Aquat. Res. 39(2): 75–81. https://doi.org/10.1016/j.ejar.2013.06.001

Hasan, H.M.I. and Islam, M. 2010. The concentrations of some heavy metals of Al-Gabal Al-Akhdar Coast Sediment. Arch. Appl. Sci. Res. 2(6): 59–67.

Herust, B., Hornung, H., Krom, M.D., Kress, N. and Cohen, Y. 1993. Trace metals in shallow sediments from the Mediterranean coastal region of Israel. Mar. Pollut. Bull. 26(12): 675–682. https://doi.org/10.1016/0025-326X(93)90550-4.

Herut, B., Hornung, H., Kress, N. and Cohen, Y. 1996. Environmental relaxation in response to reduced contaminant input: The case of mercury pollution in Haifa Bay, Israel. Mar. Pollut. Bull. 32(4): 366–373. https://doi.org/10.1016/0025-326X(95)00206-3

Ibrahim, A. 2009. Impacts of urban activities on the coastal and marine ecosystems of Syria, and the adaptative measures. Impact of Large Coastal Mediterranean Cities on Marine Ecosystems-Alexandria, Egypt. 1–4.

IUCN. 2011. Towards a representative network of Marine Protected areas in Libya. Gland, Switzerland. Retrieved from www.iucn.org

Izdar, E., Muezzinoglu, A. and Cihangir, B. 2015. Oceonographic and Pollution Monitoring Studies at the Eighties in the Aegean Sea. pp. 537–547. *In:* Katagan, T. Tokac, A., Besiktepe, S. and Ozturk, B. (eds). The Aegean Sea: Marine Biodiversity, Fisheries, Conservation and Governance. Istanbul, Turkey: Turkish Marine Research Foundation.

Joksimović, A., Đurović, M., Zonn, I.S., Kostianoy, A.G., Semenov, A.V. 2021. Introduction. pp. 1–13. *In:* Joksimović, A., Đurović, M., Zonn, I.S., Kostianoy, A.G. and Semenov, A.V. (eds). The Montenegrin Adriatic Coast. The Handbook of Environmental Chemistry, vol 109. Springer, Cham. https://doi.org/10.1007/698_2020_725

Kalantzi, I., Shimmield, T.M., Pergantis, S.A., Papageorgiou, N., Black, K.D. and Karakassis, I. 2013. Heavy metals, trace elements and sediment geochemistry at four Mediterranean fish farms. Sci. Total Environ. 444: 128–137. https://doi.org/10.1016/j.scitotenv.2012.11.082

Kalantzi, I., Rico, A., Mylona, K., Pergantis, S.A. and Tsapakis, M. 2021. Fish farming, metals and antibiotics in the eastern Mediterranean Sea: Is there a threat to sediment wildlife? Sci. Total Environ. 764: 142843. https://doi.org/10.1016/j.scitotenv.2020.142843

Karageorgis, A.P., Botsou, F., Kaberi, H. and Iliakis, S. 2020. Geochemistry of major and trace elements in surface sediments of the Saronikos Gulf (Greece): Assessment of contamination between 1999 and 2018. Sci. Total Environ. 717: 137046. https://doi.org/10.1016/j.scitotenv.2020.137046

Karditsa, A., Poulos, S.E., Botsou, F., Alexakis, D. and Stamatakis, M. 2014. Investigation of major and trace element distribution patterns and pollution status of the surficial sediments of a microtidal inner shelf influenced by a transboundary river. The case of the Alexandroupolis Gulf (northeastern Aegean Sea, Greece). J. Geochem. Explor. 146: 105–118. https://doi.org/10.1016/j.gexplo.2014.08.004

Katagan, T., Tokac, A., Besiktepe, S. and Ozturk, B. (eds). 2015. The Aegean Sea Marine Biodiversity, Fisheries, Conservation and Governance. Turkish Marine Research Foundation (TUDAV). Türk Deniz Araştırmaları Vakfı. Istanbul.

Keshta, A.E., Shaltout, K.H., Baldwin, A.H. and Sharaf El-Din, A.A. 2020. Sediment clays are trapping heavy metals in urban lakes: An indicator for severe industrial and agricultural influence on coastal wetlands at the Mediterranean coast of Egypt. Mar. Pollut. Bull. 151: 110816. https://doi.org/10.1016/j.marpolbul.2019.110816

Khaled, A., Ahdy, H.H.H., Hamed, E.S.A.E., Ahmed, H.O., Abdel Razek, F.A. and Fahmy, M.A. 2021. Spatial distribution and potential risk assessment of heavy metals in sediment along Alexandria Coast, Mediterranean Sea, Egypt. Egypt. J. Aquat. Res. 47(1): 37–43. https://doi.org/10.1016/j.ejar.2020.08.006

Kletou, D., Kleitou, P., Savva, I., Attrill, M.J., Charalambous, S., Loucaides, A., et al. 2020. "Seagrass of Vasiliko Bay, Eastern Mediterranean: Lost Cause or Priority Conservation Habitat?" Journal of Marine Science and Engineering 8 (9). https://doi.org/10.3390/JMSE8090717.

Kontaş, A., Uluturhan, E., Akçalı, İ., Darılmaz, E. and Altay, O. 2015. Spatial distribution patterns, sources of heavy metals, and relation to ecological risk of surface sediments of the Cyprus Northern Shelf (Eastern Mediterranean). Environ. Forensics 16(3): 264–274. https://doi.org/10.1080/15275922.2015.1059386

Kontas, A., Uluturhan, E., Alyuruk, H., Darilmaz, E., Bilgin, M. and Altay, O. 2020. Metal contamination in surficial sediments of Edremit Bay (Aegean Sea): Spatial distribution, source identification and ecological risk assessment. Reg. Stud. Mar. Sci. 40: 101487. https://doi.org/10.1016/j.rsma.2020.101487

Korkmaz, C., Ay, Ö., Ersoysal, Y., Köroğlu, M.A. and Erdem, C. 2019. Heavy metal levels in muscle tissues of some fish species caught from north-east Mediterranean: Evaluation of their effects on human health. J. Food Compos. Anal. 81: 1–9. https://doi. org/10.1016/j.jfca.2019.04.005

Kovač, N., Glavaš, N., Ramšak, T., Dolenec, M. and Rogan Šmuc, N. 2018. Metal(oid) mobility in a hypersaline salt marsh sediment (Sečovlje Salina, northern Adriatic, Slovenia). Sci. Total Environ. 644: 350–359. https://doi.org/10.1016/j.scitotenv.2018.06.252

Kucuksezgin, F., Kontas, A. and Uluturhan, E. 2011. Evaluations of heavy metal pollution in sediment and *Mullus barbatus* from the Izmir Bay (Eastern Aegean) during 1997–2009. Mar. Pollut. Bull. 62(7): 1562–1571. https://doi.org/10.1016/j.marpolbul.2011.05.012

Lopes-Rocha, M., Langone, L., Miserocchi, S., Giordano, P. and Guerra, R. 2017. Detecting long-term temporal trends in sediment-bound metals in the western Adriatic (Mediterranean Sea). Mar. Pollut. Bull. 124(1): 270–285. https://doi.org/10.1016/j marpolbul.2017.07.026

Maggi, C., Berducci, M.T., Di Lorenzo, B., Dattolo, M., Cozzolino, A., Mariotti, S., et al. 2017. Temporal evolution of the environmental quality of the Vallona Lagoon (Northern Mediterranean, Adriatic Sea). Mar. Pollut. Bull. 125(1–2): 45–55. https://doi. org/10.1016/j.marpolbul.2017.07.046

Malster, I. 2019. Pollution Balance in the Sea 2017 (Hebrew). Retrieved from https://www. sviva.gov.il/infoservices/reservoirinfo/doclib2/publications/p0801-p0900/p0875.pdf

Marziali, L., Valsecchi, L., Schiavon, A., Mastroianni, D. and Viganò, L. 2021. Vertical profiles of trace elements in a sediment core from the Lambro River (northern Italy): Historical trends and pollutant transport to the Adriatic Sea. Sci. Total Environ. 782: 146766. https://doi.org/10.1016/j.scitotenv.2021.146766

Merhaby, D., Ouddane, B., Net, S. and Halwani, J. 2018. Assessment of trace metals contamination in surficial sediments along Lebanese Coastal Zone. Mar. Pollut. Bull. 133: 881–890. https://doi.org/10.1016/j.marpolbul.2018.06.031

Naifar, I., Pereira, F., Zmemla, R., Bouaziz, M., Elleuch, B. and Garcia, D. 2018. Spatial distribution and contamination assessment of heavy metals in marine sediments of the southern coast of Sfax, Gabes Gulf, Tunisia. Mar. Pollut. Bull. 131: 53–62. https://doi. org/10.1016/j.marpolbul.2018.03.048

Nassif, N., and Saade, Z. 2010. Studying heavy metals in sediments layers along selected sites on the Lebanese Coast. J. Water Resour. Prot. 02(01): 48–60. https://doi.org/10.4236/ jwarp.2010.21006

Nicos, Neocleous. 2018. Cyprus Water Management Case Experience with Desalination and Water Reuse. Malta. Retrieved from https://www.gwp.org/contentassets/ aa500f6c8cb749d7ac324a4065395386/202.the-cyprus-experience_neocleous.pdf

Nour, H.E. and El-Sorogy, A.S. 2017. Distribution and enrichment of heavy metals in Sabratha coastal sediments, Mediterranean Sea, Libya. J. Afr. Earth Sci. 134: 222–229. https://doi.org/10.1016/j.jafrearsci.2017.06.019

Obhodaš, J. and Valković, V. 2010. Contamination of the coastal sea sediments by heavy metals. Applied Radiation and Isotopes 68(4–5): 807–811. https://doi.org/10.1016/j. apradiso.2009.12.026

Oral, N. 2015. Maritime Boundaries in the Aegean Sea and Protection of Biodiversity. pp. 705–713. *In:* Katagan, T., Tokac, A., Besiktepe, S. and Ozturk, B. (eds). The Aegean Sea: Marine Biodiversity, Fisheries, Conservation and Governance. Istanbul, Turkey: Turkish Marine Research Foundation.

Otansev, P., Taşkin, H., Başsari, A. and Varinlioğlu, A. 2016. Distribution and environmental impacts of heavy metals and radioactivity in sediment and seawater samples of the Marmara Sea. Chemosphere 154: 266–275. https://doi.org/10.1016/j.chemosphere.2016.03.122

Othman, I., Al-Masri, M.S. and Al-Rayyes, A.H. 2000. Sedimentation rates and pollution history of the eastern Mediterranean Sea: Syrian coast. Sci. Total Environ. 248(1): 27–35. https://doi.org/10.1016/S0048-9697(99)00473-8

Ozsoy, E., Cagatay, N., Balkis, N., Balkis, N. and Ozturk, B. 2016 (eds). The Sea of Marmara Marine Biodiversity, Fisheries, Conservation and Governance. Turkish Marine Research Foundation, Istanbul, Turkey.

Panayiotis, P., Sotiris, O., Vasilis, G. and Vasilis, P. 2020. Coastal habitats in the Aegean Sea: Soft bottom, seagrasses, and hard bottom. pp. 1–18. *In*: The Handbook of Environmental Chemistry. Springer, Berlin, Heidelberg. https://doi.org/10.1007/698_2020_678

PAP/REP. 2005. Coastal Area Management in Turkey. Split, Croatia. https://www.medcoast.net/uploads/documents/Coastal_Area_Management_in_Turkey.pdf.

Papakostidis, G., Grimanis, A.P., Zafiropoulos, D., Griggs, G.B. and Hopkins, T.S. 1975. Heavy metals in sediments from the Athens sewage outfall area. Mar. Pollut. Bull. 6(9): 136–139. https://doi.org/10.1016/0025-326X(75)90170-8

Pinardi, N., Arneri, E., Crise, A., Ravaioli, M. and M. Zavatarelli. 2016. The physical, sedimentary and ecological structure and variability of shelf areas in the Mediterranean sea (27). The Sea 14: 1243–1272.

Ramon, D., Morick, D., Croot, P., Berzak, R., Scheinin, A., Tchernov, D., et al. 2021. A survey of arsenic, mercury, cadmium, and lead residues in seafood (fish, crustaceans, and cephalopods) from the south-eastern Mediterranean Sea. J. Food Sci. 86(3): 1153–1161. https://doi.org/10.1111/1750-3841.15627

Regional Activity Centre for Specially Protected Areas 2016. National Monitoring Programme for Marine Biodiversity in Libya. Tunis. https://www.rac-spa.org/sites/default/files/ecap/imap_libya/imap_libya.pdf.

Rivaro, P., Ianni, C., Massolo, S., Ruggieri, N. and Frache, R. 2004. Heavy metals in Albanian coastal sediments. Toxicol. Environ. Chem. 86(1–4): 85–97. https://doi.org/10.1080/02772240410001688260

Romano, E., Bergamin, L., Croudace, I.W., Pierfranceschi, G., Sesta, G. and Ausili, A. 2021. Measuring anthropogenic impacts on an industrialised coastal marine area using chemical and textural signatures in sediments: A case study of Augusta Harbour (Sicily, Italy). Sci. Total Environ. 755: 1–13. https://doi.org/10.1016/j.scitotenv.2020.142683

Sarı, E., Çağatay, M.N., Acar, D., Belivermiş, M., Kılıç, Ö., Arslan, T.N. et al. 2018. Geochronology and sources of heavy metal pollution in sediments of Istanbul Strait (Bosporus) outlet area, SW Black Sea, Turkey. Chemosphere 205: 387–395. https://doi.org/10.1016/j.chemosphere.2018.04.096.

Scheinin, A., Tsemel, A., Barnea, O., Edelist, D., Klass, K., Gefen-Gilad, Anat., et al., 2013. Report on the Environmental State of the Mediterranean Sea 2013 (Hebrew). Jerusalem.

Shoham-Frider, E., Gertner, Y., Guy-Haim, T., Herut, B., Kress, N., Shefer, E. and Silverman, J. 2020. Legacy groundwater pollution as a source of mercury enrichment in marine food web, Haifa Bay, Israel. Sci. Total Environ. 714: 136711. https://doi.org/10.1016/j.scitotenv.2020.136711

Shoham-Frider, E., Shelef, G. and Kress, N. 2007. Mercury speciation in sediments at a municipal sewage sludge marine disposal site. Mar. Environ. Res. 64(5): 601–615. https://doi.org/10.1016/j.marenvres.2007.06.003

Signa, G., Mazzola, A., Di Leonardo, R. and Vizzini, S. 2017. Element-specific behaviour and sediment properties modulate transfer and bioaccumulation of trace elements in a highly-contaminated area (Augusta Bay, Central Mediterranean Sea). Chemosphere 187: 230–239. https://doi.org/10.1016/j.chemosphere.2017.08.099

Simboura, N., Maragou, P., Paximadis, G., Kapiris, K., Papadopoulos, V.P., Sakellariou, D., et al. 2018. Greece. World Seas: An Environmental Evaluation, Volume I: Europe, the Americas and West Africa, 2nd Ed. Elsevier Ltd. https://doi.org/10.1016/B978-0-12-805068-2.00012-7

Spada, L., Annicchiarico, C., Cardellicchio, N., Giandomenico, S. and Di Leo, A. 2012. Mercury and methylmercury concentrations in Mediterranean seafood and surface sediments, intake evaluation and risk for consumers. Int. J. Hyg. Environ. Health 215(3): 418–426. https://doi.org/10.1016/j.ijheh.2011.09.003

Stamatis, N., Ioannidou, D., Christoforidis, A. and Koutrakis, E. 2002. Sediment pollution by heavy metals in the Strymonikos and Ierissos Gulfs, North Aegean Sea, Greece. Environ. Monit. Assess. 80(1): 33–49. https://doi.org/10.1023/ A:1020382011145

Stamatis, N., Kamidis, N., Pigada, P., Sylaios, G. and Koutrakis, E. 2019. Quality indicators and possible ecological risks of heavy metals in the sediments of three semi-closed East Mediterranean Gulfs. Toxics 7(2): 30. https://doi.org/10.3390/toxics7020030

Sustainable Development Turkey. 2012. Turkey's Sustainable Development Report: Claiming the Future. Retrieved from https://sustainabledevelopment.un.org/content/documents/853turkey.pdf

Tanhua, T., Hainbucher, D., Schroeder, K., Cardin, V., Alvarez, M. and Civitarese, G. 2013. The Mediterranean Sea system : A review and an introduction to the special issue. Ocean Sci. 9: 789–803. https://doi.org/10.5194/os-9-789-2013.

Telesca, L., Belluscio, A., Criscoli, A., Ardizzone, G., Apostolaki, E.T., Fraschetti, S., et al. 2015. Seagrass meadows (*Posidonia oceanica*) distribution and trajectories of change. Sci. Rep. 5: 1–14. https://doi.org/10.1038/srep12505

Tranchida, G., Bellanca, A., Angelone, M., Bonanno, A., Langone, L., Mazzola, S., et al. 2010. Chronological records of metal deposition in sediments from the Strait of Sicily, central Mediterranean: Assessing natural fluxes and anthropogenic alteration. J. Mar. Syst. 79(1–2): 157–172. https://doi.org/10.1016/j.jmarsys.2009.08.001

Türkmen, A., Türkmen, M., Tepe, Y. and Akyurt, I. 2005. Heavy metals in three commercially valuable fish species from İskenderun Bay, Northern East Mediterranean Sea, Turkey. Food Chem. 91(1): 167–172. https://doi.org/10.1016/j.foodchem.2004.08.008

Ubeid, K.F., Al-Agha, M.R. and Almeshal, W.I. 2018. Assessment of heavy metals pollution in marine surface sediments of Gaza Strip, southeast Mediterranean Sea. J. Mediterr. Earth Sci. 10: 109–121. https://doi.org/10.3304/JMES.2018.001

Uluturhan, E. 2010. Heavy metal concentrations in surface sediments from two regions (Saros and Gökova Gulfs) of the Eastern Aegean Sea. Environ. Monit. Assess. 165(1–4): 675–684. https://doi.org/10.1007/s10661-009-0978-6

Uluturhan, E., Kontas, A. and Can, E. 2011. Sediment concentrations of heavy metals in the Homa Lagoon (Eastern Aegean Sea): Assessment of contamination and ecological risks. Mar. Pollut. Bull. 62(9): 1989–1997. https://doi.org/10.1016/j.marpolbul.2011.06.019

UNEP/MAP. 2012. State of the marine and coastal mediterranean. UNEP/MAP – Barcelona Convention. Athens. https://www.unep.org/unepmap/.

Wafi, H.N. 2015. Assessment of Heavy Metals Contamination in the Mediterranean Sea along Gaza Coast—A case Study of Gaza Fishing Harbor. Al-Azhar University Gaza.

Zaqoot, H.A., Aish, A.M. and Wafi, H.N. 2017. Baseline concentration of heavy metals in fish collected from gaza fishing harbor in the Mediterranean Sea along Gaza Coast, Palestine. Turk. J. Fish. Aquat. Sci. 17: 101–109. https://doi.org/10.4194/1303-2712-v17

Zonn, I.S., Kostianoy, A.G., Semenov, A.V., Joksimovic, A. and Durovic, M. 2021. The Adriatic Sea Encyclopedia. Springer, Cham. https://doi.org/https://doi.org/10.1007/978-3-030-50032-0

Chapter **6**

Cadmium, Mercury and Lead in the Bones of Marine Mammals

Violeta Evtimova[1]*, Atanas Grozdanov[1] and Dimitar Parvanov[2]

[1]Department of Zoology and Anthropology,
Faculty of Biology, Sofia University, Sofia, Bulgaria.
[2]Research Department, Nadezhda, Women's Health Hospital, Sofia, Bulgaria.

INTRODUCTION

Biosphere metal loads have significantly increased with industrial development, especially in the last decades of the 20th century. This problem is becoming even more serious because, in general, breakdown mechanisms for heavy metals do not exist, as these contaminants simply move from one natural reservoir to another, interacting with different categories of living organisms thus, resulting in serious negative effects worldwide (Georgiev et al., 2011).

Metals accumulate mostly in the soil and sediments of water bodies. Bacterial processes extract metal ions from the bottom mass to the solution, which are then included in the biogeochemical cycle. In these processes, due to the absence of chemical or biochemical decomposition processes, their amount slowly but steadily increases throughout the cycle flow (Georgiev et al., 2011).

Metals adversely affect the development of marine mammals through their accumulation. The biological reactions resulting from the influence of these contaminants are very diverse. Their effects are manifested not only as a direct toxicity to a particular individual, but also result in indirect effects at the population level (Honda et al., 1982). Depending on the conditions, changes in the density, diversity, group structure and species composition of the populations may occur.

*Corresponding authors: vilka@abv.bg

The nature and extent of these changes depend on the content and chemical form of the respective metals in the water and sediments. Lead (Pb), cadmium (Cd) and mercury (Hg), for example, reduce the vital functions of marine mammals and may lead to suppression of the immune system in various species (Zelikoff and Tomas, 1998). Not all metals are toxic or harmful, however, and many of them are vital for marine mammals, as they are part of many enzymes and cell microstructures paramount for the maintenance of various biochemical and physiological functions in living organisms, as long as those metals are at concentrations below toxic thresholds. However, the toxicity thresholds for each marine mammal vary significantly and are difficult to be determined (Lavery et al., 2008).

Following contaminant absorption, bone tissue certainly provides more information than soft tissues, such as liver and kidney. With the exception of young animals, this can be explained by the substantially long life of bones that have been just synthetized, which allows for the accumulation of many different compounds (Triffit, 1985). Analysing compact bones, which have low turnover rates, indicates this accumulation throughout individual life history. Therefore, bone tissue acts as an accurate record of contaminant exposure history and as a bioindicator for elemental exposure in a given environment (Gdula-Argasinska et al., 2004).

Environmental pollution caused by anthropogenic activities may directly result in marine mammal death or greatly affect their immune systems, thus making the development of diseases more likely (Geraci et al., 1999). Overall toxicity is, however, subject to the exact chemical compound, the physiological status of the animal, its age, status, size, gender, reproductive status and affected organs (Reijnders et al., 1993).

LEAD (Pb)

Lead is a major global polluter, as it has been mined for centuries by humans, and the use of lead-based products in industry has increased significantly (Lantzy and Mackenzie, 1979). It enters the environment and, in particular, marine ecosystems mainly from mining, rechargeable batteries and chemicals (Cossa et al., 1993). Some marine mammal species living near anthropogenic sources show relatively high concentrations of lead (Law et al., 1991, 1992).

This metal displays the ability to form organic and inorganic ligands in marine environments, affecting the toxic action of this element (Cossa et al., 1993). Lead effects vary, young individuals and foetuses are most sensitive to this metal (Núñez-Nogueira et al., 2019). Table 6.1 displays the bone tissue Pb concentrations in different marine mammal species worldwide.

A high variability in metal accumulation among different tissues is noted (Frank et al., 1992; Frodello and Marchand, 2001; Garcia-Fernandez et al., 2009). Overall, investigations on Pb bioaccumulation in cetacean bones are rare, but virtually all indicate that Pb concentrations in this hard tissue is usually higher than in the soft ones such as liver, kidney or muscle (Fujise et al., 1988; Honda et al., 1982).

In one assessment, Pb bone concentrations were investigated in three seal species from different areas of Swedish marine waters. No differences were noted between Ringed seal and Gray seal bone Pb concentrations (6.9 µg g^{-1} dry weight), while Harbor seal bones contained much higher concentrations (20.5 µg g^{-1} dry weight) (Enhus et al., 2011). While fish is the main food of seals, the Pb concentrations in that study were higher than concentrations found in fish in 2010 in the same areas (Bignert et al., 2010). Differences in Pb concentrations in different bone types (humerus, femur and rib) in California sea lions (*Zalophus californianus*) of age between 1 and 4 years from the coast of San Luis Obispo Country, California have also been reported (Braham, 1973). In humerus the mean Pb concentration is the highest—34.2 µg g^{-1} dry weight, in femur—20.6 µg g^{-1} dry weight and in the rib is the lowest value – 8.7 µg g^{-1} dry weight. In that study, lead is calculated to be significantly higher in females than in males.

In another study, this time on small cetaceans stranded along the Dutch, Irish, French and Spanish coasts, average bone Pb concentrations were lower than 1 µg g^{-1} dry weight, except for French common dolphins, which exhibited mean concentration of 1.471 µg g^{-1} dry weight (Caurant et al., 2006). Mean Pb concentrations found in the bones of nine common dolphin foetuses in the same area were recorded as 0.081 ± 0.040 µg g^{-1} dry weight, about 10-fold lower when compared to adult values (1.024 ± 0.267 µg g^{-1}). Pb accumulation with growth could not be studied throughout the foetal period, because foetus length only ranged between 37 and 60 cm. Furthermore, Pb concentrations in foetuses were not correlated to Pb concentrations in their mothers. A placental transfer of Pb occurs as early as the early embryonic stage. After birth, Pb accumulation in hard tissues was age-related. However, it is known that Pb levels increase significantly during the nursing period and is probably transmitted from the mother to the baby through milk (Honda et al., 1986; Vighi et al., 2017). Pb is also usually lower in female individuals than in the male, probably due to parturition and lactation events, which eliminate Pb from the body (Honda et al., 1986).

Concerning age differences in Pb levels, concentrations in the bone of striped and common dolphins tend to increase with age after birth (Caurant et al., 2006; Honda et al., 1982, 1986). Differences in Pb concentration in bones with age have also been observed in Antillean manatees (*Trichechus manatus manatus*) found stranded in lagoons and rivers in the Tabasco and Campeche areas, in the Gulf of Mexico, and in Chetumal Bay, in the Caribbean region (Romero-Calderon et al., 2015). Geographically significant differences between the two sampling areas were also noted, with higher levels found in the Gulf of Mexico group compared to the Mexican Caribbean group (p < 0.05). The highest Pb value in the bones of *T. m. manatus*, and, in fact, in Sirenians in general, has been reported in Quintana Roo, Mexico, of 128 mg kg^{-1} (Rojas-Mingüer et al., 1997). One study on fin whales (*Balaenoptera physalus*) inhabiting the waters of north-western Spain and western Iceland also reported a sharp increase in Pb concentrations with age in individuals from the Spanish region. This is probably related to the high level of industrialization observed in Spain when compared to Iceland, implying contamination of the waters off the Spanish coast (Vighi et al., 2017).

Harbour porpoises (*Phocoena phocoena*) are coastal species and, thus, are highly affected by anthropogenic activities, such as by-catch in fishing nets, noise and chemical pollution. One study regarding Pb concentrations in dead harbour porpoises found on the shores of the Bulgarian Black Sea coast reported significantly higher levels in animals found on the northern Black Sea coast than on the southern Black Sea coast (Evtimova et al., 2019). This may be due to the main Black Sea polluter, the Dunav River, which is located in the north and imports industrial waters into the Black Sea. When comparing the results with other assessments, studies for metals in bones, higher Pb levels are noted in the Black Sea specimens (13.80 mg kg^{-1}, 33 individuals) compared to the same species along the coasts of France (0.315 mg kg^{-1}, 7 individuals), Ireland (0.337 mg kg^{-1}, 3 individuals) and Spain (0.143 mg kg^{-1}, 1 individual) (Caurant et al., 2006).

Differences in Pb values in the bones of the same species in different parts of the world have also been reported in other studies. For example, common dolphin (*Delphinus delphis*) Pb bone levels in the Mediterranean was detected at 4.2 µg g^{-1} dry weight (Frodello and Marchand, 2001), while the same species in Western European waters presented 0.66 µg g^{-1} dry weight Pb levels (Caurant et al., 2006).

The lowest Pb value reported for marine mammal bones was reported for Weddell seals (*Leptonychotes weddellii*) around the coast of Antarctica (0.01 µg g^{-1} wet weight) by Yamamoto et al. (1987) and by Dall's porpoise (*Phocoenoides dalli*) in the North-western Pacifi, at the same value (Fujise et al., 1988).

CADMIUM (Cd)

Cadmium is released into marine waters directly or through rivers mainly from mining and smelting of sulfide ores, fuel combustion, application of phosphate fertilizers and sewage (Briffa et al., 2020). Table 6.1 displays Cd concentrations in bone tissue from different marine mammal species worldwide.

Cadmium is considered as one of the most toxic metals for aquatic animals, including marine mammals (Borgmann et al., 2005). Small concentrations could lead to birth defects, genetic mutations and cancer. However, the presence of this element does not necessarily mean lethal effects or severe damage to the marine mammals (O'Shea et al., 1984). Its concentration and adverse effects are strongly associated with bones, especially dental bones in some species, such as the dugong (*Dugong dugon*), and results in direct toxic effects (Yoshiki et al. 1975), such as bone deformations and decreas in bone density (Kido et al., 1989), bone weakening and fracture (Bansal and Asthana, 2018; O'Shea et al., 1999). Relatively high levels of this metal may also possibly result in delayed sexual and physical maturity in marine mammals (Kemper and Gibbs, 1997). However, as per one study on ringed seals (*Phoca hispida*) in Greenland, no relationship between Cd, degree of bone mineralisation, and nephropathy was observed, and the authors suggested that high dietary intakes of vitamin D and calcium may protect the seals from cadmium toxicity (Sonne-Hansen et al., 2002).

Usually, Cd accumulates to highest levels first in marine mammal kidneys, followed by the liver (Honda et al., 1983; Meador et al., 1993) or even the skin in some species, such as the striped dolphin *Stenella coeruleoalba* (Frodello and Marchand, 2001). Furthermore, levels in the bones in contrast to other metals such as Pb and zinc are relatively lower compared to other tissues and organs, such as lung, liver, skin, kidney, and muscle tissue (Frodello and Marchand, 2001; Watanabe et al., 1996). Recently, Cd concentrations in bones have been reported in harbor porpoises along the Bulgarian Black Sea coast (Evtimova et al., 2019) and in skulls, ribs and flipper bones from manatees from Mexican waters (Núñez-Nogueira et al., 2019), as well as in and also in tusks from dugongs in Australia (Nganvongpanit et al., 2017). The concentrations in bones ranged from 0.01 mg kg^{-1} to over 5 mg kg^{-1} wet weight depending on the species characteristics and geographical region (Escobar-Sánchez, 2010; Lavery et al., 2008). In manatees (*Trichechus manatus*), for example, concentrations have been reported between 3 and 5 mg kg^{-1} dry weight, similar to Harbour porpoises (*Phocoena phocoena*), with mean values of 2.45 ± 0.64 mg kg^{-1} dry weight (Evtimova et al., 2019). However, concentrations in the tusks of dugongs (*Dugong dugon*) were less than $0.02 \pm 0.01\%$. In addition, the Cd content in South Australian adult bottlenose dolphins (*Tursiops aduncus*) ranged between 0.005 and 0.33 mg kg^{-1} wet weight (Lavery et al., 2008). Cadmium concentrations are also significantly different in various bones. For example, a detailed analysis in the bones of striped dolphin (*Stenella coeruleoalba*) revealed higher contents in the vertebra compared to the ribs and the skull (Honda et al., 1984). An explanation for this could be the observed positive correlation between Cd concentration and the different content of protein and moisture in these specific bones.

The Cd content in the bones of certain marine mammals such as *Phocoena phocoena* and *Trichechus manatus* have displayed significant positive correlations with the concentrations of other metals such as Pb (Evtimova et al., 2019; Rojas-Mingüer and Morales-Vela, 2002), Cr and Mn (Rojas-Mingüer and Morales-Vela, 2002). The observed association could be explained by the nature of metal pollution and the specificity of their deposition and bioaccumulation.

The amount of bioaccumulated Cd is also associated with marine mammal age and sex, with higher concentrations usually observed in older individuals, as expected (Das et al., 2003; Takeuchi et al., 2016; Vighi et al., 2017), and in females compared to males (Honda et al., 1986). Honda et al., for example, reported found relatively higher Cd concentrations in pregnant striped dolphins (*Stenella coeruleoalba*) compared to mature males or foetuses (Honda et al., 1982).

MERCURY (Hg)

The origin of the Hg that accumulates in marine mammals may be natural, human-related or a combination of both. This element exists in both organic and inorganic forms, resulting in clear absorption rate differences. Both forms exist in different ratios in the marine mammals diet and indicate differences concerning metabolism, target organs and susceptibility during different life stages (O'Hara et al., 2003). Some studies of mammals suggest that Hg can cause toxic effect at concentrations in hepatic tissues higher than 100 μg g^{-1} (Wagemann and Muir, 1984). Table 6.1 displays Hg concentrations in bone tissue from different marine mammal species worldwide.

As they occupy higher marine food web levels, marine mammals are often exposed to high risks regarding Hg accumulation, due to biomagnification processes, which makes them important biomonitoring subjects. However, Hg accumulates and retains in bones to a lesser degree in comparison to other metals (Ewers and Schlipköter, 1991). In addition, despite their generally lower Hg concentrations, bones probably release the accumulated Hg during their long remodelling process (Outridge et al., 2000).

The most widespread form of Hg, which accumulates in fish and, thus, in many marine mammals, is monomethyl mercury (MeHg) (Bloom, 1992). Due to their fish-eating habits, carnivorous sea mammals like the harbour seal (*Phoca vitulina richardii*) become a target for the biomagnification process of the highly neurotoxic MeHg, which concentrates and absorbs across the gut (Tiffini et al., 2008; Wiener et al., 2003). In one study on harbour seal cubs, the highest concentrations of total Hg were determined in the fur and liver, the highest total Hg burden was identified in the pelt and muscle tissue, and both indicators were reported at their lowest in bones. In addition, the same research identified a correlation between total Hg bone concentrations with heart, liver, kidney, muscle, blubber, pelt concentrations (Tiffini et al., 2008).

Mercury has also been reported at high concentrations in the liver of many other marine mammals, such as False orcas (*Pseudorca crassidens*) at the eastern coasts of Australia and the Short-finned pilot whale (*Globicephala macrorhynchus*) in the West Indies and Georgia, reaching highly significant levels of 454 mg kg^{-1} on wet weight basis (Kemper et al., 1994; Wagemann and Muir, 1984). Similarly, Hg was also identified in high levels in the liver and in low levels in the bones of Dall's porpoise (*Phocoenoides dalli*) in a study conducted off the Japanese coast (Yang et al., 2002).

Research regarding Hg concentrations in bones and associated age variations has also been conducted for the striped dolphin (*Stenella coeruleoalba*) (Honda et al., 1986), where total and MeHg in bones were low in foetus stage. Total Hg increased with age from birth until about 20 years and later remained constant, while MeHg reached a plateau at about 10 years of age (Honda et al., 1986). Very similar total and MeHg concentrations were observed in the calf and the foetus, indicating limited Hg transfer via milk due the very low Hg concentrations detected in milk (0.002 ± 0.002 μg g^{-1} wet weight) (Itano et al., 1984).

Concerning Sirenians (dugongs and manatees), Hg has been rarely detected or is absent from bone samples, with higher muscle and liver concentrations reported. In general, these herbivorous species exhibit lower Hg concentrations in comparison with cetaceans and other carnivorous sea mammals, associated with different diets and, thus, differences in Hg biomagnification (Núñez-Nogueira et al., 2019).

Table 6.1 Mean Cd, Hg and Pb concentrations in bone tissue from different marine mammal species worldwide

Species	N	Region	Sex	Age	Pb	Hg	Cd	Unit	References
Ringed seal (*Pusa hispida*)	12	Swedish waters	NA	NA	6.92 ± 6.29	0.19 ± 0.17	0.08 ± 0.06	µg g⁻¹ dry weight	Enhus et al., 2011
Gray seal (*Halichoerus grypus*)	15	Swedish waters	NA	NA	6.99 ± 10.7	0.12 ± 0.08	0.09 ± 0.05	µg g⁻¹ dry weight	
Harbor seal (*Phoca vitulina*)	17	Swedish waters	NA	NA	20.5 ± 63.7	0.22 ± 0.30	0.14 ± 0.15	µg g⁻¹ dry weight	
Striped dolphin (*Stenella coeruleoalba*)	11	Kii Peninsula, Japan	NA	Foetus	0.09 ± 0.04	0.09 ± 0.03	0.003 ± 0.002	µg g⁻¹ wet weight	Honda et al., 1986
Striped dolphin (*Stenella coeruleoalba*)	13	Kii Peninsula, Japan	Male	0–1.5	0.27 ± 0.06	0.06 ± 0.01	0.04 ± 0.03	µg g⁻¹ wet weight	
Striped dolphin (*Stenella coeruleoalba*)	6	Kii Peninsula, Japan	Male	1.5–7.5	0.35 ± 0.06	0.62 ± 0.83	0.12 ± 0.02	µg g⁻¹ wet weight	
Striped dolphin (*Stenella coeruleoalba*)	5	Kii Peninsula, Japan	Male	7.5–36.5	0.74 ± 0.26	1.44 ± 0.63	0.09 ± 0.02	µg g⁻¹ wet weight	
Striped dolphin (*Stenella coeruleoalba*)	5	Kii Peninsula, Japan	Female	7.5–36.5	0.39 ± 0.08	1.55 ± 0.50	0.18 ± 0.05	µg g⁻¹ wet weight	
Fin whale (*Balaenoptera physalus*)	23	Iceland	NA	NA	111.91 ± 90.42			mg kg⁻¹ dry weight	Vighi et al., 2016
Fin whale (*Balaenoptera physalus*)	21	NW Spain	NA	NA	161.82 ± 59.89			mg kg⁻¹ dry weight	
Bottlenose dolphin (*Tursiops truncatus*)	7	Mediterranean Sea	NA	Juvenile to 3 years	3.6–13 (±6.4)		NS–0.37 (±0.16)	µg g⁻¹ dry weight	Frodello et al., 2001
Striped dolphin (*Stenella coeruleoalba*)	3	Mediterranean Sea	NA	Nursing to 6 years	3.2–16 (±8.1)		NS–0.05 (±0.04)	µg g⁻¹ dry weight	
Risso's dolphin (*Grampus griseus*)	3	Mediterranean Sea	NA	NA	4.5–10 (±7.2)		0.1–0.31 (±0.23)	µg g⁻¹ dry weight	
Pilot whale (*Globicephala melas*)	3	Mediterranean Sea	NA	6 to 7 years	5.1–9.2 (±7.7)		0.02–0.2 (±0.09)	µg g⁻¹ dry weight	

Species	N	Region	Sex	Age	Pb	Hg	Cd	Unit	References
Common dolphin (*Delphinus delphis*)	2	Mediterranean Sea	NA	2 and 4 years	4.2–4.5 (±4.3)		0.01–0.02 (±0.02)	$\mu g\ g^{-1}$ dry weight	
Striped dolphin (*Stenella coeruleoalba*)	3	Kii Peninsula	NA	NA	0.62 ± 0.06		0.28 ± 0.03	$\mu g\ g^{-1}$ wet weight	Honda et al., 1982
Harbor seal (*Phoca vitulina*)	10	Britain	NA	NA	3.5 ± 2.5	0.17 ± 0.22	0.41 ± 0.33	$\mu g\ g^{-1}$ wet weight	Roberts et al., 1976
Pilot whale (*Globicephala melas*)		Mediterranean Sea	Female	6		2.3 ± 0.2		$\mu g\ g^{-1}$ dry weight	Frodello et al., 2000
Striped dolphin (*Stenella coeruleoalba*)		Mediterranean Sea	Male	>7		2.1 ± 0.2		$\mu g\ g^{-1}$ dry weight	
Common dolphin (*Delphinus delphis*)		Mediterranean Sea	Female	2		3.4 ± 0.2		$\mu g\ g^{-1}$ dry weight	
Bottlenose dolphin (*Tursiops truncatus*)		Mediterranean Sea	Female	4		7.9 ± 0.6		$\mu g\ g^{-1}$ dry weight	
Risso's dolphin (*Grampus griseus*)		Mediterranean Sea	Female	2.5		150 ± 20		$\mu g\ g^{-1}$ dry weight	
Indo-Pacific bottlenose dolphin (*Tursiops aduncus*)	35/15	South Australia	NA	NA	2.84 ± 3.10		0.05 ± 0.08	$mg\ kg^{-1}$ wet weight	Lavery et al., 2008
Common dolphin (*Delphinus delphis*)	30	European coasts	NA	NA	0.661 ± 0.511			$\mu g\ g^{-1}$ dry weight	Caurant et al., 2006
Harbour porpoise (*Phocoena phocoena*)	21	European coasts	NA	NA	0.369 ± 0.309			$\mu g\ g^{-1}$ dry weight	
Striped dolphin (*Stenella coeruleoalba*)	6	European coasts	NA	NA	0.410 ± 0.366			$\mu g\ g^{-1}$ dry weight	
Dall's porpoise (*Phocoenoides dalli*)	17	Sanriku coast, Japan	Female/ Male	NA		0.2 ± 0.2		$\mu g\ g^{-1}$ dry weight	Yang et al., 2002
Cuvier's Beaked Whale (*Ziphius cavirostris*)	1	Mediterranean Sea	NA	NA	4.2 ± 0.06	0.34 ± 0.02	0.04 ± 0.01	$\mu g\ g^{-1}$ dry weight	Frodello et al., 2002

(Contd.)

Table 6.1 Mean Cd, Hg and Pb concentrations in bone tissue from different marine mammal species worldwide (*Contd.*)

Species	N	Region	Sex	Age	Pb	Hg	Cd	Unit	References
Indo-Pacific bottlenose dolphin (*T. aduncus*)	36	South Australia	NA	NA	2.78 ± 3.07		0.047 ± 0.081	mg kg⁻¹ wet weight	Lavery et al., 2007
Bottlenose dolphin (*Tursiops truncatus*)	6	South Australia	NA	NA	0.85 ± 0.19			mg kg⁻¹ wet weight	
Common dolphin (*Delphinus delphis*)	35	South Australia	NA	NA	1.03 ± 0.55			mg kg⁻¹ wet weight	
Antillean manatees (*Trichechus manatus manatus*)	22	Mexican Caribbean	NA	NA	11.2 ± 3.2		3.9 ± 0.5	μg g⁻¹ wet weight	Romero-Calderon et al., 2015
Antillean manatees (*Trichechus manatus manatus*)	11	The Gulf of Mexico	NA	NA	14.0 ± 2.4		4.1 ± 0.4	μg g⁻¹ wet weight	
Antillean manatees (*Trichechus manatus manatus*)		Quintana Roo, Mexico	NA	NA	128			mg kg⁻¹ waight	Rojas-Mingüer et al., 1997
West Indian manatee (*Trichechus manatus*)		Chetumal Bay, Mexico	NA	NA	41.0 ± 3.4		4 ± 1.0	μg g⁻¹ wet weight	Rojas-Minguer et al., 2002
Pacific walrus (*Odobenus rosmarus divergens*)	19	Chukchi Peninsula	NA	NA	3.46 ± 4.03		0.17 ± 0.17	μg g⁻¹ dry weight	Trukhin et al., 2014
Dall's porpoise (*Phocoenoides dalli*)		Northwestern Pacific	NA	NA	0.1 ± 0.1		0.2 ± 0.2	μg g⁻¹ wet weight	Fujise et al., 1988
Northern fur seal (*Callorhinus ursinus*)		Pribilof Islands, Alaska	NA	NA	1.6 ± 2.2		0.1 ± 0.0	μg g⁻¹ wet weight	Goldblatt et al., 1983
Striped dolphin (*Stenella coeruleoalba*)		Taiji, Japan	NA	NA	0.4 ± 0.2		0.1 ± 0.1	μg g⁻¹ wet weight	Honda et al., 1986
California sea lion (*Zalophus californianus*)		Gulf of California, Mexico	NA	NA	27.8 ± 1.4		2.8 ± 0.5	μg g⁻¹ wet weight	Szteren et al., 2013
Weddell seal (*Leptonychotes weddellii*)		Syowa Station, Antartica	NA	NA	0.1 ± 0.1		0.0 ± 0.0	μg g⁻¹ wet weight	Yamamoto et al., 1987

Species	N	Region	Sex	Age	Pb	Hg	Cd	Unit	References
Harbor seal (*Phoca vitulina*)	26	California, USA	NA	NA		0.038 ± 0.002		µg g⁻¹ wet weight	Brookens et al., 2008
Baikal seal (*Phoca sibirica*)		Lake Baikal, Russia	male	19		0.02		µg g⁻¹ wet weight	Watanabe et al., 1996
Baikal seal (*Phoca sibirica*)		Lake Baikal, Russia	female	13		0.01		µg g⁻¹ wet weight	Watanabe et al., 1997
California sea lion (*Zalophus californianus*)	13	Northern Gulf of California	NA	NA	28.25 ± 12.37	0.16 ± 0.29	2.61 ± 1.76	µg g⁻¹ dry weight	Szteren et al., 2013
California sea lion (*Zalophus californianus*)	17	Ángel de la Guarda, Gulf of California	NA	NA	28.26 ± 5.77	0.15 ± 0.14	2.63 ± 1.25	µg g⁻¹ dry weight	
California sea lion (*Zalophus californianus*)	22	Central Gulf of California	NA	NA	28.98 ± 16.67	0.02 ± 0.05	3.57 ± 2.68	µg g⁻¹ dry weight	
California sea lion (*Zalophus californianus*)	14	Southern Gulf of California	NA	NA	25.67 ± 12.87	0.04 ± 0.04	2.59 ± 1.84	µg g⁻¹ dry weight	
East Asian finless porpoise (*Neophocaena asaeorientalis sunameri*)	27	East China Sea, China	NA	NA	0.52 ± 0.31	0.61 ± 0.56	0.14 ± 0.10	mg kg⁻¹ dry weight	Hao et al., 2020
Harbour porpoise (*Phocoena phocoena*)	33	Bulgarian Black sea coast	NA	NA	13.80 ± 3.46		2.45 ± 0.64	mg kg⁻¹ dry weight	Evtimova et al., 2018

*NA – not available

CONCLUSIONS

The studies discussed herein indicate that bones may be considered as a suitable tissue to monitor metals in marine mammals.

Lead usually exhibits the highest levels compared to Hg and Cd. The highest Pb concentration was observed in West Indian manatee (*Trichechus manatus*) in Chetumal Bay, Mexico (41 µg g^{-1} w.w.). Californian seal lions (*Zalophus californianus*) in Gulf of California also exhibited high mean values (28 µg g^{-1} dry weight). The lowest value was observed in Weddell seals (*Leptonychotes weddellii*) from Antarctica (0.1 µg g^{-1} wet weight). The highest Hg concentrations were observed in several different small cetaceans from the Mediterranean Sea (concentrations up to 8 µg g^{-1} dry weight), with the exception of one individual Risso's dolphin (*Grampus griseus*), which presented high Hg levels. All other studies indicated Hg values < 1 µg g^{-1}, except for Striped dolphins (*Stenella coeruleoalba*) around Kii Peninsula, Japan, who presented 1.55 µg g^{-1} w.w. The highest Cd concentration was of 4.1 µg g^{-1} w.w. in Antillean manatees (*Trichechus manatus manatus*) from the Gulf of Mexico and the lowest was non-detected concentrations in Weddell seals (*Leptonychotes weddellii*) from Antarctica.

The results of the different studies provide a comprehensive tool to investigate a variety of aspects of marine mammal ecology and physiology, and high Pb, Hg and Cd concentrations may reveal poor physiological states in marine mammals. Among the several factors affecting individual metal variations, the diet is probably one of the most important, as most persistent contaminants are incorporated into marine mammals via the dietary route through the ingestion of contaminated food items.

REFERENCES

Bansal, S.L. and Asthana, S. 2018. Biologically essential and non essential elements causing toxicity in environment. J. Environ. Anal. Toxicol. 8: 557.

Bignert, A., Danielsson, S., Nyberg, E., Asplund, L., Eriksson, U., Nylund, K., et al. 2010. Comments Concerning the National Swedish Contaminant Monitoring Programme in Marine Biota, 2010. 1:2010. Swedish Museum of Natural History, Stockholm, Sweden.

Bloom, N.S. 1992. On the chemical form of mercury in edible fish and marine invertebrate tissue. CJFAS 49(5): 1010–1017.

Borgmann, U., Couillard, Y., Doyle, P. and Dixon, D.G. 2005. Toxicity of sixty-three metals and metalloids to Hyalella azteca at two levels of water hardness. Environ. Toxicol. Chem. 24(3): 641–652.

Braham, H.W. 1973. Lead in the California sea lion (*Zalophus californianus*). Environ. Pollut. (1970) 5(4): 253–258. doi:10.1016/0013-9327(73)90002-5.

Briffa, J., Sinagra, E. and Blundel, R. 2020. Heavy metal pollution in the environment and their toxicological effects on humans. Heliyon 6(9): e04691. https://doi.org/10.1016/j.heliyon.2020.e04691.

Caurant, F., Aubail, A., Lahaye, V., Van Canneyt, O., Rogan, E., López, A., et al. 2006. Lead contamination of small cetaceans in European waters—The use of stable isotopes for identifying the sources of lead exposure. Mar. Environ. Res. 62: 131–148.

Cossa, D., Elbaz-Poulichet, F., Gnassia-Barelli, M. and Romeò, M. 1993. Le plomb en milieu marin biogèochimie et ècotoxicologie. Repères ocèan n°3, IFREMER.

Das, K., Debacker, V., Pillet, S. and Bouquegneau, J.-M. 2003. Heavy metals in marine mammals. pp. 135–167. *In*: Vos, J.G., Bossart, G.D., Fournier, M. and O'Shea, T.J. (eds). Toxicology of Marine Mammals, Vol. 3. Taylos & Francis, New York, NY, USA.

Enhus, C., Boalt, E. and Bignert, A. 2011. A retrospective study of metals and stable isotopes in seals from Swedish waters. 5: 2011. Swedish Museum of Natural History, Stockholm, Sweden.

Escobar-Sánchez, O. 2010. Bioacumulación y Biomagnificación de Mercurio y Selenio en peces Pelágicos Mayores de la Costa Occidental de Baja California sur, México. Ph.D. Thesis, IPN-CICIMAR, La Paz, México.

Evtimova, V., Parvanov, D., Grozdanov, A., Tserkova, F., Zlatkov, B., Vergilov, V., et al. 2019. Heavy metals in bones from harbour porpoises Phocoena phocoena from the Western Black Sea Coast. ZooNotes 136: 1–4. Available onlineat: http://www. zoonotes.bio.uni-plovdiv.bg/ZooNotes_2019/ZooNotes_2019_Full.pdfFélix.

Ewers, U. and Schlipköter, H.W. 1991. Lead. In: Merian, E. (ed.), Metals and their Compounds in the Environment. VCH, Weinheim.

Frank, A., Galgan, V., Roos, A., Olsson, M., Petersson, L.R. and Bignert, A. 1992. Metal concentrations in seals from swedish waters. Ambio 21(8): 529–538.

Frodello, J., Roméo, M. and Viale, D. 2000. Distribution of mercury in the organs and tissues of five toothed-whale species of the Mediterranean. Environ. Polluti. 108(3): 447–452. doi:10.1016/s0269-7491(99)00221-3

Frodello, J.P. and Marchand, B. 2001. Cadmium, copper, lead, and zinc in five Toothed whale species of the Mediterranean Sea. Intern. Journ. of Toxicol. 20: 339–343.

Fujise, Y., Honda, K., Tatsukawa, R. and Mishima, S. 1988. Tissue distribution of heavy metals in Dall's porpoise in the Northwestern Pacific. Mar. Pollut. Bull. 19(5): 226–230.

Garcia-Fernandez, A.J., Gomes-Ramirez, P., Martinez-Lopez, E., Hernandez-Garcia, A., Maria-Mojica, P., Romero, D., et al. 2009. Heavy metals in tissues from loggerhead turtles (*Caretta caretta*) from the southwestern Mediterranean (Spain). Ecotoxol. and Environ. Safety 72: 557–563.

Gdula-Argasinska, J., Appleton, J., Sawicka-Kapusta, K. and Spence, B. 2004. Further investigation of the heavy metal content of the teeth of the bank vole as an exposure indicator of environmental pollution in Poland. Environ. Pollut. 131: 71–79. http:// dx.doi.org/10.1016/ j.envpol.2004.02.025

Georgiev, D., Velcheva, I., Gecheva, G., Petrova, S. and Mollov, I. 2011. Contamination of waters and its impact on the ecosystems. University of Plovdiv. Publishing House.

Geraci, J.R., Harwood, J. and Lounsbury, V.J. 1999. Marine mammal dieoffs. pp. 367–395. *In*: Twiss, J.R. and Reeves, R.R. (eds). Conservation and Management of Marine Mammals. Smithsonian Institution Press, Washington DC.

Goldblatt C.J. and Anthony, R.G. 1983. Heavy metals in northern fur seals (*Callorhinus ursinus*) from the Pribilof Islands, Alaska. J. Environ. Qual. 12: 478–482.

Hao X, Shan H, Wu C, Zhang D, Chen B. 2020. Two decades' variation of trace elements in bones of the endangered east asian finless porpoise (*Neophocaena asaeorientalis sunameri*) from the East China Sea, China. Biol. Trace. Elem. Res. 198(2): 493–504.

Hoffman, D.J., Rattner, B.A., Burton, G.A., Jr. and Cairns, J., Jr. (eds). 2003. Handbook of Ecotoxicology, 2nd Ed. Lewis Publishers, Boca Raton, FL.

Honda, K., Tatsukawa, R. and Fujiyama, T., 1982. Distribution characteristics of heavy metals in the organs and tissues of striped dolphin, *Stenella coeruleoalba*. Agricult. Biol. Chem. 46(12): 3011–3021.

Honda, K., Tatsukawa, R., Itano, K., Miyazaki, N. and Fujiyama, T. 1983. Heavy metal concentrations in muscle, liver and kidney tissue of Striped dolphin *Stenella coeruleoalba* and their variations with body, length, weight, age and sex. Agricult. Biol. Chem. 47: 1219.

Honda, K., Fujise, Y., Tatsukawa, R. and Miyazaki, N. 1984. Composition of chemical components in bone of striped dolphin, *Stenella coeruleoalba*: Distribution characteristics of major inorganic and organic components in various bones, and their age related changes. Agric. Biol. Chem., 48(2): 409–418.

Honda, K., Fujise, Y., Tatsukawa, R., Itano, K. and Miyazaki, N. 1986. Age-related accumulation of heavy metals in bone of the striped dolphin, *Stenella coeruleoalba*. Mar. Environ. Res. 20: 143–160.

Itano, K., Kawai, S., Miyazaki, N., Tatsukawa, R. and Fujiyama, T. 1984. Mercury and selenium levels at the fetal and suckling stages of striped dolphin, *Stenella coeruleoalba*. Agric. Biol. Chem., 48: 1691–8.

Kemper, C., Gibbs, Ph., Obendorf, D., Marvanek, S. and Lenghaus, C. 1994. A review of heavy metal and organochlorine levels in marine mammals in Australia. Sci. Total Environ. 154(2–3): 129–39.

Kemper, C.M. and Gibbs, S.E. 1997. A study of the life history parameters of dolphins and seals entangled in tuna farms near Port Lincoln, and comparisons with information from other South Australian dolphin carcasses. Report to Environment Australia, p. 98.

Kido, T., Nogawa, K., Yamada, Y., Honda, R., Tsuritani, I., Ishizaki, M., et al. 1989. Osteopenia in inhabitants with renal dysfunction induced by exposure to environmental cadmium. Int. Arch. Occup. Environ. Health 61: 271–276.

Lantzy, R.J. and Mackenzie, F.T. 1979. Atmospheric trace metals: Global cycles and assessment of man's impact. Geochimica et Cosmochimica Acta 43(4): 511–525.

Lavery, T.J., Butterfield, N., Kemper, C.M., Reid, R.J. and Sanderson, K. 2008. Metals and selenium in the liver and bone of three dolphin species from South Australia, 1988–2004. Sci. Total Environ. 390: 77–85.

Lavery, T.J., Kemper, C.M., Sanderson, K., Schultz, C.G., Coyle, P., Mitchell, J.G., et al. 2009. Heavy metal toxicity of kidney and bone tissues in South Australian adult bottlenose dolphins (*Tursiops aduncus*). Mar. Environ. Res. 67(1): 1–7. doi: 10.1016/j.marenvres.2008.09.005. Epub 2008 Oct. 10. PMID: 19012959.

Law, R.J., Fileman, C.F., Hopkins, A.D., Baker, J.R., Harwood, J., Jackson, D.B., et al. 1991. Concentrations of trace metals in the livers of marine mammals (seals, porpoises and dolphins) from waters around the British Isles. Mar. Pollut. Bull. 22: 183–191.

Law, R.J., Jones, B.R., Baker, J.R., Kennedy, S., Milne, R. and Morris, R.J. 1992. Trace metals in the livers of marine mammals from the Welsh coast and the Irish Sea. Mar. Pollut. Bull. 24: 296–304.

Meador, J.P., Varanasi, U., Robisch, P.A. and Chan, S.L. 1993. Toxic metals in Pilot Whales (*Globicephala melaena*) from strandings in 1986 and 1990 on Cape Cod, Massachusetts. Can. J. Fish. Aquat. Sci. 50: 2698–2706.

Nganvongpanit, K., Buddhachat, K., Piboon, P., Euppayo, T., Kaewmong, P., Cherdsukjai, P., et al. 2017. Elemental classification of the tusks of dugong (Dugong dugong) by HHXRF analysis and comparison with other species. Sci. Rep. 7: 1 1.

Núñez-Nogueira, G., Pérez-López, A. and Santos-Córdova, J.M. 2019. As, Cr, Hg, Pb, and Cd Concentrations and bioaccumulation in the dugong dugong dugon and manatee trichechus manatus: A review of body burdens and distribution. Int. J. Environ. Res. Public Health 16(3): 404. doi:10.3390/ijerph16030404

O'Hara, T., Woshner, V. and Bratton, G. 2003. Inorganic pollutants in arctic marine mammals. pp. 206–246. *In*: Vos, J.G., Bossart, G.D., Fournier, M. and O'Shea, T.J. (eds). Toxicology of Marine Mammals, Vol. 3. Taylos & Francis, New York, NY, USA.

O'Shea, T.J., Moore, J.F. and Kochman, H.I. 1948. Contaminant concentrations in manatees in Florida. J. Wildl. Manag. 3: 741–748.

O'Shea, T.J. 1999. Environmental contaminants and marine mammals. pp. 485–566. *In*: Reynols, J.E., III, and Rommel, S.A. (eds). Biology of Marine Mammals. Smithsonian Institute Press: Washington DC, USA.

Outridge, P.M., Wagemann, R. and McNeely, R. 2000. Teeth as biomonitors of soft tissue mercury concentrations in beluga, *Delphinapterus leucas*. Environ. Toxicol. Chem. 19: 1517–1522.

Reijnders, P., Brasseur, S., van der Toorn, J., van der Wolf, P., Boyd, I., Harwood, J., et al. 1993. Seals, fur seals, sea lions and walrus. Status Survey and Conservation Action Plan. Gland, Switzerland, IUCN The World Conservation Union, 88 p.

Rojas-Mingüer, A., Morales-Vela, B. and Rosiles-Martínez, R. 1997. Metals in Bone and Blood of manatees (Trichechus manatus manatus) from Chetumal Bay, Quintana Roo, México. ECOSUR, México. p. 287.

Rojas-Mingüer, A. and Morales-Vela, B. 2002. Metales en hueso y sangre de manatíes de (*Trichechus manatus manatus*) de la Bahía de Chetumal, Quintana Roo, Mexico. pp. 133–142. *In*: Rosado-May, F.J., Romero Mayo, R. and De Jesus Navarrete, A. (eds). Contribuciones de la ciencia al manejo costero integrado de la Bahía de Chetumal y su area de influencia. Universidad de Quintana Roo, Mexico.

Romero-Calderón, A.G., Morales-Vela, B., Rosíles-Martínez, R., Olivera-Gómez, L.D. and Delgado-Estrella, A. 2015. Metals in Bone Tissue of Antillean Manatees from the Gulf of Mexico and Chetumal Bay, Mexico. Bull. Environ. Contam. Toxicol. 96(1): 9–14. doi:10.1007/s00128-015-1674-6

Sonne-Hansen, C., Dietz, R., Leifsson, P.S., Hyldstrup, L. and Riget, F.F. 2002. Cadmium toxicity to ringed seals (*Phoca hispida*)–an epidemiological study of possible cadmium induced nephropathy and osteodystrophy in ringed seals (*Phoca hispida*) from Qaanaaq in Northwest Greenland. Sci. Total Environ. 295: 167–181.

Szteren, D. and Aurioles-Gamboa, D. 2013. Trace elements in bone of Zalophus californianus from the Gulf of California: A comparative assessment of potentially polluted areas. Cienc. Mar. 39: 303–315.

Takeuchi, N.Y., Walsh, M.T., Bonde, R.K., Powell, J.A., Bass, D.A., Gaspard, J.C., et al. 2016. Baseline reference range for trace metal concentrations in whole blood of wild and managed west Indian manatees (*Trichechus manatus*) in Florida and Belize. Aquat. Mamm. 42: 440–453.

Tiffini, J., Brookens, T.J., O'Hara, T.M., Taylor, R.J., Bratton, G.R. and Harvey, J.T. 2008. Total mercury body burden in Pacific harbor seal, Phoca vitulina richardii, pups from central California. Mar. Pollut. Bull. 56(1): 27–41.

Triffitt, J.T. 1985. Receptor molecules, coprecipitation and ion exchange process in the deposition of metal ions in bone. pp. 3–20. *In*: Priest, N.D. (ed.). Metals in Bone. Springer, Dordrecht. https://doi.org/10.1007/978-94-009-4920-1_1.

Trukhin, A.M., Kolosova, L.F. and Slin'ko, E.N. 2015. Heavy metals in the organs of Pacific walruses (*Odobenus rosmarus divergens*) from the Chukchi Peninsula. Russian J. Ecol. 46(6): 585–588. doi:10.1134/s1067413615060211

Vighi, M., Borrell, A. and Aguilar, A. 2017. Bone as a surrogate tissue to monitor metals in baleen whales, Chemosphere 171: 81–88. doi: 10.1016/j.chemosphere.2016.12.036

Wagemann, R. and Muir, D.C.G. 1984. Concentrations of heavy metals and organochlorines in marine mammals of northern waters: overview and evaluation. Western Region, Department of Fisheries and Ocean, Canada.

Watanabe, I., Ichihashi, H., Tanabe, S., Amano, M., Miyazaki, N., Petrov, E.A., et al. 1996. Trace element accumulation in Baikal seal (*Phoca sibirica*) from the Lake Baikal. Environ. Pollut. 94(2): 169–179. doi:10.1016/s0269-7491(96)00079-6.

Wiener, J.G., Krabbenhoft, D.P., Heinz, G.H. and Scheuhammer, A.M. 2003. Ecotoxicology of mercury. pp. 415–465. *In*: Hoffman, D.J., Rattner, B.A., Burton, G.A., Jr. and Cairns, J., Jr. (eds.), Handbook of Ecotoxicology, 2nd Ed. Lewis Publishers, Boca Raton, FL.

Yamamoto, Y., Honda, K., Hidaka, H. and Tatsukawa, R. 1987. Tissue distribution of heavy metals in Weddell seals (*Leptonychotes weddellii*). Mar. Pollut. Bull. 18: 164–169.

Yang, Jian and Kunito, Takashi and Tanabe, Shinsuke and Miyazaki, Nobuyuki. 2002. Mercury in tissues of Dall's porpoise (*Phocoenoidesdalli*) collected off Sanriku coast of Japan. Fisheries Science 68: 256–259. 10.2331/fishsci.68.sup1_256.

Yoshiki, S., Yanagisawa, T., Kimura, M., Otaki, N. and Suzuki, M. 1975. Bone and kidney lesions in experimental cadmium intoxication. Arch. Environ. Health 30: 559–562.

Zelikoff, J.T. and Thomas, P.T. 1998. Immunotoxicology of Environmental and Occupational Metals. Taylor & Francis, London.

Chapter 7

Physicochemical Water Parameters Affecting Cadmium, Lead and Mercury Speciation, Bioavailability and Toxicity in the Aquatic Environment

Paloma de Almeida Rodrigues[1,2,3*], Rafaela Gomes Ferrari[1,2,4]
and Carlos Adam Conte Junior[1,2,3,5,6,7]

[1]Center for Food Analysis (NAL), Technological Development Support Laboratory (LADETEC), Federal University of Rio de Janeiro (UFRJ), Cidade Universitária, Rio de Janeiro 21941-598, RJ, Brazil.

[2]Laboratory of Advanced Analysis in Biochemistry and Molecular Biology (LAABBM), Department of Biochemistry, Federal University of Rio de Janeiro (UFRJ), Cidade Universitária, Rio de Janeiro 21941-909, RJ, Brazil.

[3]Graduate Program in Veterinary Hygiene (PPGHV), Faculty of Veterinary Medicine, Fluminense Federal University (UFF), Vital Brazil Filho, Niterói 24220-000, RJ, Brazil.

[4]Agrarian Sciences Center, Department of Zootechnics, Federal University of Paraiba,Areias 51171-900, PB, Brazil.

[5]Graduate Program in Sanitary Surveillance (PPGVS), National Institute of Health Quality Control (INCQS), Oswaldo Cruz Foundation (FIOCRUZ), Rio de Janeiro 21040-900, RJ, Brazil.

[6]Graduate Program in Food Science (PPGCAL), Institute of Chemistry (IQ), Federal University of Rio de Janeiro (UFRJ), Cidade Universitária, Rio de Janeiro 21941-909, RJ, Brazil.

[7]Graduate Program in Chemistry (PGQu), Institute of Chemistry (IQ), Federal University of Rio de Janeiro (UFRJ), Cidade Universitária, Rio de Janeiro 21941-909, RJ, Brazil.

*Corresponding author: paloma_almeida@id.uff.br

INTRODUCTION

The aquatic environment is a target for several chemical pollutants, including many toxic metals, such as mercury (Hg), cadmium (Cd), and lead (Pb). These are three important toxic metals widely studied by the scientific community. All three are emitted by natural and anthropogenic sources, finally entering aquatic environment (Jan et al., 2015). In this environmental compartment, these elements present similar dynamics, remaining in suspension or deposited in sediment. Physicochemical water parameters are significantly affected by biogeochemical cycles, influencing elemental speciation, resulting in greater or lesser toxicity, or even influencing metal bioavailability, among other aspects (Li et al., 2020; Van Ael et al., 2017).

In this regard, pH, salinity, dissolved oxygen, organic matter, and temperature are noteworthy, in addition to the presence of organic and inorganic compounds, which may complex with these metals. These elements, as mentioned previously, sediment in greater amounts and become available for absorption by the biota when resuspended, either by marine currents or by animal movement itself (Li et al., 2020; Owens et al., 2009). Absorption then takes place through the gills, due to the process of respiration or feeding (Li et al., 2020; Raknuzzaman et al., 2016). Such metals accumulate in the animal organism and can reach humans through seafood consumption, resulting in environmental and human health concerns. Different elements display different toxicity mechanisms, leading to deleterious effects on the respiratory, reproductive, immunological, renal, hepatic, and nervous systems (Das et al., 2019; Li et al., 2016; Xu et al., 2016; Zhao et al., 2010; Zhu et al., 2018). In this context, this chapter will address the main abiotic variables responsible for affecting the behavior of Cd, Hg and Pb, exploring their chemical forms in different aquatic environments, contaminant sources, the main physical-chemical parameters that influence speciation, toxicity, and bioavailability and their toxic effects on aquatic biota and human health.

CADMIUM (Cd)

Sources

This metal is found naturally in soil and can reach the aquatic environment. Natural Cd sources to the aquatic environment comprise of rock erosion, dissolution of Cd present in minerals, deposition of atmospheric Cd, and volcanic emanations. The most common minerals that contain cadmium are forms of CdS (greenockite and hawleyite), cadmoselite (CdSe), monteponite (CdO), otavite ($CdCO_3$), and cadmian metacinnabar ((HgCd)S) (Cullen and Maldonado, 2012). Anthropogenic activities are, however, the main source of water contamination. These include industries, such as the Ni-Cd battery industry, pigment and plastic production, gasoline additives, anticorrosion agents and mining activities (Järup, 2003; Zhang and Reynolds, 2019). The irregular discharge of industrial and domestic sewage into water is the main Cd input to the aquatic environment. When considering domestic sewage, inappropriate battery disposal stands out, as these are rarely recycled. These batteries contaminate the aquatic environment due to the lixiviation of the discarded in landfills by the action of rain or waste incineration, which promotes a greater release of Cd into the atmosphere. Another important source of Cd emission to the environment is smoking (Jarup, 2003). Some sources, such as industrial activity, volcanic activity, biomass combustion, mining, and cigarette smoke, are responsible for emitting dust and Cd vapor into the air. These tiny particles of atmospheric Cd are deposited in water. Upon reaching the water, as well as other elements, it can be deposited in the sediment. When resuspended, it becomes available for absorption or even associated with suspended particles, also becoming available for absorption by the biota, accumulating in the body, in addition to the potential for biomagnification between benthic chains (Jarup, 2003; Zhang and Reynolds, 2019). This is because some articles point out that Cd, as well Cu, have a better biomagnification capacity in benthic chains, as compared to other elements, such as Hg, which can biomagnify throughout the entire chain (Costa et al., 2016; Saadati et al., 2019; Tchounwou et al., 2012; Wiech et al., 2018; Zhang and Reynolds, 2019). Figure 7.1 summarizes the dynamics of cadmium as it reaches the aquatic environment.

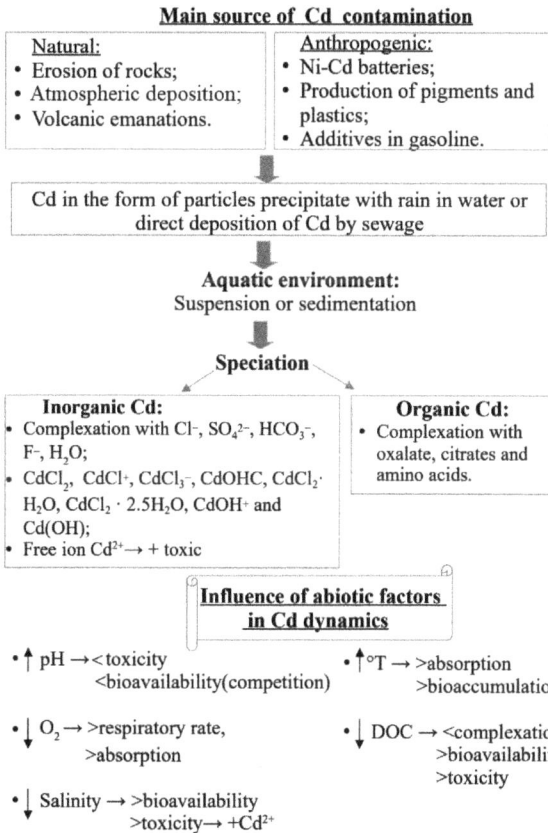

Figure 7.1   Cd dynamics and contamination in water media.

Speciation and Influence of Abiotic Parameters on Cd Dynamics in Water

Cd can exist in several phases depending on the environment: water, sediment, and air. In addition, this element can be found in different forms, being in smaller proportions in the form of ion, due to low solubility, and the majority complexed to inorganic ions (e.g., Cl^-, SO_4^{2-}, HCO_3^-, F^-) and organic ligands (e.g., oxalate, citrates, and amino acids). For example, in these cases, the cadmium chloride ($CdCl_2$) complex, which is soluble, can dissolve in water to form Cd^{2+}, $CdCl^+$, $CdCl_3^-$ and $CdOHCl$. It is still possible to complex with water and form $CdCl_2 \cdot H_2O$ and $CdCl_2 \cdot 2.5H_2O$. They can also form complexes such as $CdOH^+$ and $Cd(OH)_2$ (Mason, 2013; Zhang and Reynolds, 2019).

Cd species and metal behavior may vary according to the type of aquatic environment. In freshwater environments, the element is quite mobile compared to the marine environment. Another issue is that Cd's availability and toxicity in freshwater are strictly related to the chemical species. Organic matter is normally complexed in freshwater systems. How complexed Cd will be is related not only to dissolved organic carbon (DOC) concentrations but also to the pH (tending to acid) of the water and the existence of other cations promoting competition between the ligands. In these environments, there is a predominance of complexed forms and a lower prevalence of the free form Cd^{2+}. In an estuarine environment, where there is a transition between fresh and saltwater, this speciation is altered by the inflow of saline water. In these regions, there is the formation of a complex mainly with chlorine. In the ocean, the main natural source of Cd is through river flow. In this location, the element is mainly in dissolved form. The concentration of Cd in the marine environment is also related to depth, having lower concentrations at the surface, intermediate values in the middle of the water column, and higher concentrations as it approaches the bottom. In the coastal region, the surface water region has higher concentrations of Cd when compared to the ocean region. This is due to the greater influence of river flow, terrestrial emissions, and the phenomenon of upwelling towards the coast, which inverts the deep waters to the surface, enriching this region with nutrients and pollutants that were submerged. In this environment, a higher percentage of dissolved Cd is found to form a complex with organic ligands, especially in surface water. In deep waters are mainly $CdCl^+$ and $CdCl_2$ species (Cullen and Maldonado, 2012).

The speciation of Cd can be influenced by physicochemical parameters of water, which in turn affect the toxicity and bioavailability of the element. One of the main physicochemical parameters of water reported in the literature as influencing the behavior of metals is salinity. Thwala et al. (2011) identified that the toxicity of Cd is increased in a lower salinity environment and justifies that this in fact may be related to the form of free ion (Cd^{2+}), which is more abundant in this environment, is the most bioavailable and most toxic, in addition to the reduction of the chloride ion, which consequently reduces the complexation of the Cd and an increase in the free form. In the marine environment, the main form is complex and with less toxic potential (Ma et al., 2019; Piazza et al., 2016, Thwala et al., 2011). Another issue related to salinity and toxicity is linked to osmoregulatory organisms, such as crustaceans. For these animals, a lower Cd uptake is identified in waters with salinity close to the isosmotic point due to the possible influence of the osmoregulatory ion-exchange mechanism, responsible for the absorption of the metal. That is to say that, in an isosmotic situation, the balance of salinity allows less uptake of the metal by the animal. In environments with salinity below the isosmotic point, the organism tends to absorb water and, consequently, the elements there, such as Cd. This one, in particular, is easily absorbed by the calcium transporting channels in the branchial epithelium, as they have the same radius ionic and charge (Rainbow, 1995), becoming indistinguishable from transporters. Thus, it is also possible to identify that the calcium concentration in the water will influence the absorption of Cd by the biota (Thwala et al., 2011; USEPA, 2016). Not only would the calcium present in the water be considered a competitor for absorption and consequent toxicity of Cd, but also other cations such as magnesium (Mn), while sulfate (SO_4^{2-}) and chloride (Cl^-), which are anions, can be connected to Cd, generating forms of less toxicity (Bielmyer-Fraser et al., 2018; Ma et al., 2019). Thus, the hardness of the water, that is to say, the concentration of ions such as calcium and magnesium in the form of carbonates is also an important factor related to the dynamics of elements such as Cd. The increase in hardness reduces the acute toxicity of Cd in aquatic animals (Sprague, 1985; USEPA, 2016).

The relationship between water pH and Cd is explored in different studies and remains complex. The acidic pH, despite providing greater solubility and bioavailability, leads to reduced toxicity of metals such as Cd. This is because the increase in hydrogen ions promotes competition between the H^+ and Cd ions for the binding sites to be absorbed by the biota. A reduced competition in alkaline water would then be expected. However, it is important to emphasize that alkalinity is caused by the presence of carbonates, bicarbonates, and hydroxides, which can also compete for absorption with Cd ions or even complex with this metal, generating insoluble forms, promoting the reduction of toxicity, mainly in brackish waters where there is a higher concentration of carbonates (Cornelis et al., 2007; Sandrin et al., 2002).

The temperature of the water can also be considered a factor that influences the behavior of Cd. Studies show a positive correlation between the accumulation of the element and the temperature. The increase in temperature increases the accumulation of the element in aquatic organisms, as already demonstrated in oysters, due to the increased permeability of the membrane, leading to better absorption of metal ions as seen with Cd. In addition, the increase in water temperature is related to the increase in the respiratory rate and consequently the greater uptake via breathing of metals such as Cd, as has been demonstrated in several studies related to this metal (Lan et al., 2019; Piazza et al., 2016; Prato et al., 2008; Quevedo et al., 2018; Vellinger et al., 2012; Vouyer and Modica, 1990). This means that water temperature and animal metabolism are directly related. Another related factor is the concentration of DOC in the water. Studies show that the increase in DOC leads to a reduction in Cd toxicity in aquatic animals and a reduction in bioavailability due to the complexation of DOC with Cd

(Clifford and McGeer, 2010; Giesy and Wiener, 1977; Niyogi et al., 2008; USEPA, 2016). The table below (Table 7.1) presents some recent studies that measured related environmental variables and Cd concentrations in seafood and/or water.

Table 7.1 Different studies that correlate abiotic variables with Cd concentration

Salinity	pH	T °C	DOC (mg L^{-1})	DO (mg L^{-1})	Cd concentration	References
–	–	20 24 28 32	–	–	0.88 in fish 1.19 in fish 2.13 in fish 2.59 in fish	Abdel–Tawwab and Wafeek (2014)[1]
–	8.23 8.14	28.05 28.10		6.35 6.05	0.25 in water 0.31 in water	Jacaúna et al. (2020)[2]
32.2 32.2	7.92 7.89	29.5 24.6	–	–	0.07 in water 0.39 in oysters 0.05 in water 0.22 in oysters	Lan et al. (2019)[3]
–	7.9 to 9.3	–	–	–	<0.004–1 in water	Qu et al. (2019)[4]
–	7.01 6.69	23 24.3	–	–	0.19 in water <0.1 in water	Samayamanthula et al. (2021)[5]
–	7.5 6.29	17.3 19.4	–	8.1 4.5	0.9 in water 0.3 in fish 0.9 in water 0.1 in fish	Weber et al. (2013)[6]

Water samples were measured in µg L^{-1} and animal samples were measured or converted in mg kg^{-1} d.w. DOC: Dissolved organic carbon; DO: Dissolved oxygen.

[1] Fish species: *Oreochromis niloticus*. Local: Egypt;

[2] Values obtained from the two collection points. Local: Igarapé do Quarenta, Manaus, Brazil;

[3] Values were obtained from two collection points (A and G). The animal sample was in wet weight. Oysters specie: *Crassostrea gigas*. Local: Xiamen Bay, China;

[4] The values presented refer to the lowest and highest value of Cd detected in the water samples, from Yarlung Tsangpo and Shule He rivers. Local: Tibetan Plateau;

[5] Average of three-day values at two collection points in 2018. Local: P1 Qortuba and P2 Jaber Al Ahmed, Kuwait;

[6] Results for the two collection points: upper Sinos River and lower Sinos River. Fish species *Oligosarcus gender*. Local: Rio Grande do Sul-Brazil.

Toxicity

Cd is considered a non-essential metal with toxic potential, affecting animal and human health. Animals are exposed to the contaminant through absorption via food or through the gills, which are more vulnerable due to the large surface area and the thin thickness of the epithelium. Gills histological changes are observed, such as epithelial wrinkling, which can compromise osmoregulatory activity. Furthermore, Cd in aquatic organisms can also promote the inhibition of energy metabolism;

reduction of antioxidant enzymes; altered sperm quality, and cardiac involvement, such as myocardial edema and inflammatory cell infiltrate in addition to impaired liver function (Jaishankar et al., 2014; Lei et al., 2011; Ma et al., 2019; Wu et al., 2015; Zhu et al., 2018).

In humans, in addition to the route of contamination by food consumption, contamination can occur through the inhalation of Cd particles from industrial emissions or even from tobacco, especially for smokers (with a lower risk ratio for passive smokers) (Jaishankar et al., 2014; Zhang and Reynolds, 2019). Among the foods, fruit and vegetables are the ones with the highest concentrations of the element (Jaishankar et al., 2014). However, fish also stands out as a source of contamination for humans. When the particles are inhaled, damage to the respiratory system is identified (Jaishankar et al., 2014). Through the food route, Cd after being absorbed can cause damage to different tissues, especially to the kidneys, bones, liver, and reproductive system (Das et al., 2019; Jaishankar et al., 2014; Tang et al., 2020). For example, in the kidneys, tubular and glomerular damage that can progress to an irreversible nature are identified (Zhang and Reynolds, 2019). Cd is also considered a carcinogenic agent, belonging to group 1 of the IARC classification, and has been mainly related to cancer cases in the renal system (IARC, 1993).

MERCURY (Hg)

Sources

Mercury is a metal that can be found in nature associated with other elements, with HgS being the main compound found, called minerals cinnabar and metacinnabar. By heating it, followed by condensation processes, it is possible to release metallic/elemental mercury from the HgS compound (Micaroni et al., 2000). Among other natural sources of Hg emission are the emanations from volcanoes, continental degassing, weathering of rocks containing mercury salts, and the escape of Hg from the oceans. In addition to natural sources, anthropogenic sources are predominant in the contamination of the aquatic environment. These include mining, Chlor-alkali industry, acetaldehyde production, waste incinerators, paint factory, pesticides, fungicides, mercury vapor lamps, batteries, metallurgical activities, dental waste, and coal burning (Condini et al., 2017; Delgado-Alvarez et al., 2015). Hg can be released in the form of vapor and transported over long distances, reaching the ground and water. Other mercurial forms are associated with organic or inorganic compounds. They can be formed in water from elemental Hg or directly discharged into water, coming mainly from industrial and domestic sewage or through leaching from the upper portions of the soil profile or through surface runoff from the soil to the water. It is important to emphasize that despite having natural sources of mercury release, in addition to the transition processes between species that are also natural, the main form of environmental contamination is through human activities (Condini et al., 2017; Delgado-Alvarez et al., 2014; Díez, 2009; Harayashiki et al., 2018; Hintelmann, 2010;

Kojadinovic et al., 2006; Murphy et al., 2007; Onsanit et al., 2011). Figure 7.2 summarizes the dynamics of mercury as it reaches the aquatic environment.

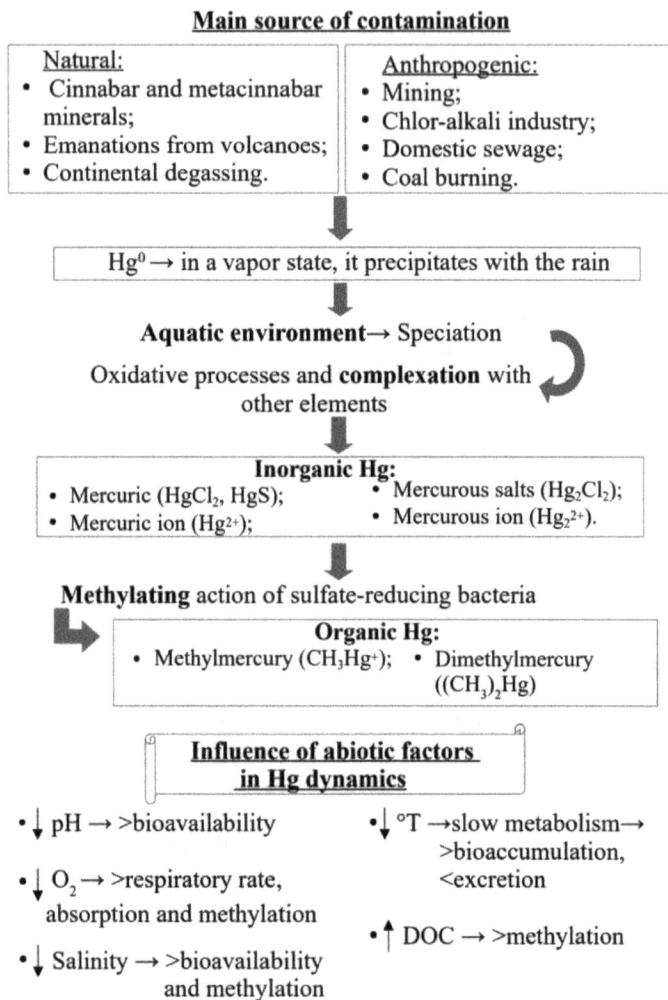

Main source of contamination

Natural:	Anthropogenic:
• Cinnabar and metacinnabar minerals; • Emanations from volcanoes; • Continental degassing.	• Mining; • Chlor-alkali industry; • Domestic sewage; • Coal burning.

$Hg^0 \rightarrow$ in a vapor state, it precipitates with the rain

Aquatic environment→ Speciation

Oxidative processes and **complexation** with other elements

Inorganic Hg:
- Mercuric ($HgCl_2$, HgS);
- Mercuric ion (Hg^{2+});
- Mercurous salts (Hg_2Cl_2);
- Mercurous ion (Hg_2^{2+}).

Methylating action of sulfate-reducing bacteria

Organic Hg:
- Methylmercury (CH_3Hg^+);
- Dimethylmercury (($CH_3)_2Hg$)

Influence of abiotic factors in Hg dynamics

• ↓ pH → >bioavailability

• ↓ O_2 → >respiratory rate, absorption and methylation

• ↓ Salinity → >bioavailability and methylation

• ↓ °T →slow metabolism→ >bioaccumulation, <excretion

• ↑ DOC → >methylation

Figure 7.2   Hg dynamics and contamination in water media.

Speciation and Influence of Abiotic Parameters on Hg Dynamics in Water

Mercury can be found in both inorganic and organic forms. In inorganic form, this metal can present three oxidation states: Elemental or metallic Hg (Hg^0) is a liquid at room temperature and has a high volatilization capacity. This, as aforementioned, can be emitted naturally through volcanic emanations and continental degassing or by evaporation when in its metallic form. Due to this property, Hg^0 can be transported over long distances. When precipitating with rain, it can reach aquatic environments, depositing itself in the sediment. Upon reaching the water, the elemental form can, through oxidative processes and by combining with other elements, give rise to other states of its inorganic form, the main ones being mercuric ($HgCl_2$, HgS) or mercurous salts (Hg_2Cl_2) and the mercuric (Hg^{2+}) or mercurous ion (Hg_2^{2+}), that can be deposited in the sediment (Bjørklund et al., 2017; Díez, 2009; Hong et al., 2012; Micaroni et al., 2000; Rice et al., 2014; Sunderland and Selin, 2013). Hg^0, just as it can reach water and undergo conversion processes, can also follow the opposite path and return to the atmosphere. For example, Hg^{2+} can, by photochemical or microbial action, be reduced to gaseous Hg^0 form and return to the atmosphere (Bowman et al., 2019). The resuspension of the sediment contaminated by Hg in its inorganic form (especially the mercuric ion) promotes the re-availability of Hg, which through the methylating action of sulfate-reducing bacteria gives rise to the organic form, mainly represented by methylmercury (CH_3Hg^+), also represented by MeHg. MeHg is the predominant form in aquatic organisms, being the most toxic, the most bioavailable for absorption by the biota, and the slowest one to be excreted, in addition to its good biomagnification capacity. This phenomenon is related to the ability that some elements have to be transferred to higher trophic level animals through their food in such a way that top-of-chain animals will have higher Hg concentration than low trophic level organisms (Mallory et al., 2018; Panichev and Panicheva, 2014; Rice et al., 2014; Ruus et al., 2015; 2017; Sevillano-Morales et al., 2015; Taylor and Calabrese, 2018).

Mercury has a variable distribution in different aquatic environments. Some species are more prevalent in some environments than others, which may be related to abiotic characteristics, which favor or disfavor certain species, or also due to the main polluting sources in each region. In colder areas, such as the Arctic and Antarctica, the main form of mercury contamination is through atmospheric deposition, whether wet or dry. In estuarine regions, the main contamination route is through direct discharge of Hg into the water and atmospheric deposition to a lesser extent. In the coastal areas, a significant reported contamination pathway, in addition to atmospheric deposition, direct discharge, and sediment acquisition, is ocean transport via upwelling.

In general, oceanic waters present mainly Hg in the species Hg^0, Hg^{2+}, MeHg, and diMeHg. In these waters, the metal, especially Hg^{2+}, has a preference for forming compounds with chlorine (e.g., $HgCl_3^-$ and $HgCl_4^{2-}$) and halides, unlike in the freshwater environment, where the compounds are mainly with oxides. Compounds with halide usually have greater difficulty in undergoing reduction and methylation processes when compared to other forms. In the portion of oceanic surface water, there is predominantly the Hg^0 species. In deep waters, under anoxia conditions, HgS, HgS_2H_2, HgS_2 H^- and HgS_2^{2-} forms are predominant, mainly in benthic sediments (Gworek et al., 2016). In lakes that have waters with more acidic pH, the form of MeHg is usually predominant, since in these places there is richness in sulfate and, consequently, the great activity of reducing sulfate bacteria carrying out the methylation process (Thomas et al., 2020).

Different abiotic factors are responsible for influencing the Hg dynamics in the aquatic environment. Among them, pH, dissolved organic matter, temperature, salinity, and oxygen saturation are parameters related to Hg established in the scientific literature (Ando et al., 2010; Boyd et al., 2017; Chen et al., 2014; Dong et al., 2016; Murphy et al., 2007; Reinhart et al., 2018; Rodrigues et al., 2021; Wang et al., 2010). Regarding pH, some studies indicate that Hg is more bioavailable for absorption in acidic pH; therefore, it becomes more suitable for being methylated. Thus, studies show higher concentrations of MeHg at acidic pH. Concerning organic matter, its presence in water favors methylation as it provides carbon for the bacteria responsible for this process (Dong et al., 2016). However, carbon can also bind to organic mercury and reduce its bioavailability and hence its absorption rate. Thus, the reduction of dissolved organic matter favors the absorption of Hg but can reduce methylation rates (Chen et al., 2014; Dong et al., 2016; Wang et al., 2010). Summer rains and storms are also responsible for increasing the mobilization of organic matter and sediments due to surface runoff, leading to increased Hg concentrations in the water (Chételat et al., 2015).

Water temperature is also an important factor in mercurial dynamics. There are two aspects that can influence Hg absorption and distribution. One aspect is that the water temperature affects the proximate composition of the animal's musculature, specifically the protein content, consequently affecting the concentration of the bioaccumulated element in the animal organism, linked to proteins. The other aspect is related to animal metabolism. Elevated temperatures accelerate the metabolism and consequently the rate of excretion of the element. Aquatic organisms subjected to an environment with varying water temperatures tend to warm up over a period of time which can limit the absorption of Hg by water compared to animals under constant temperature. This limitation of metal acquisition may be due to reduced metabolic capacity and also to increased metal excretion due to increased enzymatic biotransformation activity. Despite the possible reduction in bioaccumulation and increased excretion at higher water temperatures, it is also possible that with the change in metabolic activity, the animal organism has greater difficulty in activating its antioxidant defenses, which leads to oxidative damage, and can be translated into greater toxicity (Coppola et al., 2017; Múgica et al., 2015). Another issue is related to climate change and local precipitation dynamics, which can affect, for example, the microbial community responsible for the methylation process, especially by altering water chemistry. According to studies, there may be an increase in the DOC load in wet seasons, and with high temperatures, there will be an increase in the pH of the water (Dijkstra et al., 2013; Paranjape and Hall, 2017; Thomas et al., 2020; Ullrich et al., 2001). These changes in abiotic parameters directly interfere with the element's speciation, methylation, and toxicity.

Higher water salinity is related to findings of lower Hg concentrations in water and biota, especially due to competition during absorption. Furthermore, salinity is also related to the methylation process. Higher salinity environments compromise the survival of sulfate-reducing bacteria due to their sensitivity to salt. Moreover, the sulfide present in saline water can bind to the inorganic form of Hg, reducing its bioavailability for both absorption and methylation (Reinhart et al., 2018). Another finding is linked to a higher concentration of loaded forms of inorganic mercury in salt water, which is more difficult to be absorbed by bacteria for the methylation process (Ando et al., 2010; Boyd et al., 2017; Murphy et al., 2007; Reinhart et al., 2018). As for oxygen, lower concentrations increase the animal respiratory frequency and provide greater uptake of Hg by the biota. In addition, anoxic environments tend to favor the action of bacteria responsible for the formation of MeHg (Chen et al., 2018; Murphy et al., 2007). The table below (Table 7.2) presents some recent studies that measured related environmental variables and Hg concentrations in seafood and/or water.

Table 7.2 Different studies that correlate abiotic variables with Hg concentration

Salinity g/L	pH	T °C	DOC (mg L^{-1})	DO mg/L	Hg concentration	References
–	–	–	1.3 18.2	–	0.0013 0.0066 in water	Braaten et al. (2018)[1]
–	–	17°C 17°C→21°C	–	–	0.036 0.015 in mussel	Coppola et al. (2018)[2]
–	6.03 to 6.63	–	–	–	0.00065 to 0.02384 in water 0.07 to 3.51 in fish	Lino et al. (2019)[3]
–	7.9 to 9.3	–	–	–	1.49 to 4.99 in water	Qu et al. (2019)[4]
29.38 28.87	8.49 8.51	24.86 23.71	–	6.39 3.56	0.110 0.108 in crab	Rodrigues et al. (2020)[5]
35.62 31.17	8.66 8.58	17 25	–	6.13 4.24	0.034 0.043 in seafood	Rodrigues et al. (2021)[6]
–	7.21	–	8.03	8.18	0.67 in fish	Thomas et al. (2020)[7]

Water samples were measured in μg L^{-1} and animal samples were measured or converted in mg kg^{-1} d.w. DOC: Dissolved organic carbon; DO: Dissolved oxygen; → : temperature variation.

[1] Examples using the highest and lowest Hg values found in water samples from lakes Igletjern and Opptjern. Abiotic factor measured was total organic carbon (TOC). Local: Southeast Norway;

[2] Example of the first two experimental conditions used in the article. Mussel specie: *Mytilus galloprovincialis*. Local: Ria de Aveiro (northwest Atlantic coast of Portugal);

[3] Range of values presented in the article regarding pH and Hg concentration in water and fish species samples of importance for consumption. Local: Tapajos River basin, Brazil;

[4] The values presented refer to the lowest and highest value of Hg detected in the water samples, from Yarlung Tsangpo river. Local: Tibetan Plateau;

[5] Mean values of collection points outside the studied bay (Urca and Flamengo) and inside (Engenho and Paquetá) in 2018. Crab species: *Callinectes sapidus*, *Achelous spinimanus*, and *Achelous spinicarpus*. Local: Guanabara bay-Rio de Janeiro-Brazil;

[6] Obtainable values at two collection points (Cagarras Island and Seaport). Seafood species: Swimming crabs (*Achelous spinimanus*, *A. spnicarpus*, and *Callinectes sapidus*), Squid (*Doryteuthis sanpaulensis*, *D. plei*, and *Lolliguncula brevis*), Shrimp (*Farfantepenaeus brasiliensis*, *F. paulensis*, and *Litopenaeus schmitti*). Local: Guanabara bay-Rio de Janeiro-Brazil;

[7] Examples of results obtained with environmental and Walleye fish (*Sander vitreus*) samples. Local: Scattered lakes across Province of Ontario, Canada.

Toxicity

Most studies indicate that almost 100% of the mercury present in the animal's body is in the form of MeHg since approximately 100% of what is ingested in this way is absorbed. In contrast, only about 7 to 15% of the inorganic form is absorbed (Adams and Engel, 2014; Mallory et al., 2018; Souza-Araujo et al., 2016; Taylor and Calabrese, 2018). This element, like the others, can be acquired through food or water via gills (where it is absorbed via transport channels in membranes). Inside the body, Hg is usually distributed in different tissues, accumulating, mainly due to its lipoafinity and the presence of organic radical (in the case of MeHg), which facilitates its rapid entry into the bloodstream. Inorganic mercury, in turn, binds to components of the plasma fraction of blood and has less mobility when compared to the organic form. The latter tends to be concentrated in the tissues, especially in aquatic animals, in the gills, where they acquire stable bonds with mucoproteins, reducing the spread of the elements through the bloodstream. Animal tissues that accumulate MeHg include liver/hepatopancreas, musculature, kidneys, skin, digestive glands, heart, brain, gills, reproductive organs, and eggs (Azevedo et al., 2016; Bastos et al., 2015; Mallory et al., 2018; Nowosad et al., 2018; O'Bryhim et al., 2017; Pethybridge et al., 2010; Raimundo et al., 2010; Storelli et al., 2006). In humans, one of the main ways of metal exposure is through seafood feeding. In this case, Hg also tends to accumulate, mainly reaching the nervous system, due to its ability to cross the blood-brain barrier, renal, cardiovascular, immune system, in addition to the reproductive system, including being able to cross the placental barrier (Crowe et al., 2017; Dórea et al., 2013; Gutiérrez-Mosquera et al. 2017; Yin et al., 2017).

In general, Hg, especially as MeHg, prefers to establish bonds with albumin, glutathione, cystine, or proteins rich in cysteine, such as metallothionein, which are mainly present in the liver, an organ with detoxifying capacity. This protein contains a high percentage of amino, nitrogen, and sulfur groups that are used to sequester metals (Hosseni et al., 2013). Thus, metals can accumulate in the aforementioned tissues or even have a percentage eliminated via bile, reaching the intestine. In the organ, by enzymatic action, MeHg can be hydrolyzed and become partially demethylated, which makes it return to its inorganic form. It is then reabsorbed and eliminated in the feces. It is interesting to note that abiotic factors such as pH and microbial community can influence the metal dynamics in water, as aforementioned and within the organism. In this case, the internal pH can compromise the hydrolytic enzymatic action and the intestinal bacterial community, interfering in the demethylation and methylation process inside the organism and changing the Hg ligands (Clarkson et al., 2007; Land et al., 2018; Liao et al., 2019). At the cellular level, the element can cause several damages. During the MeHg degradation process, free radicals are generated, which are responsible for cell damage. In addition, changes in antioxidant enzymes, genotoxic and mutagenic effects, changes in Na^+/K^+ ATPase expression standards, alteration of calcium homeostasis, impairment of lymphocyte and monocyte proliferation, cytokine production and immunoglobulin secretion, induction of apoptosis of cerebellar neurons, changes in sperm indices and delay in neurological development when exposed during the gestational period are seen, both in humans, where most cases are mild and with lower percentage of severe cases or premature death, and in animals, where the involvement can also lead to death or compromise functions such as swimming performance (Bakar et al., 2017; García-Medina et al., 2017; Lackner et al., 2014; Macirella et al., 2017; Rasinger et al., 2017; Yang et al., 2019; Zhang et al., 2016).

LEAD (Pb)

Sources

Lead is a toxic contaminant of the aquatic environment but is a rare element in the continental crust. It is present in the mineral called galena (PbS) and, in an oxidation state, in the form of anglesite ($PbSO_4$), cerussite ($PbCO_3$), and pyromorphite (Pb_4 (PbCl) ($PO_4)_3$) (Cullen and McAlister, 2017). The contamination can occur through natural origins such as the erosion of rocky beds, volcanic outgassing, forest fire, radioactive decay; and anthropogenic sources such as industrial processes such as cosmetics, food, oil, mining, smelting, refining, marine leisure, battery manufacturing, protective X-ray materials, ammunition, glass and ceramic hardening, paint, weights and shielding applications, pesticide and fertilizer production (Jan et al., 2015; Li et al., 2016; Tchounwou et al., 2012). It is also used in the octane process of gasoline (Jan et al., 2015; Li et al., 2016; Tchounwou et al., 2012). Contamination can occur either by directly input of the element in water through industrial sewage discharge or atmospheric deposition. In the latter case, the small Pb particles (maybe up to in submicron size) can be transported by the wind over long distances and reach water via dry or wet deposition (rain, snow, sleet, or fog precipitation). This element can then reach water bodies directly or be first attached to plants and soil and then by river transport to the ocean. Pb in the form of particles that reach the water can take about 6 hours to dissolve (Bordon et al., 2018; Espejo et al., 2019; Jaishankar et al., 2014; Jan et al., 2015). Like the other elements, Pb can deposit in the sediment or remain suspended in water, interacting with other elements and being influenced by abiotic factors, changing its speciation and, consequently, its toxicity and bioavailability. Figure 7.3 summarizes the dynamics of lead as it reaches the aquatic environment.

Speciation and Influence of Abiotic Parameters on Pb Dynamics in Water

Pb can appear in different chemical forms depending on the abiotic factors in the water. One of the influencing factors of this speciation is the salinity of water. In a freshwater environment, where salinity is low, and pH varies between 6 to 8, Pb is mainly in the form of Pb^{2+}, $Pb(CO_3)_2^{2-}$ or even in complex with SO_4^2 (at pH tending to acidic), OH^- (at pH tending to alkaline) or organic matter in freshwater with good oxygenation. In an estuarine environment, where water salinity can be

Main source of contamination

Natural:	Anthropogenic:
• Erosion of rocky beds;	• Industrial sewage;
• Volcanic outgassing;	• Battery, paint, protective X-ray
• Florest fire;	• Ammunition;
• Radioactive decay.	• Octane process of gasoline.

⬇

Wet (rain) or dry deposition of Pb particles or direct discharge
of Pb and species by sewage

⬇

Aquatic environment:
Suspension or sedimentation

⬇

Speciation

Inorganic Pb:	**Organic Pb:**
• Complexation with CO_3^{2-}, NO^{3-}, OH$^-$ and with different oxidation states (0, I, II, IV);	• Tetraethyl lead
• Free ion Pb^{2+} → main inorganic form	• Tetramethyl lead

Influence of abiotic factors in Pb dynamics

• ↓ pH → >sediment release of Pb   • ↓ °T →slow metabolism:
      < absorption (competition)   >bioaccumulation

• ↑ O$_2$ → >sediment release of Pb   • ↓ DOC → <complexation
      > absorption   >bioavailability
                                              >toxicity

• ↓ Salinity→ >oxygen solubility, respiratory
      frequency and toxicity

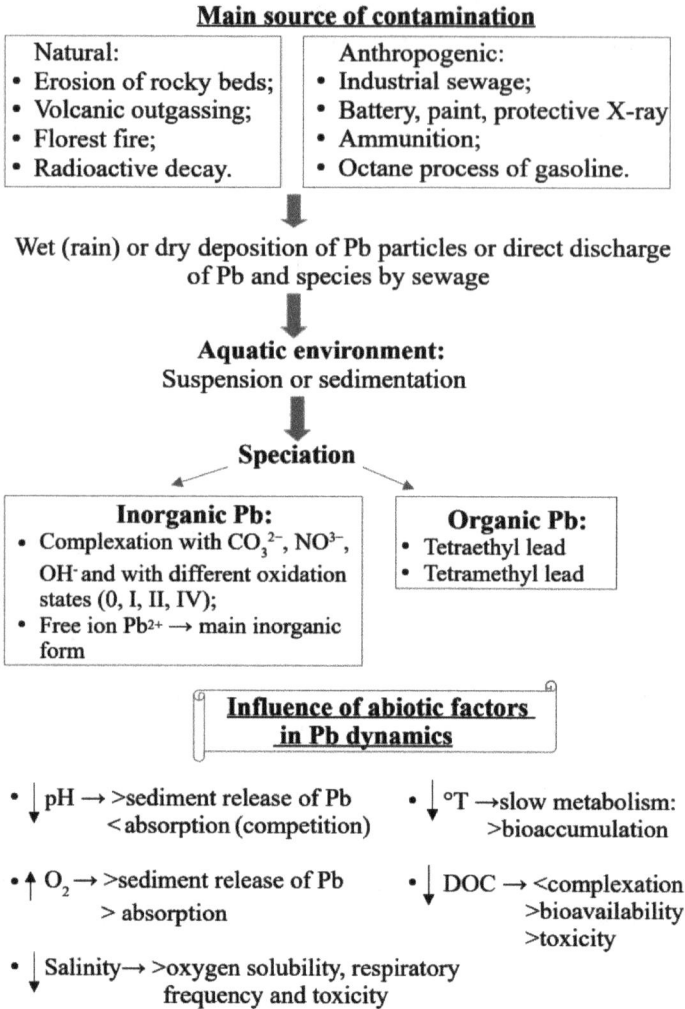

Figure 7.3 Pb dynamics and contamination in water media.

considered intermediate between fresh and saltwater, Pb tends to be complexed to particulate or colloid material, tending to increase this form as it moves towards the ocean. Furthermore, it is also possible to find this metal in solution, in inorganic form, when associated, for example, with Cl$^-$, CO and OH$^-$ or organic form when associated with organic binders. The latter case occurs in waters with a higher concentration of dissolved organic carbon. In a marine environment, there is a change in Pb speciation, in addition to the dissolution potential, which is reduced as salinity increases. In the open ocean, where there is a low concentration of DOC, Pb complexed to organic ligands decreases, and inorganic forms increase (Cullen and McAlister, 2017). In addition to salinity, other forces such as pH, ionic strength, and cation concentration will influence speciation and adsorption to suspended particles or sediment. Through evaluations of Pb speciation, most of the

Pb dissolved in water is found complexed to inorganic compounds; among them are preferentially CO_3^{2-} (25%), NO_3^- (22%), and OH^- (19%) and with different oxidation states (0, I, II, IV). Regarding the oxidation state, Pb^{2+} is considered the most stable form and generally the main inorganic form found in aquatic organisms, especially in waters with less salinity. As for the organic form, tetraethyl lead, and tetramethyl lead (mainly used in fuel), due to their lipophilic nature, are better bioaccumulated (Cullen and McAlister, 2017; Santos et al., 2014; Tonietto et al., 2015). Regarding the biota's ability to absorb and bioaccumulate the element, research indicates that low pH (acid), low salinity, and high-water temperatures are responsible for favoring better and greater absorption and bioaccumulation (Osuna-Flores et al., 2014; Santos et al., 2014).

Table 7.3 Different studies that correlate abiotic variables with Pb concentration

Salinity	pH	T °C	DOC	DO (mg L^{-1})	Pb concentration	References
17.53 ppt	8.43	16.53	0.07(%)	8.96	7.92 in water 70.74 in mussel	Baltas et al. (2017)[1]
30 g L^{-1}	7.8	17 22	–	–	57.18 in water 2.32 in mussel 55.30 in water 2.32 in mussel	Pirone et al. (2019)[2]
14.1 28.0 ppm	7.96 7.92	18.2 18.1	2.14 1.73 mg L^{-1}	7.58 6.88	0.9 7.5 in fish	Reynolds et al. (2018)[3]
–	7.5 6.29	17.3 19.4	–	8.1 4.5	8.5 in water 1.5 in fish 9.9 in water 1.3 in fish	Weber et al. (2013)[4]
32.87 g L^{-1}	8.17	29.52	–	–	0.002 in water 0.133 in fish	Wattimena et al. (2018)[5]

Water samples were measured in µg L^{-1} and animal samples were measured or converted in mg kg^{-1} d.w. DOC: Dissolved organic carbon; DO: Dissolved oxygen.

[1] Example taken from the Giresun Province collection area (average values). Mussel species *Mytilus galloprovincialis.* Local: Giresun Province-Turkish;

[2] Values obtained through the average of Pb values in four weeks of analysis in water. Mussel specie: *Mytilus galloprovinciali.* Local: Ria de Aveiro (northwest Atlantic coast of Portugal);

[3] Example taken from the results related to the collection made on site 1 (confluence of the Tietê and Piracicaba rivers) in 2007 (winter and summer). Fish species *Atherinops Affinis.* Local: São Paulo-Brazil;

[4] Results for the two collection points: upper Sinos River and lower Sinos River. Fish species *Oligosarcus gender.* Local: Rio Grande do sul-Brazil;

[5] Fish species *Caranx* sp. Local: Ambon Bay-Indonesia.

Regarding the pH, in high H^+ concentration of, that is, in acidic waters, a greater release of Pb present in the sediment occurs, increasing bioavailability. Due to the increased competition mechanism between the element and H^+ by ligands, under acidic pH the adsorption capacity of elements are initially reduced and increasing mobility ensues, which results in more soluble metals and increases the bioavailability, as takes place with Pb. Regarding the oxygen content dissolved in water, this can affect the oxidation of organic compounds, facilitating the release of Pb into the water and, consequently, its absorption by the biota. Thus, the greater the dissolved oxygen content, the greater the release of Pb from the sediment into the water (Santos et al., 2014). It is also possible to identify a difference in the tissue distribution of Pb influenced by water salinity. Saltwater animals drink plenty of water to avoid dehydration; consequently, the greatest accumulation of potentially toxic metals, such as Pb, is in the intestine, while freshwater animals actively regulate osmotic regulation via gills to maintain body ions since they are in a low osmotic pressure environment. Thus, the main bioaccumulation of Pb is in the gills (Lee et al., 2019). Another factor that favors greater toxicity at low salinity is the reduction in competition between cations (Ca^{2+}, Mg^{2+}) and Pb by biotic ligands in the gills, which favors metal absorption (Santos et al., 2014; Reynolds et al., 2018). Ca^{2+}, in particular, has an important influence on Pb bioavailability and toxicity. This ion also aids in relieving the accumulation of Pb, due to its function of adhering to Pb dissolved in the environment (Lee et al., 2019). Concerning temperature, as well as for other metals, the increase or decrease in water temperature influences animal metabolism. It was observed that higher temperatures hinder the absorption of Pb and accelerate the metabolism in order to favor the elimination of the metal (Merçon et al., 2019). The table above (Table 7.3) presents some recent studies that measured related environmental variables and Pb concentrations in seafood and/or water.

Toxicity

Due to its ability to cause toxic effects in animals, humans, and plants, lead is highlighted as a pollutant of priority attention in international bodies. In animals, the metal is bioaccumulated and consequently can be fatal at low concentrations. Absorption can occur via the gills or food. The element can affect different systems in aquatic animals, such as reproductive, immunological, hematopoietic, cardiovascular, hepatic, and renal (Lavradas et al., 2014; Lee et al., 2019; Li et al., 2016; Liu et al., 2014; Tavabe et al., 2019; Xu et al., 2019).

Pb is considered one of the elements with the greatest potential for bioaccumulation due to its affinity to form stable bonds with sulfur and oxygen molecules in proteins. When absorbed, the liver plays a crucial role in the process of detoxifying metals such as Pb. It is reported that Pb binds to bile steroids and can be reabsorbed from the intestine or eliminated in the feces (Lee et al., 2019). The main alterations identified in animals comprise the impairment of the ionic regulatory function. This metal has the ability to inhibit ionic transport in the gills, mainly calcium transport, which can lead to hypocalcemia and impaired neurotransmitter release. However, Ca^{2+} also has a protective effect against contamination of various metals, such as Pb in animals, helping to remove the metal and controlling its tissue distribution (Lee et al., 2019; Rogers et al., 2003). Pb also stimulates the antioxidant response due to the production of reactive oxygen species, leading to damage to proteins and DNA, and RNA. In the nervous system, behavioral and cognitive damage are observed, evidenced by neurodegeneration, compromised neurotransmission, morphological changes in the brain, and dysregulation of cell signaling (Lee et al., 2019; Verstraeten et al., 2008). Regarding the compromise of the immune system, a reduction in hematopoietic function, antibody production, and phagocytosis activity is observed (Lee et al., 2019). The role of Pb in the reproductive system of aquatic organisms is mainly related to impaired sperm function, for example, reducing motility and affecting the integrity of the sperm plasma membrane, among other changes that are translated in some cases into infertility (Li et al., 2016). In females, it was possible to identify the transfer of Pb to eggs (Lavradas et al., 2014).

There are two main ways of contamination in humans: inhalation of dust particles containing Pb or via food, with emphasis on fish as a carrier food source. By inhalation, about 50% of the Pb present in dust can be inhaled and absorbed by the lungs, while absorption via food is between 10 to 15% in adults and 50% in children (Jarup, 2003). When absorbed, Pb can bind to erythrocytes, thus having a good ability to reach different tissues. Its excretion is slow and occurs mainly through urine. It can accumulate in various tissues, including bone tissue, having a half-life of about 30 years (WHO, 1995). Both its inorganic and organic forms have toxic potential. However, like the nervous system, while the inorganic form cannot penetrate the blood-brain barrier in adults, only in children, the organic form has this ability in both age groups (Jarup, 2003). The symptoms of Pb intoxication are mainly related to the nervous and gastrointestinal systems. It can also cause anemia and is considered a possible carcinogen. It is possible to identify irritability, acute psychosis, mental confusion, memory loss, among other symptoms related to encephalopathy (Jarup, 2003).

Final Considerations

Cd, Hg, and Pb are the three important elements explored by the world scientific community due to the known toxicity in animal, plant, and human species. The physical-chemical parameters of water are closely related to the dynamics of each of the metals covered in this chapter. The low salinity and the low concentration of dissolved organic carbon are highlighted as the main contributors of metal toxicity, except for Hg, where a higher concentration of DOC will favor the formation of methylmercury. Temperature is most closely related to bioaccumulation, given its influence on animal metabolism. In addition to these, pH and dissolved oxygen close the list of the main explored abiotic factors associated to these element's bioavailability concerning biota absorption. It was also possible to identify that these mentioned factors directly affect the speciation of the three metals in different aquatic environments, such as oceans, coastal waters, and freshwater environments.

The toxicity of Cd, Hg, and Pb is predominantly reported in cases of chronic contamination due to the bioaccumulation capacity of the elements, either in animals or in humans via their consumption. In this sense, the different scientific works related to this theme, compiled in this chapter, seek to quantify the elements in water and animal samples, with the objective of controlling environmental and animal contamination and also evaluating the potential risk to public health. However, not all studies seek to quantify and correlate physical-chemical water parameters with elemental values in either water or animals, an important correlation that required further exploration in these types of assessments.

REFERENCES

Abdel-Tawwab, M. and Wafeek, M. 2014. Influence of water temperature and waterborne cadmium toxicity on growth performance and metallothionein–cadmium distribution in different organs of Nile tilapia, *Oreochromis niloticus* (L.). J. Thermal. Biol. 45: 157–162.

Adams, DH. and Engel, ME. 2014. Mercury, lead, and cadmium in blue crabs, *Callinectes sapidus*, from the Atlantic coast of Florida, USA: A multipredator approach. Ecotoxicol Environ Saf. 102: 196–201.

Ando, M., Seoka, M., Mukai, Y., Jye, MW., Miyashita, S. and Tsukamasa Y. 2010. Effect of water temperature on the feeding activity and the resultant mercury levels in the muscle of cultured bluefin tuna *Thunnus orientalis* (Temminck and Schlegel). Aquac. Res. 42(4): 516–524.

Azevedo, L.S., Almeida, M.G., Bastos, W.R., Suzuki, M.S., Recktenvald, M.C.N.N., Bastos, M.T.S. et al. 2016. Organotropism of methylmercury in fish of the southeastern of Brazil. Chemosphere 185: 746–753.

Bakar, A.N., Mohd, S.N.S., Ramlan, N.F., Wan Ibrahim, W.N., Zulkifli, S.Z., Che Abdullah, C.A. et al. 2017. Evaluation of the neurotoxic effects of chronic embryonic exposure with inorganic mercury on motor and anxiety-like responses in zebrafish (*Danio rerio*) larvae. Neurotoxicol. Teratol. 59: 53–61.

Baltas, H., Sirin, M., Dalgic, G., Bayrak, E.Y. and Akdeniz, A. 2017. Assessment of metal concentrations (Cu, Zn, and Pb) in seawater, sediment and biota samples in the coastal area of Eastern Black Sea, Turkey. Mar. Pollut. Bull. 122(1–2): 475–482.

Bastos W.R., Dórea J.G., Bernardi J.V.E., Lauthartte LC. and Mussy MH., Hauser M. et al. 2015. Mercury in muscle and brain of catfish from the Madeira river, Amazon, Brazil. Ecotoxicol. Environ. Saf. 118: 90–97.

Bielmyer-Fraser, G.K., Harpe,r B., Picariello, C. and Albritton-Ford, A. 2018. The influence of salinity and water chemistry on acute toxicity of cadmium to two euryhaline fish species. Comp. Biochem. Physiol. C: Toxicol. Pharmacol. 214: 23–27.

Bjørklund, G., Dadar, M., Mutter, J. and Aaseth, J. 2017. The toxicology of mercury: current research and emerging trends. Environ. Res. 159: 545–554.

Bordon, I.C., Emerenciano, A.K., Melo, J.R.C., Silva, J.R.M.C., Favaro, D.I.T., Gusso-Choueri, P.K., et al. 2018. Implications on the Pb bioaccumulation and metallothionein levels due to dietary and waterborne exposures: The *Callinectes danae* case. Ecotoxicol. Environ. Saf. 162: 415–422.

Bowman, K.L., Lamborg, C.H. and Agather, A.M. 2019. A global perspective on mercury cycling in the ocean. Sci Total Environ. 136166.

Boyd, E.S., Yu, R.Q., Barkay, T., Hamilton, T.L., Baxter, B.K., Naftaz, D.L., et al. 2017. Effect of salinity on mercury methylating benthic microbes and their activities in great salt Lake, Utah. Sci. Total. Environ. 581–582: 495–506.

Braaten, H.F.V., de Wit, H.A., Larssen, T. and Poste, A.E. 2018. Mercury in fish from Norwegian lakes: The complex influence of aqueous organic carbon. Sci. Total Environ. 627: 341–348.

Chen, C.Y., Borsuk, M.E., Bugge, D.M., Hollweg, T., Balcom, P.H., Ward, D.M. et al. 2014. Benthic and pelagic pathways of methylmercury bioaccumulation in estuarine food webs of the Northeast United States. PLoS One 9(2): e89305.

Chen, M.M., Lopez, L., Bhavsar, S.P. and Sharma, S. 2018. What's hot about mercury? Examining the influence of climate on mercury levels in Ontario top predator fishes. Environ. Res. 162: 63–73.

Chételat, J., Amyot, M., Arp, P., Blais, J.M., Depew, D. and Emmerton, C.A. et al. 2015. Mercury in freshwater ecosystems of the Canadian Arctic: Recent advances on its cycling and fate. Sci. Total Environ. 509–510: 41–66.

Clarkson, T.W., Vyas, J.B. and Ballatori, N. 2007. Mechanisms of mercury disposition in the body. Am. J. Ind. Med. 50(10): 757–764.

Clifford, M. and McGeer, J.C. 2010. Development of a biotic ligand model to predict the acute toxicity of cadmium to *Daphnia pulex*. Aquat. Toxicol. 98(1): 1–7.

Condini, M.V., Hoeinghaus, D.J., Roberts, A.P., Soulen, B.K. and Garcia, A.M. 2017. Mercury concentrations in dusky grouper *Epinephelus marginatus* in littoral and neritic habitats along the southern Brazilian coast. Mar. Pollut. Bull. 115(1–2): 266–272.

Coppola, F., Almeida, Â., Henriques, B., Soares, A.M.V.M., Figueira, E., Pereira, E., et al. 2017. Biochemical impacts of Hg in *Mytilus galloprovincialis* under present and predicted warming scenarios. Sci. Total Environ. 601–602: 1129–1138.

Coppola, F., Henriques, B., Soares, A.M.V.M., Figueira, E., Pereira, E. and Freitas, R. 2018. Influence of temperature rise on the recovery capacity of *Mytilus galloprovincialis* exposed to mercury pollution. Ecol. Indic. 93: 1060–1069.

Cornelis, R. and Nordberg, M. 2007. General chemistry, sampling, analytical methods, and speciation. pp. 197–208. *In*: Gunnar, F.N., Bruce, A.F., Monica, N. and Lars, T.F. (eds). Handbook on the Toxicology of Metals, 3rd Ed. Academic Press, Burlington.

Costa, F.N., Korn, M.G.A., Brito, G.B., Ferlin, S. and Fostier, A.H. 2016. Preliminary results of mercury levels in raw and cooked seafood and their public health impact. Food Chem. 192: 837–841.

Crowe, W., Allsopp, P.J., Watson, G.E., Magee, P.J., Strain, J.J., Armstrong, D.J., et al. 2017. Mercury as an environmental stimulus in the development of autoimmunity—A systematic review. Autoimmun Rev. 16(1): 72–80.

Cullen, J.T. and Maldonado, M.T. 2013. Biogeochemistry of cadmium and its release to the environment. pp. 31–62. *In*: Sigel, A., Sigel, H. and Sigel, R.K.O. (eds). Cadmium: From Toxicity to Essentiality. Metal Ions in Life Sciences, Vol. 11. Springer, Dordrecht.

Cullen, J.T. and McAlister, J. 2017. Biogeochemistry of lead. Its release to the environment and chemical speciation. Met. Ions Life Sci. 17: 21–48.

Das, S., Tseng, L.C., Chou, C., Wang, L., Souisssi, S. and Hwang, J.S. 2019. Effects of cadmium exposure on antioxidant enzymes and histological changes in the mud shrimp *Austinogebia edulis* (Crustacea: Decapoda). Environ. Sci. Pollut. Res. 26: 7752–7762.

Delgado-Alvarez, C.G., Ruelas-Inzunza, J., Osuna-López, J.I., Voltolina, D. and Frías-Espericueta, M.G. 2015. Mercury content and their risk assessment in farmed shrimp *Litopenaeus vannamei* from NW Mexico. Chemosphere 119: 1015–1020.

Díez, S. 2009. Human health effects of methylmercury exposure. Rev. Environ. Contam. Toxicol. 198: 111–132.

Dijkstra, J.A., Buckman, K.L., Ward, D., Evans, D.W., Dionne, M., Chen, C.Y. et al. 2013. Experimental and natural warming elevates mercury concentrations in estuarine fish. PloS One 8(3): e58401.

Dong, W., Liu, J., Wei, L., Jingfeng, Y., Chernick, M. and Hinton, D.E. 2016. Developmental toxicity from exposure to various forms of mercury compounds in medaka fish (*Oryzias latipes*) embryos. PeerJ 4: e2282.

Dórea, J.G., Farina, M. and Rocha, J.B. 2013. Toxicity of ethylmercury (and Thimerosal): A comparison with methylmercury. J. Appl. Toxicol. 33(8): 700–711.

Espejo,W., Padilha, J.A., Gonçalves, R.A., Dorneles, P.R., Barra, R., Oliveira, D., et al. 2019. Accumulation and potential sources of lead in marine organisms from coastal ecosystems of the Chilean Patagonia and Antarctic Peninsula area. Mar. Pollut. Bull. 140: 60–64.

García-Medina, S., Galar-Martínez, M., Gómez-Oliván, L.M., RuizLara, K., Islas-Flores, H. and Gasca-Pérez, E. 2017. Relationship between genotoxicity and oxidative stress induced by mercury on common carp (*Cyprinus carpio)* tissues. Aquat. Toxicol. 192: 207–215.

Giesy Jr., J.P. and Wiener, J.G. 1977. Frequency distributions of trace metal concentrations in five freshwater fishes. Trans. Am. Fish Soc. 106(4): 393–403.

Gutiérrez-Mosquera, H., Sujitha, S.B., Jonathan, M.P., Sarkar, S.K., Medina-Mosquera, F., Ayala-Mosquera, H., et al. 2017. Mercury levels in human population from a mining district in Western Colombia. J. Environ. Sci. 1–8.

Gworek, B., Bemowska-Kalabun, O., Kijenska, M. and Wrzosek-Jakubowska, J. 2016. Mercury in Marine and Oceanic Water—a Review. Water, Air Soil Pollut. 227: 371.

Harayashiki, C.A.Y., Reichelt-Brushett, A., Cowden, K. and Benkendorff, K. 2018. Effects of oral exposure to inorganic mercury on the feeding behaviour and biochemical markers in yellowfin bream (*Acanthopagrus australis*). Mar. Environ. Res. 134: 1–15.

Hintelmann, H. 2010. Organomercurials. Their formation and pathways in the environment. Met. Ions Life Sci. 7: 365–401

Hong, Y.S., Kim, Y.M. and Lee, K.E. 2012. Methylmercury exposure and health effects. J. Prev. Med. Public Health. 45(6): 353–363.

Hosseni, M., Nabavi, S.M. and Parsa, Y. 2013. Bioaccumulation of trace mercury in trophic levels of benthic, benthopelagic, pelagic fish species, and sea birds from Arvand River, Iran. Biol. Trace Elem. Res. 156(1–3): 175–180.

IARC. 1993. Summaries & evaluations: Cadmium and cadmium compounds (Group 1). Lyon, International Agency for Research on Cancer, p. 119 (IARC Monographs on the Evaluation of Carcinogenic Risks to Humans, Vol. 58.

Jacaúna, R.P., Kochhann, D., Campos, D.F. and Val, A.L. 2020. Aerobic Metabolism Impairment in Tambaqui (*Colossoma macropomum*) Juveniles Exposed to Urban Wastewater in Manaus, Amazon. Bull. Environ. Contam. Toxicol. 105(6): 853–859.

Jaishankar, M., Tseten, T., Anbalagan, N., Mathew, B.B. and Beeregowda, K.N. 2014. Toxicity, mechanism and health effects of some heavy metals. Interdiscip. Toxicol. 7(2): 60–72.

Jan, A.T., Azam, M, Siddiqui, K., Ali, A., Choi, I. and Haq, Q.M. 2015. Heavy metals and human health: Mechanistic insight into toxicity and counter defense system of antioxidants. Int. J. Mol. Sci. 16(12): 29592–630.

Järup, L. 2003. Hazards of heavy metal contamination. Br. Med. Bull. 68(1): 167–182.

Kojadinovic, J., Potier, M., Le Corre, M., Cosson, R.P. and Bustamante, P. 2006. Mercury content in commercial pelagic fish and its risk assessment in the Western Indian Ocean. Sci. Total Environ. 366(2–3): 688–700.

Lackner, J., Weiss, M., Muller-Graf, C. and Greiner, M. 2018. Disease burden of methylmercury in the German birth cohort 2014. PLoS One 13(1): ee0190409.

Lan, W.R., Huang, X.G., Lin, L., Li, S.X. and Liu, F.J. 2019. Thermal discharge influences the bioaccumulation and bioavailability of metals in oysters: Implications of ocean warming. Environ. Pollut. 113821.

Land, S.N., Rocha, R.C.C., Bordon, I.C., Saint'Pierre, T.D., Ziolli, R.L. and Hauser-Davis, R.A. 2018. Biliary and hepatic metallothionein, metals and trace elements in environmentally exposed neotropical cichlids *Geophagus brasiliensis*. J. Trace. Elem. Med. Biol. 50: 347–355.

Lavradas, R.T., Hauser-Davis, R.A., Lavandier, R.C., Rocha, R.C., São Pedro, T.D., Seixas, T., et al. 2014. Metal, metallothionein and glutathione levels in blue crab (*Callinectes* sp.) specimens from southeastern Brazil. Ecotoxicol. Environ. Saf. 107: 55–60.

Lee, J.W., Choi, H., Hwang, U.K., Kang, J.C., Kang, Y.J., Kim, K.I., et al. 2019. Toxic effects of lead exposure on bioaccumulation, oxidative stress, neurotoxicity, and immune responses in fish: A review. Environ. Toxicol. Pharmacol. 68: 101–108.

Lei, W., Wang, L., Liu, D., Xu, T. and Luo, J. 2011. Histopathological and biochemical alternations of the heart induced by acute cadmium exposure in the freshwater crab *Sinopotamon yangtsekiense*. Chemosphere 84(5): 689–694.

Li, N., Hou, Y.H., Jing, W.X., Dahms, H.U. and Wang, L. 2016. Quality decline and oxidative damage in sperm of freshwater crab *Sinopotamon henanense* exposed to lead. Ecotoxicol. Environ. Saf. 130: 193–198.

Li, R., Tang, X., Guo, W., Lin, L. and Zhao, L. 2020. Spatiotemporal distribution dynamics of heavy metals in water, sediment, and zoobenthos in mainstream sections of the middle and lower Changjiang River. Sci. Total. Environ. 714: 136779.

Liao, W., Wang, G., Zhao, W., Zhang, M., Wu, Y., Liu, X. et al. 2019. Change in mercury speciation in seafood after cooking and gastrointestinal digestion. J. Hazard Mat. 375: 130–137.

Lino, A.S., Kasper, D., Guida, Y.S., Thomaz, J.R. and Malm, O. 2019. Total and methyl mercury distribution in water, sediment, plankton and fish along the Tapajós River basin in the Brazilian Amazon. Chemosphere 235: 690–700.

Liu, N., Wang, L., Yan, B., Li, Y., Ye, F., Li, J. et al. 2014. Assessment of antioxidant defense system responses in the hepatopancreas of the freshwater crab Sinopotamon henanense exposed to lead. Hydrobiologia 741: 3–12.

Ma, J., Zhou, B., Duan, D. and Pan, K. 2019. Salinity-dependent nanostructures and composition of cell surface and its relation to Cd toxicity in an estuarine diatom. Chemosphere 215: 807–814.

Macirella, R. and Brunelli, E. 2017. Morphofunctional alterations in Zebrafish (*Danio rerio*) gills after exposure to mercury chloride. Int. J. Mol. Sci. 18(4): 824.

Mallory, M.L., O'Driscoll, N.J., Klapstein, S., Varela, JL., Ceapa, C. and Stokesbury, M.J. 2018. Methylmercury in tissues of Atlantic sturgeon (*Acipenser oxyrhynchus*) from the Saint John River, New Brunswick, Canada. Mar. Pollut. Bull. 126: 250–254.

Mason, R.P. 2013. Trace Metals in Aquatic Systems. Wiley-Blackwell.

Merçon, J., Pereira, TM., Passos, LS., Lopes, TO., Coppo, G., Barbosa, B., et al. 2019. Temperature affects the toxicity of lead-contaminated food in *Geophagus brasiliensis* (QUOY & GAIMARD, 1824). Environ. Toxicol. Pharmacol. 66: 75–82.

162 *Lead, Mercury and Cadmium in the Aquatic Environment*

Micaroni, R.C.C.M., Bueno, M.I.M.S. and Jardim, W.F. 2000. Mercury compounds, review on determination, tratament and disposal methods. Quim. Nova. 23(4): 487–495.

Múgica, M., Izagirre, U. and Marigómez, I. 2015: Lysosomal responses to heat-shock of seasonal temperature extremes in Cd-exposed mussels. Aquat. Toxicol. 16499–107.

Murphy, G.W., Newcomb, T.J. and Orth, D.J. 2007. Sexual and seasonal variations of mercury in smallmouth Bass. J. Freshwater Ecol. 22(1): 135–143.

Niyogi, S., Kent, R. and Wood, C.M. 2008. Effects of water chemistry variables on gill binding and acute toxicity of cadmium in rainbow trout (*Oncorhynchus mykiss*): a biotic ligand model (BLM) approach. Comp. Biochem. Physiol., C. 48: 305–314.

Nowosad, J., Kucharczyk, D. and Łuczyńska, J. 2018. Changes in mercury concentration in muscles, ovaries and eggs of European eel during maturation under controlled conditions. Ecotoxicol. Environ. Saf. 148: 857–861.

O'Bryhim, J.R., Adams, D.H., Spaet, J.L.Y., Mills, G. and Lance, S.L. 2017. Relationships of mercury concentrations across tissue types, muscle regions and fins for two shark species. Environ. Pollut. 223: 323–333.

Onsanit, S. and Wang, W.X. 2011 Sequestration of total and methyl mercury in different subcellular pools in marine caged fish. J. Hazard Mater. 198: 113–122.

Osuna Flores, I., Meyer-Willerer, A.O., Olivos-Ortiz, A., Vázquez, F.J.B. and Marmolejo-Rodríguez, A.J. 2014. Lead in Shrimp Litopenaeus vannamei Boone in Sublethal Concentrations. J. Toxicol. Environ. Health, Part A 77(18):1084–1090.

Owens, E., Effer, S.W., Bookman, R., Driscoll, C.T., Matthews, D.A. and Effier, A.J.P. 2009. Resuspension of mercury-contaminated sediments from an in-lake industrial waste deposit. J. Environ. Eng. 135: 526–534.

Panichev, N.A. and Panicheva, S.E. 2014. Determination of total mercury in fish and sea products by direct thermal decomposition atomic absorption spectrometry. Food. Chem. 166: 432–441.

Paranjape, A.R. and Hall, B.D. 2017. Recent advances in the study of mercury methylation in aquatic systems. Facets 2(1): 85–119.

Pethybridge, H., Cossa, D. and Butler, E.C. 2010. Mercury in 16 demersal sharks from Southeast Australia: Biotic and abiotic sources of variation and consumer health implications. Mar. Environ. Res. 69(1): 18–26.

Piazza, V., Gambardella, C., Canepa, S., Costa, E., Faimali, M. and Garaventa, F. 2016. Temperature and salinity effects on cadmium toxicity on lethal and sublethal responses of *Amphibalanus amphitrite nauplii*. Ecotoxicol. Environ. Saf. 123: 8–17.

Pirone, G., Coppola, F., Pretti, C., Soares, AMVM., Solé, M. and Freitas, R. 2019. The effect of temperature on Triclosan and Lead exposed mussels. Comp. Biochem. Physiol. Part B, Biochem. Mol. Biol. 232: 42–50.

Prato, E., Scardicchio, C. and Biandolino, F. 2008. Effects of temperature on the acute toxicity of cadmium to *Corophium insidiosum*. Environ. Monit. Assess. 136: 161–166.

Qu, B., Zhang, Y., Kang, S. and Sillanpää, M. 2019. Water quality in the Tibetan Plateau: Major ions and trace elements in rivers of the "Water Tower of Asia." Sci. Total Environ. 649: 571–581.

Quevedo, L., Ibáñez, C., Caiola, N. and Mateu, D. 2018. Effects of thermal pollution on benthic, macroinvertebrate communities of a large Mediterranean river. J. Entomol. Zoo. Stud. 6: 500–507.

Raimundo, J., Vale, C., Canário, J., Branco, V. and Moura, I. 2010. Relations between mercury, methylmercury and selenium in tissues of *Octopus vulgaris* from the Portuguese coast. Environ. Pollut. 158(6): 2094–2100.

Rainbow, P.S. 1995. Biomonitoring of heavy metal availability in the marine environment. Mar. Pollut. Bull. 31: 183–192.

Raknuzzaman, M., Ahmed, M.K., Islam, M.S., Habibullah-Al-Mamun, M., Tokumura, M., Sekine, M., et al. 2016. Trace metal contamination in commercial fish and crustaceans collected from coastal area of Bangladesh and health risk assessment. Environ. Sci. Pollut. Res. 23: 17298–17310.

Rasinger, J.D., Lundebye, A.K., Penglase, S.J., Ellingsen, S. and Amlund, H. 2017. Methylmercury induced neurotoxicity and the influence of selenium in the brains of adult Zebrafish (*Danio rerio*). Int. J. Mol. Sci. 18(4): 725.

Reinhart. B.L., Kidd, K.A., Curry, R.A., O'driscoll, N.J. and Pavey, S.A. 2018. Mercury bioaccumulation in aquatic biota along a salinity gradient in the Saint John River estuary. J. Environ. Sci. 1–14.

Reynolds, E.J., Hoang, T.C., Smith, D.S. and Chowdhury, J.M. 2018. Chronic effects of lead exposure on Topsmelt Fish (*Atherinops Affinis*): Influence of salinity, organism age, and relative sensitivity to other Marine Fish. Environ. Toxicol. Chem. 37(10): 2705–2713.

Rice, K.M., Walker, E.M., Wu, M., Gillette, C. and Blough, E.R. 2014. Environmental mercury and its toxic effects. J. Prev. Med. Public Health 47(2): 74–83.

Rodrigues, P.A., Ferrari, R.G., Hauser-Davis, R.A., Santos, L.N. and Conte-Junior, C.A. 2020. Dredging Activities Carried Out in a Brazilian Estuary Affect Mercury Levels in Swimming Crabs. Int. J. Environ. Res. Public Health 17(12): 4396.

Rodrigues, P.A., Ferrari, R.G., Rosario, D.K.A., Hauser-Davis, R.A., Lopes, A.P., Santos, A.F.G.N., et al. 2021. Interactions between mercury and environmental factors: A chemometric assessment in seafood from an eutrophic estuary in southeastern Brazil. Aquatic. Toxicol. 236: 105844.

Rogers, J.T., Richards, J.G. and Wood, C.M. 2003. Ionoregulatory disruption as the acute toxic mechanism for lead in the rainbow trout (*Oncorhynchus mykiss*). Aquatic. Toxicol. 64: 215–234.

Ruus, A., Øverjordet, I.B., Braaten, H.F., Evenset, A., Christensen, G., Heimstad, E.S. et al. 2015. Methylmercury biomagnification in an Arctic pelagic food web. Environ. Toxicol. Chem. 34(11): 2636–2643.

Ruus, A., Hjermann, D.O., Beylich, B., Schoyen, M., Oxnevad, S. amd Green, N.W. 2017. Mercury concentration trend as a possible result of changes in cod population demography. Mar. Environ. Res. 130: 85–92.

Saadati, M., Soleimani, M., Sadeghsaba, M. and Hemami, M.R. 2019. Bioaccumulation of heavy metals (Hg, Cd and Ni) by sentinel crab (*Macrophthalmus depressus*) from sediments of Mousa Bay, Persian Gulf. Ecotoxicol. Environ. Saf. 191: 109986.

Samayamanthula, D.R., Sabarathinam, C. and Alayyadhi, N.A. 2021. Trace elements and their variation with pH in rain water in arid environment. Arch. Environ. Contam. Toxicol. 80: 331–349.

Sandrin, T.R. and Maier, R.M. 2002. Effect of pH on cadmium toxicity, speciation, and accumulation during naphthalene biodegradation. Environ. Toxicol. Chem. 21(10): 2075–2079.

Santos, D.B., Barbieri, E., Bondioli, A.C.V. and Melo, C.B. 2014. Effects of Lead in white shrimp (*Litopenaeus schmitti*) metabolism regarding salinity. O mundo da Saúde, São Paulo 38(2): 16–23.

Sevillano-Morales, J.S., Ramírez-Ojeda, A.M., Cejudo-Gómez, M. and Moreno-Rojas, R. 2015. Risk profile of methylmercury in seafood. Curr. Opin. Food Sci. 6: 56–60.

Souza-Araujo, J., Giarrizzo, T., Lima, M.O. and Souza, M.B. 2016. Mercury and methyl mercury in fishes from Bacajá River (Brazilian Amazon): evidence for bioaccumulation and biomagnification. J. Fish Biol. 89(1): 249–263.

Sprague, J.B. 1985. Factors that modify toxicity. pp. 124–163. *In*: Rand, G.M. and Petrocelli, S.R. (eds). Fundamentals of Aquatic Toxicology. Hemisphere Publishing Company, New York, NY.

Storelli, M.M., Giacominelli-Stuffler, R. and Marcotrigiano, G.O. 2006. Relationship between total mercury concentration and fish size in two pelagic fish species: Implications for consumer health. J. Food Prot. 69(6): 1402–1405.

Sunderland, E.M. and Selin, N.E. 2013. Future trends in environmental mercury concentrations: implications for prevention strategies. Environ. Health. 12(2): 2–5.

Tang, L., Hamid, Y., Zehra, A., Sahito, ZA., He, Z., Khan, M.B. et al. 2020. Mechanisms of water regime effects on uptake of cadmium and nitrate by two ecotypes of water spinach (*Ipomoea aquatica* Forsk.) in contaminated soil. Chemosphere 246: 125798.

Tavabe, K.R., Abkenar, B.P., Rafiee, G. and Frinsko, M, 2019. Effects of chronic lead and cadmium exposure on the oriental river prawn (*Macrobrachium nipponense*) in laboratory conditions. Comp. Biochem. Physiol., C. Toxicol. Pharmacol. 221: 21–28.

Taylor, D.L. and Calabrese, N.M. 2018 Mercury content of blue crabs (*Callinectes sapidus*) from southern New England coastal habitats: Contamination in an emergent fishery and risks to human consumers. Mar. Pollut. Bull. 126: 166–178.

Tchounwou, P.B., Yedjou, C.G., Patolla, A.K. and Sutton, D.J. 2012. Heavy metal toxicity and the environment. Exp. Suppl. 101: 133–64.

Thomas, S.M., Melles, S.J., Mackereth, R.W., Tunney, T.D., Chu, C. and Oswald, C.J. 2020. Bhavsar SP., Johnston TA. Climate and landscape conditions indirectly affect fish mercury levels by altering lake water chemistry and fish size. Environ. Res. 109750.

Tonietto, A.E., Lombardi, A.T., Choueri, R.B. and Vieira, A.A.H. 2015. Chemical behavior of Cu, Zn, Cd, and Pb in a eutrophic reservoir: speciation and complexation capacity. Environ. Sci. Pollut. Res. 22(20): 15920–15930.

Thwala, M., Newman, B. and Cyrus, D. 2011. Influence of salinity and cadmium on the survival and osmoregulation of *Callianassa kraussiand* and *Chiromantes eulimene* (Crustacea: Decapoda). Afr. J. Aquat. Sci. 36(2): 181–189.

Ullrich, S.M., Tanton, T.W. and Abdrashitova, S.A. 2001. Mercury in the aquatic environment: A review of factors affecting methylation. Crit. Rev. Environ. Sci. Technol. 31(3): 241–293.

United States Environmental Protection Agency (USEPA). 2016. Aquatic life ambient water quality criteria Cadmium. Office of Water Office of Science and Technology Health and Ecological Criteria Division, Washington DC.

Van Ael, E., Blust, R. and Bervoets, L. 2017. Metals in the Scheldt estuary: from environmental concentrations to bioaccumulation. Environ. Pollut. 228: 82–91.

Vellinger, C., Felten, V., Sornom, P., Rousselle, P. and Beisel, J.N. 2012. Usseglio-Polatera P. Behavioural and physiological responses of *Gammarus pulex* at three temperature: individual and combined effects. PLoS One 7(6): 39–53.

Verstraeten, S.V., Aimo, L., Oteiza, P.I. 2008. Aluminium and lead: Molecular mechanisms of brain toxicity. Arch. Toxicol. 82: 789–802.

Vouyer, R.A. amd Modica, G. 1990. Influence of salinity and temperature on acute toxicity of cadmium to Mysidopsis bahia Molenock. Arch. Environ. Contam. Toxicol. 19(1): 124–131.

Wang, R., Wong, M.H. and Wang, W.X. 2010. Mercury exposure in the freshwater tilapia *Oreochromis niloticus*. Environ. Pollut. 15(8): 2694–2701.

Wattimena, R.L., Selanno, D.A.J., Tuhumury, S.F. and Tuahatu, J.W. 2018. Analysis of heavy metal content (Pb) on waters and fish at the floating cages BPPP Ambon. e3s Web of Conferences. 31: 04001.

Weber, P., Behr, E.R., Knor, C.D.L., Vendruscolo, D.S., Flores, E.M.M., Dressler, V.L., et al. 2013. Metals in the water, sediment, and tissues of two fish species from different trophic levels in a subtropical Brazilian river. Microchemical J. 106: 61–66.

WHO. Lead. 1995. Environmental Health Criteria, Vol. 165. Geneva: World Health Organization.

Wiech, M., Amlund, H., Jensen, K.A., Aldemberg, T., Duinker, A. and Maage, A. 2018. Tracing simultaneous cadmium accumulation from different uptake routes in brown crab *Cancer pagurus* by the use of stable isotopes. Aquat. Toxicol. 201: 198–206.

Wu, H., Li, Y., Lang, X. and Wang, L. 2015. Bioaccumulation, morphological changes, and induction of metallothionein gene expression in the digestive system of the freshwater crab *Sinopotamon henanense* after exposure to cadmium. Environ. Sci. Pollut. Res. Int. 22(15): 11585–11594.

Xu, P., Chen, H., Xi, Y., Mao, X. and Wang, L. 2016. Oxidative stress induced by acute and subchronic cadmium exposure in the ovaries of the freshwater crab *Sinopotamon henanense* (DAI, 1975). Crustaceana 89(9): 1041–1055.

Xu, X.H., Meng, X., Gan, H.T., Liu, T.H., Yao, H.Y., Zhu, X.Y., et al. 2019. Immune response, MT and HSP70 gene expression, and bioaccumulation induced by lead exposure of the marine crab, *Charybdis japonica*. Aquat. Toxicol. 210: 98–105.

Yang, L., Zhang, Y., Wang, F., Luo, Z., Guo, S. and Strähle, U. 2019. Toxicity of Mercury: Molecular Evidence. Chemosphere 125586.

Yin, R., Zhang, W., Sun, G., Feng, Z., Hurley, J.P., Yang, L., et al. 2017. Mercury risk in poultry in the Wanshan mercury mine, China. Environ. Pollut. 230: 810–816.

Zhao, Y., Wang, X., Qin, Y. and Zheng, B. 2010. Mercury (Hg^{2+}) effect on enzyme activities and hepatopancreas histostructures of juvenile Chinese mitten crab *Eriocheir sinensis*. Chin. J. Oceanol. Limnol. 28(3): 427–434.

Zhang, Q.F., Li, Y.W., Liu, Z.H. and Chen, Q.L. 2016. Reproductive toxicity of inorganic mercury exposure in adult zebrafish: Histological damage, oxidative stress, and alterations of sex hormone and gene expression in the hypothalamic-pituitary-gonadal axis. Aquat. Toxicol. 177: 417–424.

Zhang, H. and Reynolds, M. 2019. Cadmium exposure in living organisms: A short review. Sci. Total Environ. 678: 761–767.

Zhu, Q.H., Zhou, Z.K., Tu, D.D., Zhou, Y.L., Wang, C., Liu, Z.P., et al. 2018. Effect of cadmium exposure on hepatopancreas and gills of the estuary mud crab (*Scylla paramamosain*): histopathological changes and expression characterization of stress response genes. Aquat. Toxicol. 195: 1–7.

Chapter **8**

Uptake, Bioaccumulation, Partitioning of Lead (Pb) and Cadmium (Cd) in Aquatic Organisms in Contaminated Environments

Francis Orata* and Fred Sifuna

Department of Pure and Applied Chemistry, Masinde Muliro
University of Science and Technology, P.O. Box 190-50100, Kakamega, Kenya
Email: *fomoto@mmust.ac.ke*; Tel: +254718282462

INTRODUCTION

Aquatic contamination is largely caused by rapid urbanization, increased agricultural activities (Young et al., 2008; Zhang et al., 2014) and industrialization practices. These anthropogenic activities and processes introduce influents into aquatic ecosystems, consequently resulting in aquatic ecosystem contamination. Contaminated aquatic environments lead to characteristic pollutant uptake and partitioning in various ecosystem compartments, including biota, such as benthic organisms. Contaminant loads may increases nutrients and microbial activities, altering the physicochemical properties of aquatic ecosystems. Such changes include decreased dissolved oxygen concentrations, alkalinity, and pH, which generally alter the structure of biotic communities (Pascoal et al., 2005; Suozzo, 2005).

---
*Corresponding authors: fomoto@mmust.ac.ke

Depending on the level of contamination, pharmacokinetic processes such as absorption, elimination, distribution, metabolism, and excretion of exposed organisms within the ecological system will change and affect biota pollutant uptake and partitioning, such as metals and nutrients. Among the pollutants found in water, metals are of particular concern, as they are non-biodegradable, persist in wastewater treatment plants (Enos et al., 2018) and, hence, remain in the environment long after the original source of pollution has ceased, accumulating in the environment, from where they easily translate through food chains to humans (Obike et al., 2018). Metal inputs to the environment pose a threat to human and other organisms that dwell within the contaminated environments. Contamination is enhanced due to food chain bioaccumulation as a result of the non-degradable state of metals such as Pb and Cd. Organisms are able to control metal concentrations in certain tissues of their body to minimize damage of reactive forms of trace metals and to control selective utilization of essential metals (Vijver et al., 2005), but once the metabolically available forms passed a threshold concentration, the organism suffers from toxic effects, initially sublethal but eventually lethal. Enzymatic activities, normal cell functioning and changing cell signaling, are some of the major disruptions in biological systems that are caused by Pb (Counter et al., 2009; Dietert and Piepenbrink, 2006). Pb has been shown to have effects on hemoglobin synthesis and anemia has been observed in children at Pb blood levels above 40 ug/dl (Abadin et al., 2007). Some of the effects of Pb poisoning are reversible, whereas chronic exposure to high levels of Pb may result in decreased kidney function and possible renal failure. The evidence for the carcinogenicity of Pb and several inorganic Pb compounds in humans is inadequate (Abadin et al., 2007). However, the classification of the International Agency for Research on Cancer (IARC) places Pb in class 2B and therefore it is regarded as possibly carcinogen to humans. Cd causes toxic effects on the liver, kidney, and long-term exposure to this toxic metal resulted in inflammatory infiltration, necrosis of hepatocytes, degenerative changes in testis tissues, reduction in spermatocytes, degeneration in renal tubules, and hypertrophy of renal epithelium (Khafaga et al., 2019).

LEAD (Pb) SPECIATION AND BIOAVAILABILITY IN THE AQUATIC ENVIRONMENT

Lead is one of the prime toxins among metals which is discharged into the environment by different anthropogenic activities. In the environment, Pb binds strongly to particles, such as soil, sediment, and sewage sludge. Typical speciation of Pb in freshwater ecosystem will vary with water pH. In acidic aquatic environments, Pb will exist mainly as free ion (Pb^{2+}), which is highly mobile and bioavailable to biota. However, traces of its carbonate, chloride, hydroxide and sulphate forms will co-exist. In alkaline aquatic ecosystems of approximately pH 9.0, Pb carbonate ($PbCO_3$) and to a small extent, the hydroxide ($Pb(OH)_2$) form predominate. In saline (saltwater) environments, the dominant species are the carbonate and the chloride forms. This implies that Pb can occur in four different forms, characterised by; (i) the highly mobile and bioavailable (ionic form), (ii) in a bound form with limited mobility and bioavailability (organic complexes with dissolved humus materials), (iii) the strongly bound form, with limited mobility (attached colloidal particles such as iron oxide), or (iv) very limited mobility and availability (attached to solid clay particles or dead organism remains). Pb exchanges between water and sediment segments are prominent and this element can therefore be transported to various regions within the aquatic ecosystem as soluble complexes or ions. Sediments constitute a sink for Pb in aquatic environments, with long residence time and limited mobility. In sediments, Pb forms insoluble complexes with other chemical species through adsorption onto Fe and Mn minerals and clay minerals, or through the formation of lead sulphides. In surface waters, average residence times of biological particles containing Pb have been estimated at two to five years (UNEP, 2010). This residence time is long enough to enable Pb to be absorbed or ingested by organisms. In sediments, Pb deposits display high potential to cause harm to aquatic organisms (Thornton et al., 2001), especially to the bottom feeder fish.

CADMIUM (Cd) SPECIATION AND BIOAVAILABILITY IN THE AQUATIC ENVIRONMENT

Cadmium (Cd) levels in the earth's crust fluctuates from 0.1 to 0.5 mg kg^{-1}. Cd and its compounds are relatively more water-soluble, as compared to other metals. They are therefore more mobile, generally more bioavailable, and tend to bio-accumulate to a higher extent in aquatic organisms. For example, the bioconcentration factors in microorganisms and mollusks are in the order of thousands (Maheshwari et al., 2015). In aquatic systems, Cd is most readily absorbed by organisms directly from the water in its free ionic form Cd^{2+}. The acute toxicity of Cd to aquatic organisms varies even between closely related species and is related to interactions with the calcium metabolism of animals (Maheshwari et al., 2015). The comparatively high mobility of Cd^{2+} along with its tendency to form strong complexes with sulphur-containing proteins makes Cd a potential threat to a wide range of biota. In fish, it causes hypocalcaemia, probably by inhibiting calcium uptake from the water. The toxicity of Cd to aquatic species depends on speciation, with the free ion, Cd^{2+} being the most bioavailable. Toxicity is reduced via complexation of Cd^{2+} by inorganic and organic anions and through competitive interactions between Ca^{2+} and Cd^{2+} for uptake sites (McGeer et al., 2011). Cd^{2+} ion acute toxicity involves disruption of ion homeostasis, particularly Ca, but also Na and Mg. The effects of long-term exposure include larval mortality and temporary reductions in fish growth (Enos et al., 2018; Sabry et al., 2012). Cd has been considered as one of the factors likely responsible for the decline in population of freshwater mussels due to its high toxicity, bio-accumulation potential, and ability to transfer through food chains (Ngo, 2008).

CADMIUM AND LEAD PLANT UPTAKE AND PARTITIONING IN AQUATIC AND WETLAND PLANTS

Various aquatic and wetland plants in different plants categories such as emergent, surface floating, free floating, rooted leaves, submerged macrophytes, and trees, display different ways of taking up contaminants. Pb and Cd are absorbed in their ionic forms. Example, it was reported that Pb^{+2} and Cd^{+2} uptake in *Brassica juncea,* commonly known as Indian Mustard is through the ion exchange mechanism that involves Ca^{+2} ions or protons that are released to the solution (Crist et al., 2004). Heavy metals enter the plants cells through extracellular and intracellular routes (Singh et al., 2020). The extracellular way of the root is easily permeable to solutes, where metals are absorbed in the root cell wall's negatively charged sites (Lasat, 2002). Subsequently, transporter proteins in the cells mediate the ion transport by a transporter protein such as the CPx-ATPases (Lasat, 2002; Williams et al., 2000) among others. The transporter proteins aid the metals uptake and homeostasis. Thomine et al. (2000) reported that metal ions such as Cd enter the plant through the transporters for cations such as Fe^{2+}, which are essential elements to plants. Since Cd is one of the most readily absorbed and rapidly translocated metals, it has been applied in Cd phytoremediation efforts in polluted aqueous environments using aquatic macrophytes, which are capable of up-taking higher Cd concentrations as compared to terrestrial land plants. For example, *Eichhornia crassipes* can take up to 9.13 mg kg^{-1}, *Panicum antidotale* can take up to 7.99 mg kg^{-1}, *Lemna minor* can take up to 8.35 mg kg^{-1}, and *Potamogeton crispus* up to 7.15 mg kg^{-1}. Cd can be easily transported from roots to shoots in plants (Sterckeman et al., 2015). A study by Outa et al. (2020) on metals uptake by Lake Victoria aquatic plants, for example, reported higher metals concentrations in *Vatica cuspidata* roots, than in *Ceratophyllum demersum.* A significant correlation between Cd concentration in roots and in shoots was established. In that study, Lake Victoria was reported to be contaminated with Ni, Cu, Zn, Ag, Cd and Pb. Metal partitioning, with higher accumulation in the roots than stems, agrees with the findings reported in other studies on emergent macrophytes (Brankovic et al., 2015; Galal et al., 2017). Emergent macrophytes take up and accumulate metals in the roots but restrict their translocation to the shoots, thus they are referred to as metal excluders. Such metal excluders maintain low shoot metal concentrations over a wide range of metals concentrations (Baker, 1981), as a protective measure. For instance, plants trap arsenate in the roots below the ground to prevent access of the arsenate to the shoot and leaves above-ground that hold productive reproductive tissues in order to prevent possible mutagenic consequences (Dhankher, 2005).

Lead also accumulates in the shoots and roots of submerged aquatic macrophytes. Positive correlations have been noted between the metal concentrations of submerged shoots and in underlying sediments, indicating that Pb transfer from the sediment matrix to the roots of submerged macrophytes. According to a study by Welsh and Denny (1980) on the uptake of Pb and Cu by submerged aquatic macrophytes in two English Lakes, the enrichment of Pb in the shoots of the investigated plants was largely derived from the high metal concentrations in the sediments. The pathways for the transfer of metals from the sediments to the shoots are apparent. Since Pb precipitates out upon the formation of complexes, it is unlikely that its accumulation in aquatic plants is as a result of adsorption from the water. Arreghini et al. (2017) conducted a study to assess the ability of *Schoenoplectus californicus* growing in natural marsh sediments, to tolerate and accumulate Pb, taking into account the metal distribution in sediment fractions. In that study, Pb was found to exist in the bound (to organic matter) forms, and its absorption by the plants was, therefore, likely to be through sediments. Pb was largely retained in roots (translocation factor < 1). Pb rhizome concentrations only increased significantly in treatments with high doses. These findings are interesting because they indicate that Pb can be absorbed by plants in the bound form.

CADMIUM AND LEAD UPTAKE AND PARTITIONING IN BENTHIC MACROINVERTEBRATES AND FISH

Heavy metals enter the aquatic food chain by direct consumption of water and food and subsequently into body organs through the digestive tract. The other route of entry into body organs is mainly through membranes such as the muscle and gills. Consequently, fish accumulate heavy metals in their tissues to higher levels than environmental concentration (Annabi et al., 2013). To understand the uptake mechanism of Pb and Cd in aquatic organisms such as fish, several investigators have used Ca^{2+} uptake by fish to establish the probable uptake of Cd and Pd by fish. Cd and Pb are analogous to Ca in the process of uptake and accumulation in the fish. A study by Baldisserotto et al. (2005) showed a complex interactions between waterborne and dietary effects of Ca^{2+} and Cd in juvenile Rainbow trouts, where elevated dietary Ca^{2+} protects against both dietary and waterborne Cd uptake, whereas both waterborne and dietary Cd elevations cause reduced waterborne Ca^{2+} uptake. Ca^{2+} is adsorbed through the apical membrane Ca^{2+}-channel in the gills, and is transported across the basolateral membrane with the Ca^{2+}-ATPase (Griffith, 2017). Because of its strong similarity to Ca, Pb penetrates the organism and gets involved in metabolic processes. In previous studies, dietary metal uptake has been increasingly recognized as the dominant route of metal intake and bioaccumulation in most aquatic animals (Rainbow, 2007; Wang and Rainbow, 2008). The gastrointestinal tracts uptake of metals has been found to take place via metal-specific carriers, substitution on nutrient ion transporters and simple diffusion, which are similar to uptake of metals through the fish gills (Wood et al., 2012). However, the metal-specific transporters in the gastrointestinal tracts usually function at much higher substrate levels and have much lower affinities in relation to the substrate levels compared to uptake through the gills. The relative values for gastrointestinal tract in fish are in mmol L^{-1} rather than in μmol L^{-1} range for the fish gills (Ojo and Wood, 2008). Guo et al. (2017) studied water and diet derived Cd uptake in the fish tissues by use of stable isotopes [113]Cd and [111]Cd respectively. In the study, it was found that aquatic animals under waterborne metal exposure are also very likely exposed to elevated dietary metals. The findings by Guo et al. (2017) revealed that the dietary Cd uptake mainly occurred within 12 hours after feeding. Overall, the study findings demonstrated that there was interaction between dietary and waterborne Cd uptake in marine fish. For the fish, this simultaneous uptake of metal from two routes is far more complex than the situation of a single route of metal uptake, which should be evaluated in determining metal bioaccumulation and toxicity in both laboratory and field metal exposure (Guo et al., 2017).

Benthic organisms play an important role, as food for fish, amphibians and aquatic birds in aquatic ecosystems. Benthos are also involved in the breakdown of organic matter and nutrients within the aqueous ecosystem (Mohiuddin et al., 2011). Normally, benthic macro-invertebrates exhibit the highest bio-concentration factor for pollutants, such as metals, in the food webs. Therefore, Pb and Cd being non-essential metal pollutants, result in adverse health impacts, potentially harmful to the organisms involved in the food chain above benthic organisms.

Many concerns are raised due to metal bioaccumulation through the food chain and related human health hazards (Agah et al., 2009). Oremo et al. (2019), assessed heavy metals in benthic macroinvertebrates, water and sediments in River Isiukhu, in Kenya. The study obtained below quantification levels concentrations of Cd and Pb, within the study matrices except for Pb levels in sediments. However, other metals gave positive correlations for concentrations in sediments and benthic macroinvertebrates. For example, significant correlations were observed for Zn ($r = 0.655$, $p = 0.029$) and Cu ($r = 0.641$, $p = 0.034$). This observation supports other findings that indicate sediments as sink for metals, i.e., Pb. The investigated benthic macroinvertebrate families were *Unionidae, Baetidae* and *Gerridae*. Metal concentration profiles along the river indicated an influx of pollutants from anthropogenic sources due to rapid urbanization. A similar observation was reported by Schwantes et al. (2021) in a study on the distribution of heavy metals in sediments and their bioaccumulation on benthic macroinvertebrates in a tropical Brazilian watershed. The study reported that on average, Pb levels in benthic organisms were 14.94-fold higher than the bioavailable fraction in sediments.

In the study by Oremo et al. (2019), benthic macroinverterbrates were observed to be spread in the sediments and water of the studied river. It can be concluded that sediment benthic macroinverterbrate dwellers mainly feed on algae and other substances within the sediments, while the water dwelling benthos feed on food substances within the water matrix. This makes it more likely for bottom feeder fish to ingest more Cd and Pb that deposited within the sediment matrix. The extent of metal accumulation in biota depends on their uptake and their elimination rate. Different metals display different half-lives in different species (Hazrat et al., 2019). Metals may enter the fish body directly from water or sediments through the fish gills/skin or from the fish food/prey through the dietary route. Metal bioaccumulation and subsequent distribution in fish organs is greatly interspecies specific, and is influenced by many factors like sex, age, size, reproductive cycle, swimming patterns, feeding behaviour, and geographical location (Ekundayo et al., 2014; Zhao et al., 2012). Metal retention in the body depends on many factors such as the speciation of the concerned metal and the physiological mechanisms developed by the organism for metal regulation, homeostasis, and detoxification. In a study performed by Orata and Birgen (2016) on fish tissue bio-concentration and interspecies metals uptake in Wastewater Lagoons of Kisumu in Kenya, for example, all metals (Zn, Cu, Cr, Cd and Pb) analyzed in the study were present in *Orechromis niloticus, Clarius gariespinus* and *Protopterus aethipicus* commonly known as Tilapia, African catfish and Marbled lungfish respectively. In that study, Cd was only detected in *P. aethiopicus* skin and gills at 6.20 mg kg^{-1}and 1.60 mg kg^{-1}, respectively. Cd is a non-essential metal and does not play a biochemical role in fish, and its concentrations in skin is of interest, as this could indicate Cd ion depuration processes through this organ. Pb was detected in all fish species tissues and organs. Scales and skin were observed to contain more Pb than muscles and liver, in both fish species (Figure 8.2). Uptake of metals in fish gills occurs apically via a lanthanum-sensitive voltage-independent epithelial Ca^{2+} channel located in mitochondria-rich chloride cells and basolaterally via Ca-ATPase and an Na/Ca exchanger (McGeer et al., 2011). In the aforementioned study, it was observed that there is no positive correlation among the metal ion concentrations in the various organ tissues studied. Previous studies have indicated that metal variability in different fish species depends on ecological needs, metabolism (Canlı and Kalay, 1998), age, size and length (Al-Yousuf et al., 2000), feeding habits (Watanabe et al., 2003) and habitats (Canli and Atli, 2003).

Fish scales and skin are expected to contain elevated concentrations of Pb and Cd more than those obtained in muscles and liver which suggested depuration process, considering that Pb and Cd are non-essential metal ions to these fish species. Evidence suggests that Pb uptake in fish is localised mostly in non-edible tissues including bones and scales, so that the availability in edible portions does not generally pose a human health risk. According to EU Regulation 1881/2006/ EU, the established maximum concentration limits in fish tissues for Cd and Pb are 0.05 mg/kg and 0.30 mg/kg, respectively.

BIOACCUMULATION AND BIOMAGNIFICATION OF LEAD AND CADMIUM

Bioaccumulation which subsequently leads to biomagnification within the food chains is mainly dependent on the concentration levels of the pollutants in question within a particular ecosystem/environment. Low concentrations may not significantly cause bioaccumulation and consequently biomagnification. Accumulation of metals in the tissues of aquatic organisms via the surrounding aquatic environments also depends on whether these metals are essential to metabolism of the organism or not. The bioaccumulated concentrations are divided into two components namely; the metabolically available form and the stored detoxified form. Metals in the detoxified form can be excreted or bound to proteins or other particular molecules displaying high metal affinity (Rainbow, 2002). To varied extents, both available and the detoxified forms can accumulate throughout the food chain by transfer to other organisms. High Cd concentrations in sediments revealed a high availability of these metal ions or high uptake rates from sediment to inverterbrates such as Crustaceans (Rainbow, 2002). Cd can subsequently be transferred to inverterbrate predators such as *Anadromous* fishes (those that spend most of their lives in the sea but migrate to fresh water to spawn), as observed by Syvaranta et al. (2009), between anadromous fish to the diet of European catfish (*Silurus glanis*) in a large river system (Garonne, southwestern France). In the study, it was also observed that other metals such as Cr and Cu were diluted in the food chain. A study by M'kandawire et al. (2012) observed higher Cd and Cr accumulations in shrimps. The ability of fish to bioaccumulate metals in contaminated aquatic environment was studied by Orata and Birgen (2016), by approximating the bioconcentration factors of metals that included Cd and Pb. The following equation was used in the BCFs calculations:

$$ \text{BCF} = \frac{\text{Concentration in fish at steady state (mg/kg wet fish)}}{\text{Concentration on water (or sediments) at steady state (mg/L or mg/kg)}} $$

The obtained BCFs for Pb and Cd in fish tissues based on aqueous contact and sediments concentration are shown in Table 8.1, and the transfer of Pb and Cd to the various body organs of the three studied fish species were noted, and are shown in Figure 8.1.

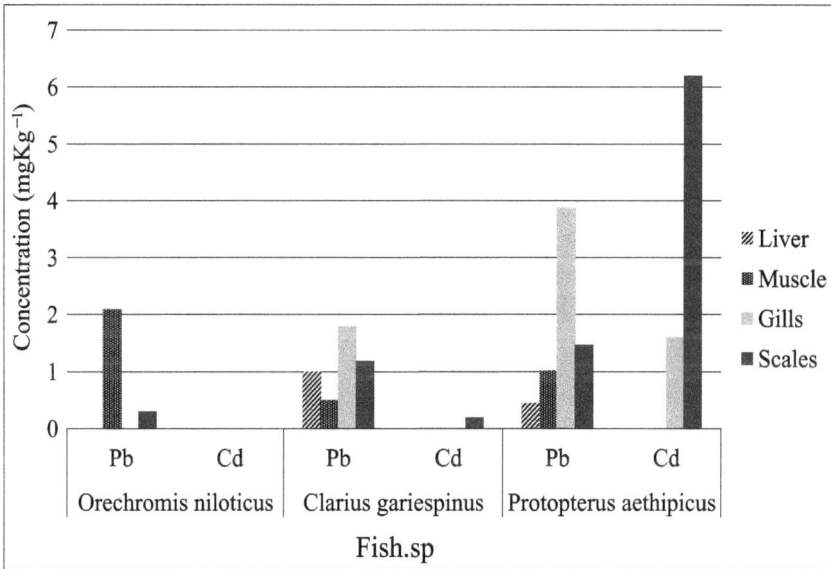

Figure 8.1   Metal tissue concentrations and distribution in *Oreochromis niloticus, Clarius gariespinus* and *Protopterus aethiopicus*. Adapted from Orata and Birgen, 2016.

Metal bioaccumulation is dependent on the rate at which organisms ingest it from the water or diet. If the rate of ingestion is faster than that of metabolism and excretion, then bioaccumulation is likely to occur. However, some organisms are tolerant to naturally accumulated Pb or other metals at least to some degree, without deleterious effects. In comparison, Pb bioaccumulation in fish is generally less than the bioaccumulation to other benthos like water fleas, shellfish,

Table 8.1 BCFs for Zn, Cu, Pb, Cd, Cr in *Orechromis niloticus, Clarius gariespinus* and *Protopterus aethiopicus* tissues based on sediments concentration. ND–non-detected

Base Media	Fish sp.	Metal	Fish Tissue BCF			
			Liver	Muscle	Gills	Scales
Aqueous media						
	Orechromis niloticus	Pb	0.24	0.3	ND	2.03
		Cd	ND	ND	ND	ND
	Clarius gariespinus	Pb	1.01	0.57	1.8	1.19
		Cd	ND	ND	ND	0.2
	Protopteru aethipicus	Pb	0.44	1.01	3.88	1.47
		Cd	ND	ND	2.91	11.27
Sediments	*Clarius gariespinus*	Pb	0.04	0.02	0.06	0.04
		Cd	ND	ND	ND	0.01
	Protopterus aethiopicus	Pb	0.02	0.04	0.14	0.05
		Cd	ND	ND	0.12	0.47

snails or aquatic insects (Thornton et al., 2001) which are bottom dwellers. Pb biomagnification is also not prominent because of natural lower concentrations, which may not magnify up the food chain (Thornton et al., 2001). In addition, and unlike the Cd compounds, most Pb compounds have low solubility in water and undergo changes to form complexes that are not bioavailable. In this regard Pb can accumulate more in biota such as mussels and worms through feeding primarily within the sediment's particles. Different fish species however will show varied concentrations of Pb and Cd depending on the feeding habits of the particular fish. For example, *O. niloticus* will likely ingest Cd and Pb bioavailable in water media. However, *O. niloticus* is not likely to ingest Pb and Cd through feeding on benthic macroinverterbrates that are found within sediments. On the other hand, bottom feeder fish, for example, *C. gariespinus*, and *P. aethiopicus* will ingest the metals through feeding on benthic macroinvertabrates that dwell mostly in sediments, as demonstrated in Figure 8.2. Cadmium is slightly soluble in water, although its chloride and sulphate salts are freely soluble. Particulate matter and dissolved organic matter may bind a substantial portion of cadmium in freshwater. Although Cd tends to bind to sediments, it is mostly readily available as Cd^{2+} and can be bioaccumulated by aquatic organisms. Bioconcentration factors (BCFs) for Cd in freshwater fish were observed to be as high as 12,400 and for a saltwater polychaete were as high as 3,160 (USEPA, 1985). Figure 8.2 shows the most likely pathways for cadmium (Cd) and Lead (Pb) transfer among different aquatic macrophytes (floating, submerged, emergent), fish (bottom feeders and non-bottom feeders) and benthic macroinvertebrates (sediments dwellers and floaters) in relation to sediments and water contamination in aqueous ecosystem. Bottom feeder fish are likely to accumulate both Cd and Pb, since bound Cd and especially Pb complexes will likely be precipitated in the sediments. In a similar reasoning, floating macrophytes will likely not accumulate Pb which partitions more in sediments due to complexing and precipitating into sediments.

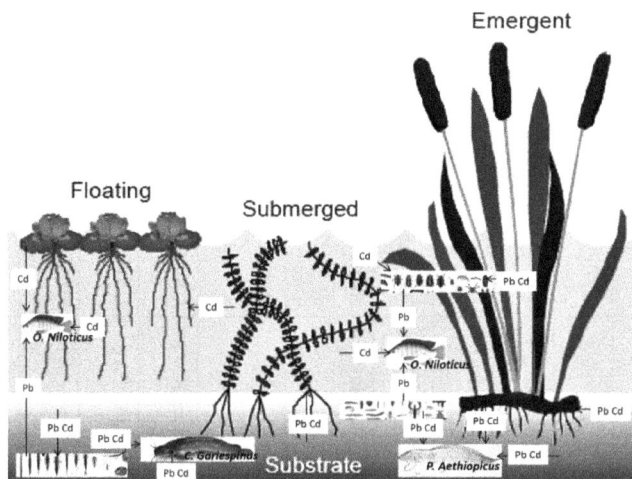

Figure 8.2 Likely pathways for cadmium (Cd) and Lead (Pb) transfer among different aquatic organisms in a contaminated aquatic system. https://sswm.info/factsheet/free-water-surface-cw. Original *Source SA'AT (2006).*

Cadmium exists in a variety of bound forms including associated with metallothioneins and/or granules (Rainbow, 2002). Cd and Pb toxicity is related to a threshold concentration of metabolically available metal to the organism rather than to total accumulated metal concentration. The onset of toxic effects depends only on the concentration of accumulated metal in a metabolically available form. It follows, therefore, that toxicity occurs when the rate of metal uptake into the body exceeds the combined rate of excretion plus detoxification of metabolically available form.

Aquatic ecosystem provides many possible pathways in which toxic substances can enter the tissues of organisms that arise from intoxicated or contaminated environment. When these toxic substances are consumed by organisms in the lower trophic levels such as phytoplankton or zooplankton, they can be passed up the trophic level. The increase in toxicity is mainly attributed to persistence pollutants such as metals. Both Pb and Cd can enter the food chain, and consequently lead to their progressive increase in concentration up the trophic levels. The general term for this process is biomagnification. Biomagnification therefore can be regarded as an ever-growing threat to aquatic ecosystem health and beyond.

Biomagnification Factor (BMF), which is described as the ratio of the chemical concentrations in the organism and the diet of the organism, can be expressed as

$$BMF = \frac{\text{Chemical concentrations in the organism}}{\text{Chemical concentrations in the diet of the organism}}$$

BMF are based on the comparison of chemicals concentration in predators and preys and are usually expressed in units of mass (i.e., kg). Biomagnification has been reported by various studies around the world (see Table 8.2) and high BMF reported for example by Chan et al. (2021). However other studies reported BMF < 1 as observed in many trophic food web relations (Madgett et al., 2021).

Table 8.2 Biomagnification factor (BMF) for various classes of organisms obtain from various studies around the world.

Heavy metal	Predator	Prey	BMF	Study area (season)	Country	References
Cd	Fish (*Coilia ectenes taihuensis*)	Zooplankton	1.73	Taihu lake	China	Zuo et al. (2018)
Cd	Fish (*Coilia ectenes taihuensis*)	Phytoplankton	1.15	Taihu lake	China	Zuo et al. (2018)
Cd	*Pseudobagrus fulvidraco*	*Carassius aumtus*	464	Hengshi and Wengjiang Rivers near Dabaoshan mine	China	Chan et al. (2021)
Cd	Juvenile mullet	Zooplankton	4.2	SE Gulf of California (dry)	USA	Jara-Marini et al. (2009)
Cd	Juvenile mullet	Zooplankton	3.4	SE Gulf of California (wet)	USA	Jara-Marini et al. (2009)
Cd	Demersal shark, fish: pelagic, demersal and flatfish	Invertebrates and zooplankton	0.4–0.7	Urbanized and industrialized estuarine locations	Scotland	Madgett et al. (2021)
Cd	Fish: *Coilia sp, Awaous banana*	Mixed zooplankton (*Copepods, Tintinnida, Chaetognatha, Meroplankton*), Crustacean, bivalve	0–1.22	Muthupet mangrove ecosystem	India	Arumugam et al. (2017)
Pb	*Pseudobagrus fulvidraco*	*Carassius aumtus*	644	Hengshi and engjiang Rivers near Dabaoshan mine	China	Chan et al. (2021)
Pb	Juvenile mullet	Zooplankton	1.3	SE Gulf of California (dry)	USA	Jara-Marini et al. (2009
Pb	Juvenile mullet	Zooplankton	2.8	SE Gulf of California (wet)	USA	Jara-Marini et al. (2009)
Pb	Fish: *Coilia sp, Awaous banana*	Mixed zooplankton (*Copepods, Tintinnida, Chaetognatha, and Meroplankton*), Crustacean, bivalve	1.08–1.27	Muthupet mangrove ecosystem	India	Arumugam et al. (2017)
Pb	*Barnacle*	Phytoplankton	2.3	SE Gulf of California (dry)	USA	Jara-Marini et al. (2009

CONCLUSIONS

Like many other metals, Pb and Cd are found naturally in the environment. Their unique chemical properties, versatile uses and economic values have made them an integral part of our modern society. From natural sources or human activities origins, Pb and Cd may enter aquatic environments, and ultimately large amounts are deposited in river sediments, estuaries and other aquatic ecosystems. The availability of Cd and Pb in these aquatic ecosystems can cause adverse health effects to aquatic organisms. Speciation forms of Pb and Cd can cause acute or chronic toxicity to benthic organisms and fish, also greatly determining their bioaccumulation potential in aquatic food chains. Moreover, speciation determines the pathways in which these metals will move from different organisms and segments of the aquatic ecosystems. Processes such as bioconcentration and biomagnification have the potential to cause adverse environmental effects to aquatic ecosystems and beyond. Aquatic ecosystem provides various possibilities in which pollutants such as Cd and Pb can be transferred to many organisms within the food web. In a contaminated aqueous environment, organisms show some levels of tolerance to various Cd and/or Pb exposure and consequently, many of the adverse effects are reversible once exposure levels decrease. Fishes should be carefully screened to ensure that high levels of these toxic trace metals are not transferred to humans.

ACKNOWLEDGEMENT

The authors would like to acknowledge the following colleagues who over time have worked on metals contamination; Prof. John Onyari, Prof. Shem Wandiga, Faith Birgen and the late Prof. Geofrey Kamau. We also acknowledge the chemistry department at University of Nairobi and the Department of Pure and applied chemistry Masinde Muliro University of Science and Technology.

REFERENCES

Abadin, H., Ashizawa, A., Stevens, Y.W., Llados, F., Diamond, G., Sage, G., et al. 2007. Toxicological Profile for Lead. Atlanta (GA): Agency for Toxic Substances and Disease Registry (US). PMID: 24049859.

Agah, H., Leermakers, M., Elskens, M., Fatemi, S.M. and Baeyens, W. 2009. Accumulation of trace metals in the muscles and liver of five fish species from the Persian Gulf. Environ. Monit. Assess. 157: 499–514.

Al-Yousuf, M., El-Shahawi, M.S. and Al-Ghais, S.M. 2000. Trace metals in liver, skin and muscle of Lethrinus lentjan fish species in relation to body length and sex. Sci. Total Environ. 256: 87–94.

Arumugam, G., Rajendran, R., Shanmugam, V., Sethu, R. and Krishnamurthi, M. 2017. Flow of toxic metals in food-web components of tropical mangrove ecosystem, Southern India. Hum. Ecol. Risk. Assess: Int. J. 24: 1367–1387. doi: 10.1080/10807039.2017.1412819

Annabi, A., Said, K. and Messaoudi, I. 2013. Cadmium: bioaccumulation, histopathology and detoxifying mechanisms in fish. Am. J. Res. Commun. 1: 60–79.

Arreghini, S., de Cabo, L., Serafini, R. and De Lario, A.F. 2017. Effect of the combined addition of Zn and Pb on partitioning in sediments and their accumulation by the emergent macrophyte *Schoenoplectus californicus*. Environ. Sci. Pollut. Res. 24: 8098–8107. https://doi.org/10.1007/s11356-017-8478-7

Baker, A. 1981 Accumulators and excluders strategies in response of plants to heavy metals. J. Plant Nutr. 3: 643–654. https://doi.org/10.1080/01904168109362867.

Baldisserotto, B., Chowdhury, M.J. and Wood, C.M. 2005. Effects of dietary calcium and cadmium on cadmium accumulation, calcium and cadmium uptake from the water and their interactions in juvenile rainbow trout. Aquat. Toxicol. 72: 99–117.

Brankovic, S., Glišić, R., Topuzovic, M. and Marin, M. 2015. Uptake of seven metals by two macrophytes species: potential for phytoaccumulation and phytoremediation. Chem. Ecol. 31: 583–593. https://doi.org/10.1080/02757540.2015.1077812

Canli, M. and Atli, G. 2003 The relationships between heavy metal (Cd, Cr, Cu, Fe, Pb, Zn) levels and the size of six Mediterranean fish species. Environ. Pollut. 121: 129–136.

Canli, M. and Kalay, M. 1998. Levels of heavy metals (Cd, Pb, Cu, Cr and Ni) in tissue of Cyprinus carpio, Barbus capito and Chondrostoma regium from the Seyhan River, Turkey. Tr. J. of Zoology 22: 149–157.

Chan, W.S., Routh, J., Luo, C., Dario, M., Miao,Y., Luo, D., et al. 2021. Metal accumulations in aquatic organisms and health risks in an acid mine-affected site in South China. Environ. Geochem. Health. 43: 4415–4440. https://doi.org/10.1007/s10653-021-00923-0

Crist, R.H., Martin, J.R. and Crist, D.R. 2004. Ion-exchange aspects of toxic metal uptake by Indian mustard. Int. J. Phytorem. 6: 85–94. DOI: 10.1080/16226510490440006.

Counter, S.A., Buchanan, L.H. and Ortega, F. 2009. Neurocognitive screening of lead-exposed Andean adolescents and young adults. J. Toxicol. Environ. Health, Part A 72: 625–632.

Dhankher, O.P. 2005. Arsenic metabolism in plants: an inside story. New Phytol. 168: 503–505. https://doi.org/10.1111/j.1469-8137.2005.01598.x.

Dietert, R.R. and Piepenbrink, M.S. 2006. Lead and immune function. Crit. Rev. Toxicol. 36: 359–385. 10.1080/10408440050053429716809103.

Ekundayo, T.M., Sogbesan, O.A. and Haruna, A.B. 2014. Study of fish exploitation pattern of lake Gerio, Yola, Adamawa State, Nigeria. J. Surv. Fish. Sci. 1(1): 9–20.

Enos, W., Wambu, Stephen A., Paul, M.S. and John, Wabomba. 2018. Removal of heavy metals from wastewater using anhydrous alumino- silicate mineral from Kenya. Chem. Soc. Ethiop. 32(1): 32–51.

Galal, T.M., Gharib, F.A., Ghazi, S.M. and Mansour, K.H. 2017. Phytostabilization of heavy metals by the emergent macrophyte *Vossia cuspidata* (Roxb.) Griff.: A phytoremediation approach. Int. J. Phytoremediation 19: 992–999. https://doi.org/10.1080/15226514.2017.1303816.

Griffith, M.B. 2017. Toxicological perspective on the osmoregulation and ionoregulation physiology of major ions by freshwater animals: teleost fish, crustacea, aquatic insects, and Mollusca Environ. Toxicol. Chem. 36: 576–600

Guo, Zhiqiang, Gao, Na; Wu, Yun and Zhang, Li. 2017. The simultaneous uptake of dietary and waterborne Cd in gastrointestinal tracts of marine yellowstripe goby Mugilogobius chulae. Environ. Pollut. 223: 31–41. doi:10.1016/j.envpol.2016.12.007.

Hazrat Ali, Ezzat Khan and Ikram Ilahi. 2019. Environmental Chemistry andEcotoxicology of Hazardous Heavy Metals: Environmental Persistence, Toxicity, and Bioaccumulation. J. Chem. 2019: Article ID 6730305. https://doi.org/10.1155/2019/6730305.

Jara-Marini, M.E., Soto-Jiménez, M.F. and Páez-Osuna, F. 2009. Trophic relationships and transference of cadmium, copper, lead and zinc in a subtropical coastal lagoon food web from SE Gulf of California, Chemosphere 77: 1366–1373. https://doi.org/10.1016/j. chemosphere.2009.09.025.

Jose, J. and Gobas, F.A.P.C. 2012. Assessing biomagnification and trophic transport of persistent organic pollutants in the food chain of the galapagos sea lion (*Zalophus wollebaeki*): Conservation and management implications. *In*: New Approaches to the Study of Marine Mammals. InTech. https://doi.org/10.5772/51725

Khafaga, A.F., Abd El-Hack, M.E., Taha, A.E., Shaaban, S., Elnesr, S.S. and Alagawany, A. 2019. The potential modulatory role of herbal additives against Cd toxicity in human, animal, and poultry: a review. Environ. Sci. Pollut. Res. 26: 4588–4604. https://doi. org/10.1007/s11356-018-4037-0

Lasat, M.M. 2002. Phytoextraction of toxic metals. J. Environ. Qual. 31(1): 109120.

Madgett, A.S., Yates, K., Webster, L., McKenzie, C. and Moffat, C.F. 2021. The concentration and biomagnification of trace metals and metalloids across four trophic levels in a marine food web. Mar. Pollut. Bull. 173(Pt. A): 112929. doi: 10.1016/j. marpolbul. 2021.112929. Epub 2021 Sep 14. PMID: 34534935.

Maheshwari, U., Mathesan, B. and Gupta, S. 2015. Efficient adsorbent for simultaneous removal of Cu (II), Zn (II) and Cr (VI): kinetic, thermodynamics and mass transfer mechanism, Process. Saf. Environ. 98: 198–210.

McGeer, J.C., Niyogi, S. and Smith, D.S. 2011. Cadmium. pp. 125–184. *In*: Wood, C.M., Farrell, A.P. and Brauner, C.J. (eds). Homeostasis and Toxicology of Non-Essential Metals: Fish Physiology, Volume 31B. Elsevier.

M'kandawire, E., Syakalima, M., Muzandu, K., Pandey, G., Simuunza, M., Nakayama, S.M.M. et al. 2012. The nucleotide sequence of metallothioneins (MT) in liver of the Kafue lechwe (*Kobus leche kafuensis*) and their potential as biomarkers of heavy metal pollution of the Kafue river. Gene. 506(2): 310–316.

Mohiuddin, K.M., Ogaway, Z.H., Otomo, K. and Shikazono, N. 2011. Heavy metals contamination in water and sediments of an urban river in a developing country. Int. J. Environ. Sci. Tech. 8: 723–736.

Ngo, H.T.T. 2008. Effects of cadmium on calcium homeostasis and physiological conditions of the freshwater mussel *Anodonta anatina*, Ph.D. thesis. Faculty of Biology, Chemistry and Geosciences, University of Bayreuth, Bayreuth, Germany.

Obike, A., Igwe, J., Emeruwa, C., Uwakwe, K. and Aghalibe, C. 2018. Diffusion-Chemisorption and Pseudo-Second Order Kinetic Models for Heavy Metal Removal from Aqueous Solutions Using Modified and Unmodified Oil Palm Fruit Fibre. Chem. Sci. Int. J. 23: 1–13.

Ojo, A.A. and Wood, C.M. 2008. In vitro characterization of cadmium and zinc uptake via the gastro-intestinal tract of the rainbow trout (*Oncorhynchus mykiss*): interactive effects and the influence of calcium. Aquat. Toxicol. 89: 55–64.

Orata, F. and Birgen, F. 2016. Fish Tissue Bio-concentration and Interspecies Uptake of Heavy Metals from Waste Water Lagoons. J. Pollut. Eff. Cont. 4: 157. doi:10.4172/2375-4397.1000157

Oremo, J., Orata, F., Owino, J. and Shivoga, W. 2019. Assessment of heavy metals in benthic macroinvertebrates, water and sediments in River Isiukhu, Kenya. Environ. Monit. Assess. 191: 646. https://doi.org/10.1007/s10661-019-7858-5

Outa, J.O., Kowenje, C.O., Plessl, C. and Jirsa, F. 2020. Distribution of arsenic, silver, cadmium, lead and other trace elements in water, sediment and macrophytes in the Kenyan part of Lake Victoria: spatial, temporal and bioindicative aspects. Environ. Sci. Pollut. Res. 27: 1485–1498.

Pascoal, C., Cassio, F., Marcotegui, A., Sanz, B. and Gomes, P. 2005. Role of fungi, bacteria, and invertebrates in leaf litter breakdown in a polluted river. J. North Am. Benthol. Soc. 24: 784–797.

Rainbow, P.S. 2002. Trace metal concentrations in aquatic invertebrates: Why and so what? Environ. Pollut. 120: 497–507.

Rainbow, P.S. 2007. Trace metal bioaccumulation: Models, metabolic availability and toxicity. Environ. Int. 33: 576–582.

Sabry, M. Shaheen, Derbalah, Aly S. and Moghanm, Farahat S. 2012. Removal of heavy metals from aqueous solution by Zeolite in competitive sorption system. J. Environ. Sci. Dev. 3: 362–367.

Singh, S., Yadav, V., Arif, N., Singh, V.P., Dubey, N.K., Ramawat, N., et al. 2020. Heavy metal stress and plant life: Uptake mechanisms, toxicity, and alleviation. Plant Life Under Changing Environment 271–287.

Sterckeman, T., Goderniaux, M., Sirguey, C., Cornu, J.-Y. and Nguyen, C. 2015. Do roots or shoots control cadmium accumulation in the hyperaccumulator *Noccaea caerulescens*? Plant Soil 392: 87–99.

Suozzo, K. 2005. The use of aquatic insects and benthic macroinvertebrate communities to assess water quality upstream and down-stream of the village of Stamford wastewater treatment facility. pp. 141–151 *In*: Proceedings of the 38th Annual Report of the Suny Oneonta Biological Field Station, Biological Field Station, Cooperstown.

Syvaranta. J., Cucherousset, J., Kopp, D., Martino, A., Cereghino, R. and Santoul, F. 2009. Contribution of anadromous fish to the diet of European catfish in a large river system. Naturwissenschaften 96(5): 631–635.

Thomine, S., Wang, R., Ward, J.M., Crawford, N.M. and Schroeder, J.I. 2000. Cadmium and iron transport by members of a plant metal transporter family in Arabidopsis with homology to Nramp genes. Proc. Natl. Acad. Sci. 97(9): 49914996.

Thornton, I., Rautiu, R. and Brush, S. 2001 Lead: The Facts. Ian Allan Printing, Hersham, Surrey, U.K.

UNEP. 2010. Final Review of Scientific Information on Lead .United Nations Environment Program, Chemical Branch, DTIE.

USEPA. (U.S. Environmental Protection Agency) 1985. Ambient aquatic life criteria for cadmium. EPA 440/5–84–032. Washington, DC.

Vijver, M.G., Vink, J.P.M., Jager, T., Wolterbeek, H. Th., Van Straalen, N.M. and Van Gestel, C.A.M. 2005 BBiphasic elimination and uptake kinetics of Zn and Cd in the earthworm Lumbricus rubellus exposed to contaminated floodplain soil. Soil Biol. Biochem. 37(10): 1843-1851.

Wang, W.-X. and Rainbow, P.S. 2008. Comparative approaches to understand metal bioaccumulation in aquatic animals. Comp. Biochem. Physiol. C: Toxicol. Pharmacol. 148(4): 315–323.

Watanabe, K.H., Fesimone, F.W., Thiyagarajah, A., Hartley, W.R. and Hindrichs, A.E. 2003. Fish tissue quality in the lower Mississippi River and health risks from fish consumption. Sci. Total Environ. 302: 109–126.

Welsh, R.P.H. and Denny, P. 1980. The uptake of lead and copper by submerged aquatic macrophytes in two english lakes. J. Ecol. 68(2): 443–455.

Williams, L.E., Pittman, J.K. and Hall, J.L. 2000. Emerging mechanisms for heavy metal transport in plants. Biochim. Biophys. Acta (BBA)-Biomembranes. 1465(12): 104126.

Wood, C.M., Farrell, A.P. and Brauner, C.J. 2012. Homeostasis and Toxicology of Essential Metals, Vol. 31A. Academic Press. Elsevier, London.

Young, R.G., Matthaei, C.D. and Townsend, C.R. 2008. Organic matter break-down and ecosystem metabolism: functional indicators for assessing river ecosystem health. J. North Am. Benthol. Soc. 27: 605–625.

Zhang, Y., Liu, L., Cheng, L., Cai, Y., Yin, H., Gao, J. and Gao, Y. 2014. Macroinver- tebrate assemblages in streams and rivers of a highly developed region (Lake Taihu basin, china). Aquat. Biol. 23: 15–28.

Zhao, S., Feng, C., Quan, W., Chen, X., Niu, J. and Shen, Z. 2012. Role of living environments in the accumulation characteristics of heavy metals in fishes and crabs in the Yangtze River Estuary, China. Mar. Pollut. Bull. 64(6): 1163–1171.

Zuo, J., Fan, W., Wang, X., Ren, J., Zhang, Y., Wang, X., et al. 2018. Trophic transfer of Cu, Zn, Cd, and Cr, and biomarker response for food webs in Taihu Lake, China. RSC Advances 8: 3410–3417.

Chapter 9

Cadmium in Mexican Sharks: Presence, Accumulation, and Public Health Risks

Gabriel Núñez-Nogueira[1*], Melina Uribe-López[1],
Mórvila Cruz-Ascencio[1], Juanita María Santos-Córdova[1],
Eva López-Dobrusin[1] and Alicia Cruz-Martínez[2]

[1]Laboratorio de Hidrobiología y Contaminación Acuática,
División Académica de Ciencias Biológicas,
Universidad Juárez Autónoma de Tabasco,
Carretera Villahermosa-Cárdenas Km. 0.5.,
Villahermosa, Tabasco 86039, México.

[2]Departamento de Ciencias Ambientales, Universidad Autónoma Metropolitana,
Unidad Lerma. Av. de las Garzas No. 10, El Panteón, C. P. 52005; Lerma de Villada, México.

INTRODUCTION

The presence of metals as pollutants in marine environments is widely recognized in various waters worldwide (Haynes and Michalek-Wagner, 2000; Klaine et al., 2008; Windom et al., 1973; Wise et al., 2014; Yu et al., 2020), including Latin America (De Marco et al., 2006; Di Marzio et al., 2019; Monteiro-Neto et al., 2003) and Mexico (Benítez et al., 2012; Flores and Albert, 2005; Pérez-Moreno et al., 2016; Ruelas-Inzunza et al., 2013). Highly toxic metals such as mercury (Hg), cadmium (Cd), lead (Pb), chromium (Cr), or metalloids such as arsenic (As), directly affecting the aquatic biota (Corrill and Huff, 1976; Jonathan et al., 2015; Núñez-Nogueira et al., 2019;

*Corresponding author: gabriel.nunez@ujat.mx

Ramírez-Ayala et al., 2020; Seixas et al., 2009). The natural occurrence of metals in the environment as well as those generated from anthropogenic activities classify metals as important pollutants widely distributed and present globally. Unfortunately, despite being a recognized type of pollution for many decades (Bryan et al., 2007; Corrill and Huff, 1976; De Marco et al., 2006; García-Hernández et al., 2007), significant discharges and inputs of metals still occur due to urban and industrial mainland activities, including soil washing, erosion, atmospheric transport, and oil spills, among others (Corrill and Huff, 1976; Lambert et al., 2000; Morais et al., 2012).

From a biological perspective, Cd is non-essential in animals, so its presence in aquatic fauna is mainly related to metal pollution. Several toxic effects have been reported for this element, although in some rare instances, it can become essential. For example, under poor zinc availability in the environment, certain marine diatoms (known as *Thalassiosira weissflogii*) exhibit Cd incorporation as a significant structural and functional constituent of the enzyme carbonic anhydrase, thus conferring an essential catalytic role in carbon metabolism in diatoms (Roberts et al., 1997; Xu et al., 2008). However, the non-essentiality of Cd is also reinforced by adverse effects observed when this element accumulates in organisms that, even at low concentrations, result in cellular, metabolic, or physiological alterations. Effects include oxidative stress, lipid membrane damage, enzyme function alterations, ionic instability, apoptosis, tissue, and organ damage (Das et al., 2019; Filippini et al., 2020; Jaishankar et al., 2014). These biological alterations at the cellular level are due to the capacity of Cd, as well as lead, metal ions capable to substitute other essential mono and divalent ions, such as Na^+, K^+, Fe^{2+}, Mg^{2+} and Ca^{2+}, becoming structural components of certain biomolecules. Cadmium and lead ions can modify proper enzyme and proteins functions and generate carcinogenicity and genotoxicity (Faroon et al., 2012; Flores-Galván et al., 2020; Klaine et al., 2008). At the same time, metal accumulation in predated species can also pose a risk to their predators, resulting in higher exposure to toxic metals that may accumulate along the food chain due to bioaccumulation and biomagnification processes (De Marco et al., 2006; Pancaldi et al., 2021a; Pinheiro et al., 2012).

As top predators in the marine food chain, sharks commonly biomagnify metals, as reported for several species, such as the blue shark *Prionace glauca* (Fricke et al., 2022) for Pb (Reátegui-Quispe and Pariona-Velarde, 2019), Caribbean reef shark *Carcharhinus perezii* (Poey 1876) for Cr, Cu (copper), Pb and Hg (Shipley et al., 2021), blacktip shark *Carcharhinus limbatus* (Valenciennes 1839) and whale shark *Rhincodon typus* (Smith 1828) for mercury (Pancaldi et al., 2019a; Norris et al., 2021).

Sharks are important fish and invertebrate predators (Ebert et al., 2021). By consuming other species, they play an essential ecological role, regulating marine organism populations, connecting the matter and energy flow in the aquatic environment (Heithaus et al., 2010). Sharks consume different prey, depending on their genus or group. Members of the Carcharhinidae and Sphyrnidae families, as well as lamnids, have a mainly piscivorous diet, as they consume teleost fish (Castro, 1993; Stevens and McLoughlin, 1991). Some sharks may even prey on other larger sharks belonging to the Carcharhinidae family (Cliff and Dudley, 1991). Other families such as Heterodontidae and Triakidae feed on molluscs (Compagno, 1990). Pelagic sharks also feed on cephalopods, such as squid, while demersal sharks consume octopus (Compagno, 1990). At the species level, particular feeding patterns are noted, such as crustaceans, which form a significant part of sharpnose shark *Rhizoprionodon longurio's* diet (Castillo-Géniz et al., 1998), or the case of the whale shark *Rhincodon typus*, a plankton-feeding organism (Compagno, 1990). Regardless of feeding habits, the transfer of metallic elements through predatory behaviour makes sharks vulnerable to metal contamination previously bioaccumulated and bio-concentrated by their prey (Matulik et al., 2017; Maz-Courrau et al., 2012; Shipley et al., 2021).

These characteristics, alongside other biological characteristics such as longevity, reproductive rate, growth rate, and overfishing, among others, makes sharks one of the most vulnerable groups concerning anthropogenic impacts, including metal pollution especially as many shark and elasmobranch species are categorized as threatened (Consales and Marsili, 2021). Recently, Consales and Marsili (2021) highlighted risks to seventeen sharks species that inhabit contaminated sites, including Mexican waters, such as the Pacific Sharpnose shark *Rhizoprionodon longurio* (Jordan and Gilbert 1882), endemic to the tropical Pacific Ocean.

SHARKS IN MEXICO

Cartilaginous fishes are the oldest group of jawed vertebrates (Wourms and Demski, 1993), and include the Subclass Elasmobranchii, comprising sharks and rays with 1,139 described species to date (Weigmann, 2016). These cartilaginous fishes inhabit a wide diversity of ecosystems, from rivers, lakes, estuaries, coastal waters, reefs, open ocean to the deep sea, though tropical to polar latitudes (Kyne and Simpfendorfer, 2007). In addition, it has been pointed out that chondrichthyans (sharks, rays, and chimaeras) do not usually live in waters deeper than 3,000 m due to the low food availability and high energy costs required to maintain their vital functions (Priede et al., 2006). About 85% of elasmobranch species are found at depths shallower than 2,000 m, and 50% are distributed above 200 m (Helfman et al., 2009). Cartilaginous fish richness in Mexico consists of approximately 214 species, 8 chimaeras, 95 batoids and 111 sharks, representing 18% of the total number of species worldwide (Del Moral-Flores et al., 2015). Mexican sharks, specifically, represent 21.5% of the total species reported worldwide (Castillo Géniz and Tovar-Ávila, 2021).

Sharks have been employed as a resource in Mexico for hundreds of years. Their use and exploitation date back to pre-Columbian times (before 1516), as evidenced by the discovery of fossil teeth in the south-southeast region of the country. Species such as white sharks, giant sharks or megalodons, as well as various other carcharhinids, were used as religious offerings in Mayan communities, although teeth may also have been used in armaments and as ornamental pieces (De Borhegyi, 1961). Concerning their use as food items, sharks are one of the most important fishery resources along the Mexican coasts. Their abundance and presence vary according to migratory patterns, mainly in the winter, when species from temperate or cold zones migrate to warmer waters, resulting in a greater diversity of species in Mexico (Fernández et al., 2011).

MEXICAN FISHERIES

Mexico is considered one of the world's main shark production countries, alongside Spain, Indonesia, India and Taiwan, where both artisanal and industrial fisheries provide this fishing resource (Castillo Géniz and Tovar-Ávila, 2021).

Artisanal fisheries are carried out near coastal areas and involve minor technological developments. This type of fishing is significant in Mexico, as it is carried out in 16 states of the Mexican Republic, namely Sonora, Sinaloa, Baja California, Baja California Sur, Nayarit, Jalisco, Colima, Michoacán, Oaxaca, Chiapas, Tamaulipas, Veracruz, Tabasco, Yucatán, Campeche and Quintana Roo (Castillo-Géniz et al., 1998) and by over 74,000 artisanal vessels (Gutiérrez-Zavala and Cabrera-Mancilla, 2019). These vessels display as main characteristics the employment of a large workforce, the use of a large number of vessels, categorized as a small fleet (less than 10 tonnes), minimal crew, and poorly technified (Flores-Hernández and Ramos-Miranda, 2004). This type of fishery extracts several species of sharks, making it one of the most important resources and fishing efforts in the country. Species such as pelagic thresher shark *Alopias pelagicus* (Nakamura 1935), silky shark *Carcharhinus falciformis* (Bibron 1839), bull shark *Carcharhinus leucas* (Valenciennes 1839), blue shark *Prionace glauca* (Linnaeus 1758), mako shark *Isurus oxyrinchus* (Rafinesque 1810), and scalloped hammerhead shark *Sphyrna lewini* (Griffith and Smith 1834) are routinely consumed by humans in Mexico, and their use is diverse, including meat, skin, and vertebrae uses (Martínez-Ortíz and García-Domínguez, 2013). Some sharks are also used in fishmeal production, mainly smaller species such as Mexican hornshark *Heterodontus mexicanus* (Taylor and Castro-Aguirre 1972).

Recently, about 3,665 artisanal fishing vessels were identified as performing shark and ray captures, although many others focused on the scale fishery also capture sharks, leading to underestimations in this regard (Fernández et al., 2011). Sharks are mainly marketed fresh (70% fresh, 22% frozen and 8% dried-salted; Cid et al., 2000). Fishery production reached 5,651 tonnes in 2003, represented by 27 captured species (Fernández et al., 2011). By 2006, production from the Mexican Pacific alone consisted in 30 species, with the pelagic thresher *C. falciformis* and oceanic thresher *A. pelagicus* and *A. vulpinus* (thresher shark) comprising the main captured species contributing with 20,960 tonnes (Seijo et al., 2013).

Between 2009 and 2018, shark catches in Mexico have increased, reaching almost 48 thousand tons in 2017 (Figure 9.1), according to official available statistics (CONAPESCA, 2018).

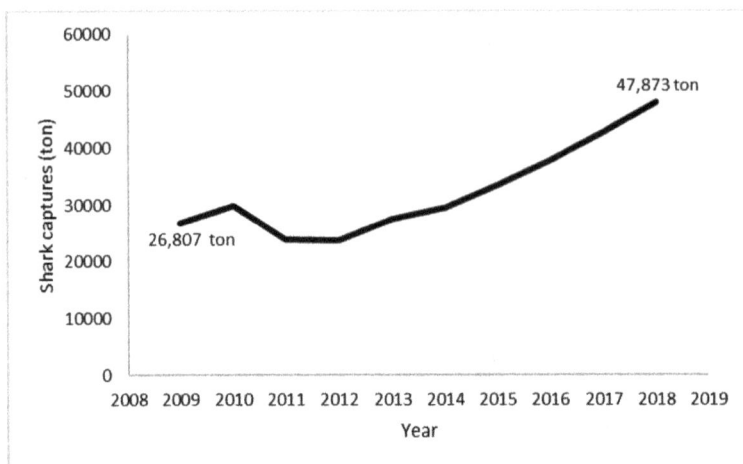

Figure 9.1   Total shark captures reported between 2009 and 2018 in Mexico, according to CONAPESCA, 2018.

SHARKS AND METALS

The presence of metals in the environment comprise an inorganic chemical hazard concerning living organisms, although some can be part of organometallic molecules and play essential roles in metabolism (Gilman and Leeper, 1951; Queen et al., 2020; Surgiewicz, 2012).

Studies have demonstrated the worldwide presence of metallic elements or metalloids in sharks, such as As, Cd, Pb, Hg, in sharks from the Brazilian Amazon (Souza-Araujo et al., 2021), Cd, Cr, Cu, Fe, Hg, and Pb from southwestern Brazil (Martins et al., 2021), Cd and Hg from Ecuador coast (Castro-Rendón et al., 2021), As, Cd, Cr, Cu, Fe, Hg, Mn, Ni, Pb, Se, and Zn from the northern Kuwait coast (Moore et al., 2015), Hg in the east coast of South Africa (McKinney et al., 2016), S, Ti, V, Cr, Mn, Fe, Co, Cu, Ni, Zn, As, Rb, Sr, Zr, Nb, Mo, Cd, Sn, Sb, Cs, Ba, La, Ce, Pr, Nd, Sm, Eu, Gd, Tb, Dy, Ho, Er, Tm, Yb, Lu, Pb, Th, U in Mediterranean Sea sharks (Bevacqua et al., 2021), As, Cd, Cu, Fe, Se, and Zn in southwestern Australian Waters sharks (Gilbert et al., 2015), Cd, Cr, Cu, Fe, Hg, and Pb in southwest Brazil (Martins et al., 2021), Cd and Hg in sharks from the Ecuadorian coast (Castro-Rendón et al., 2021), and Hg in the east coast of South Africa.

The main metal acquisition pathway in sharks is through the dietary route (Mathews and Fisher, 2009). Accumulation occurs when physiological mechanisms in place for regulating body levels are not equalized through excretion, mainly when metal and metalloid incorporation is higher, resulting in bioaccumulation (Frías-Espericueta et al., 2014; Nunez-Nogueira et al., 2006; Núñez-Nogueira, 2005).

Although harmful metal contamination impacts on sharks are not yet well studied, various alterations have been postulated including embryo malformation, as in the case of blue shark *Prionace glauca* pups caught off the coasts of Baja

California Sur and the Gulf of California (Galván-Magaña et al., 2011; Rodriguez-Romero et al., 2019), due to maternal-fetal metal transference, reported for some shark species (Dutton and Venuti, 2019; Frías-Espericueta et al., 2014). This indicates significant reproductive and population implications.

Due to their important role in the oceans and as a natural and fishery resource and significant consumption in Mexico, public health implications due to shark meat consumption must be considered. Mercury, however, is the most frequently studied element in Mexican sharks (Álvaro-Berlanga et al., 2021; Hurtado-Banda et al., 2012; Medina-Morales et al., 2020; Mendoza-Díaz et al., 2013; Murillo-Cisneros et al., 2021; Núñez et al., 1998; Núñez-Nogueira 2005; Ruelas-Inzunza et al., 2011a, b), followed by rare Cd and Pb assessments (Frías-Espericueta et al., 2014; Jonathan et al., 2015; Lara et al., 2020; Núñez-Nogueira, 2005; Ruelas-Inzunza et al., 2020; Terrazas-López et al., 2016).

Sharks fisheries in Mexico are under stress due to several reasons, including aquatic pollution caused by metals. Páez-Osuna et al. (2017) reported the frequent presence of Cd in Mexican sharks. Due to the scarce assessments concerning Cd in sharks from Mexico, studies in this regard were compiled to identify which shark species have been studied, what is their geographical distribution, which target organs have been identified, what differences exist concerning other studies worldwide and what are the potential human health risks due to their consumption under the established limits for ingestion at both the national and international levels. Due to increasing scientific interest in the levels of this metal in shark species that inhabit and are caught off the coast of Mexico, this study will be useful in indicating niches for future studies assessments.

For comparative purposes, all Cd values reported in the literature for sharks captured in Mexico on a dry basis were multiplied by a conversion factor of 0.245 to transform them into wet weight. This conversion factor has been established as the mean value for sharks (McKinney et al., 2016) and elasmobranchs in general considering a body moisture percentage ranging from 70% to 80% (Hauser-Davis et al., 2021). Thus, all concentrations are reported in milligram per kilogram (mg kg^{-1}) on a wet basis, unless otherwise noted.

CADMIUM

Cadmium is present in the environment due to natural processes such as rock and soil erosion or volcanic activity, as well as the waste derived from its use as a coating factor for metals and plastics and in the manufacture of batteries, and various technological products, as well as in the mining and petroleum industry (Simpson, 1978). Contributions from urban discharges, effluents and wastewater, as well as by the industries associated with pesticides, metallurgy, mining and atmospheric emissions as a result of fossil combustion, has increased the level of this metal in aquatic environments, resulting in higher aquatic flora and fauna exposure and at high levels throughout trophic chains worldwide (Bhat et al., 2019; Pacyna and Pacyna, 2001; Ruelas-Inzunza et al., 2020; Seixas et al., 2009; Vosylienė, 2007).

According to the Agency of Toxic Substances and Diseases Registry of the US Department of Health and Human Services (Faroon et al., 2012), Cd is considered one of the most dangerous elements to human health, alongside other elements such as As, Pb and Hg. Health effects due to Cd exposure range from bone malformation due to high bone accumulation to cancer and genotoxic effects (Kumar and Singh, 2010). Cadmium intake also leads to high accumulation in kidneys, preventing normal blood filtration and resulting in functional alterations (Faroon et al., 2012). Likewise, its presence is also associated with the liver (Faroon et al., 2012; Kumar and Singh, 2010). When exposure occurs at relatively low levels, Cd is generally neutralized in the kidney and liver through binding with regulatory cytoplasmic proteins known as metallothioneins. However, when their complexation capacity is exceeded, bioaccumulation takes place and toxic effects develop (Endo et al., 2002; Faroon et al., 2012; Nunez-Nogueira et al., 2006; Seixas et al., 2009). In fish, Cd exposure leads to different structural and functional damage in the liver, intestine, kidney and gills, as well as thyroid function alterations and bone malformations, among others (Kumar and Singh, 2010). In fish intestines, it is mainly associated with endocytosis processes, while at the gill level it is associated with absorption by chlorinated cells through calcium channels by passive diffusion (Kumar and Singh, 2010).

CADMIUM IN MEXICAN SHARKS

A review on Cd contamination in sharks captured in Mexican waters between 1994 and 2018 was conducted. A total of ten publications were detected in the last 24 years, covering a total of eleven species. The whale shark *Rhincodon typus* was the most representative, with 133 dermal and 20 kidney samples analyzed, obtained from different areas of the Baja California Peninsula (Pancaldi et al., 2019b; Vélez-Alavez et al., 2013), followed by the brown smooth-hound *Mustelus henlei* (Gill 1863) from Baja California Sur, with 83 muscle samples (Pantoja-Echevarría et al., 2020).

From a geographical distribution perspective, the higher number of species and assessments are from the Mexican Pacific, particularly from the northwest and central Pacific (Figure 9.2). The most represented geographical areas are those adjacent to the Baja California Peninsula, with reports from Nayarit, Colima, Jalisco, Sinaloa, Baja California Sur and Baja California in the Pacific. In the case of the Gulf of Mexico, samples were obtained from the coast of Veracruz.

At the species level, of the ten reports obtained in this assessment, only two are from the Gulf of Mexico, while eight are from the Mexican Pacific. One report for blue shark *P. glauca,* brown smooth-hound *M. henlei,* Pacific sharpnose shark *R. longurio,* Atlantic sharpnose shark *R. terraenovae* (Richardson 1836) and shortfin mako *I. oxyrinchus* each were obtained. No studies for species sampled from the South Pacific (coasts of Guerrero, Oaxaca and Chiapas) and the Mexican Caribbean were obtained.

Concerning analyzed tissues, muscle was the most studied tissue, totaling 451 samples, followed by liver (273 samples). The least studied tissues or organs were testicles, stomach, heart and filtering parches, all taken from a whale shark (Pancaldi et al., 2019b), the placenta, comprising 15 samples obtained from Pacific sharpnose shark *R. longurio* (Frías-Espericueta et al., 2014), the brain and gills, consisting of 43 samples obtained from *R. typus* (Pancaldi et al., 2019b), *C. limbatus* (blacktip shark) and Atlantic sharpnose shark *R. terraenovae* (Núñez-Nogueira 2005) and, finally, kidneys, with 56 samples obtained from *P. glauca* (Vélez-Alavez et al., 2013), *I. oxyrinchus* (Barrera-García et al., 2013) and *R. typus* (Pancaldi et al., 2019b). This indicates a significant knowledge gap and an interesting opportunity to expand Cd assessments in organs other than shark muscle and liver, as well as in many other species present in Mexican waters. Figure 9.2 indicates the maximum individual cadmium concentrations reported in muscle tissue and liver from Mexican sharks.

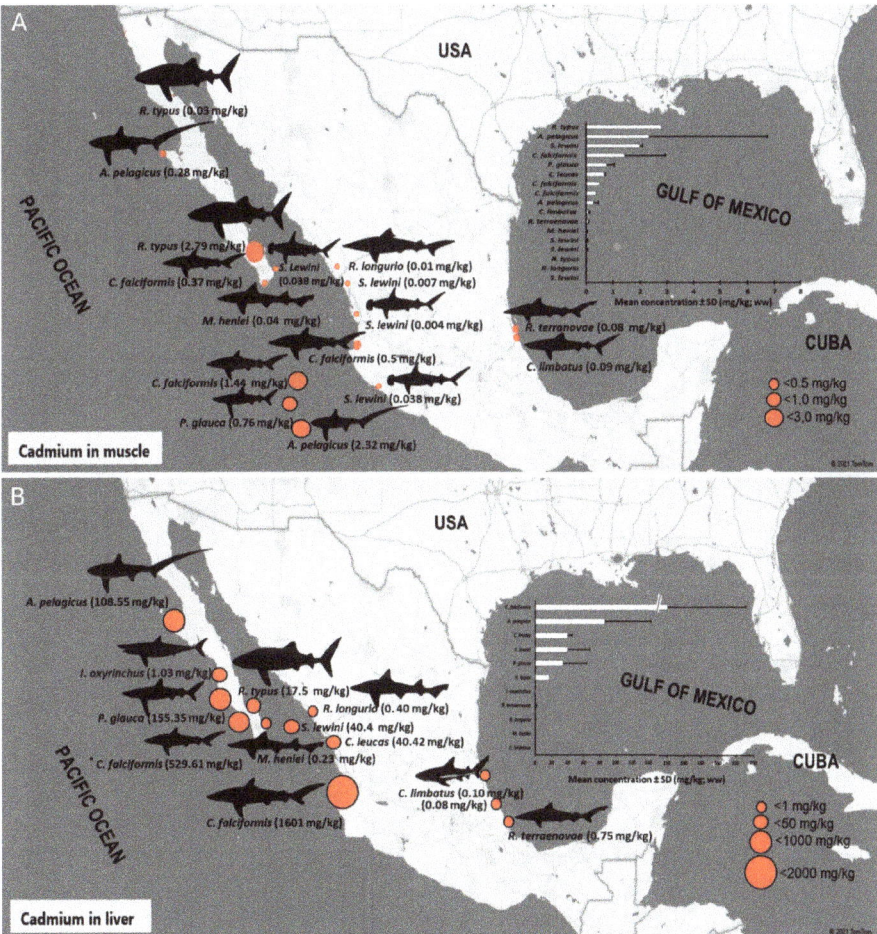

Figure 9.2 Maximum individual cadmium concentrations reported in muscle tissue (A) and (B) liver in Mexican sharks. The inner chart displays the unique or mean values per species, respectively.

Table 9.1 Cadmium concentrations (mean ± standard deviation; in mg kg⁻¹ wet weight) in the muscle of sharks from Mexican waters.

Order	Species	Mexican state	Location	Cd	n	References
Carcharhiniformes	Prionace glauca	Colima	Manzanillo	0.76 ± 0.30	30	Álvaro-Berlanga et al., 2021
		Colima	Manzanillo	1.44 ± 1.50	30	Álvaro-Berlanga et al., 2021
	Carcharhinus falciformis	Baja California	Todos Santos	0.37	6	Terrazas-López et al., 2016
		Jalisco	Bahía de Tenacatita	1.30 ± 1.14*	43	Vega-Barba, 2018
	Carcharhinus leucas	Sinaloa	Mazatlán	0.69 ± 0.03**†	1	Ruelas-Inzunza and Páez-Osuna, 2007
	Carcharhinus limbatus	Veracruz	Costa Golfo de México	0.086 ± 0.042†	21	Núñez-Nogueira, 2005
	Mustelus henlei	Baja California	Las Barrancas	0.04 ± 0.01	83	Pantoja-Echevarría et al., 2020
	Rhizoprionodon longurio	Baja California Sur	Mazatlán	0.01 ± 0.01	15	Frías-Espericueta et al., 2014
	Rhizoprionodon terraenovae	Veracruz	Costa Golfo de México	0.083 ± 0.024†	21	Núñez-Nogueira, 2005
	Sphyrna lewini	Sinaloa	Mazatlán	1.98 ± 0.10**†	1	Ruelas-Inzunza and Páez-Osuna, 2007
		Sinaloa	Las Cabras	0.007 ± 0.008	64	Ruelas-Inzunza et al., 2020
		Nayarit	Isla Isabel	0.004 ± 0.005	10	
		Colima	Bahía Navidad	0.038 ± 0.024	22	
		Baja California Sur	Cabo San Lucas	0.038 ± 0.032	20	
	Alopias pelagicus	Colima	Manzanillo	2.32 ± 4.40	30	Álvaro-Berlanga et al., 2021
Lamniformes	Isurus oxyrinchus	Baja California Sur	Bahía Tortugas	0.12 ± 0.13	13	Lara et al., 2020
		Baja California Sur	Bahía Tortugas	0.28 ± 0.18	21	
		Baja California Sur	Isla Magdalena	0.0001	20	Vélez-Alavez et al., 2013
Orectolobiformes	Rhincodon typus	Baja California Sur	Punta Bufeo	0.03	1	Pancaldi et al., 2019b
		Baja California Sur	Bahía La Paz	2.79	1	

n = sample size; *Mean ± sd calculated by adding the means from juveniles plus adults and standard deviation difference was calculated according to Williams (1993) for differences between two means. **Mean values from duplicated analyses of one tissue sample. †Modified values from dry weight to wet weight by a factor of 0.245.

CADMIUM IN SHARK MUSCLE

Specifically concerning muscle, ten studies reported the presence of Cd in three shark orders from Mexico, namely Carcharhiniformes, Lamniformes and Orectolobiformes, with Carcharhiniformes as the most represented order, comprising seven species (*Carcharhinus falciformis, C. limbatus, Mustelus henlei, Prionace glauca, Rhizoprionodon longurio, R. terraenovae* and *Sphyrna lewini*). Lamniformes and Orectolobiformes were only represented by two (*Alopias pelagicus* and *Isurus oxyrinchus*) and one (*R. typus*) species, respectively (Table 9.1). Concerning total number of muscle samples analyzed for Cd, *S. lewini* (116 muscle samples) and *M. henlei* (83 samples) were the most representative. The least studied species was *R. typus*, with only two reports available in 2019. This indicates that studies performed in different Mexico regions are still limited in terms of shark diversity, representing only 9% of the 111 present species in national waters and 19.2% of the 52 commonly caught species of fishing or commercial interest (CONAPESCA, 2018).

Cadmium concentrations ranged from 0.0001 mg kg^{-1} in *I. oxyrinchus* to 2.79 mg kg^{-1} in *R. typus* (Table 9.1). The higher Cd concentrations noted in whale sharks, which come from the waters of the Pacific Northwest, indicate the high Cd presence in those waters. In this regard, Ramírez-Ayala et al., (2020) associated the presence of Cd in areas of the Gulf of California with an upwelling enriched with this metal derived from the geotectonic activity of local hydrothermal vents. Furthermore, *P. glauca* and *A. pelagicus* from the coast of Colima presented high mean Cd concentrations, 1.44 ± 1.50 mg kg^{-1} and 2.32 ± 4.40 mg kg^{-1}, respectively. This area in the central Mexican Pacific seems to play an important role in Cd concentrations detected in sharks, as individuals sampled from this area show the highest muscle and liver values, representing an important contribution throughout the trophic chain in this marine area. Thus, research on local biomagnification and bioaccumulation processes and the monitoring of different food chain links, as well as sedimentary biogeochemistry assessments and studies on the continental contributions of this element on the coasts of Jalisco, Colima, Nayarit and Michoacán are warranted, which could aid in elucidating Cd sources and temporal-spatial distribution patterns.

It should be noted that the only two shark species studied in the Gulf of Mexico were *C. limbatus* and *R. terraenovae*, both presenting similar mean Cd levels of 0.086 ± 0.042 mg kg^{-1} and 0.083 ± 0.024 mg kg^{-1}, respectively, in muscle. These values are below the range observed for several other Pacific species, surpassed only by the Carcharhiniformes species *M. henlei* (0.04 mg kg^{-1}, Baja California Sur), *S. lewini* (0.004–0.007 mg kg^{-1}, Nayarit-Sinaloa), and Lamniformes species *I. oxyrinchus* (0.0001 mg kg^{-1}, Baja California Sur).

In general, Carcharhiniformes and Lamniformes presented Cd concentrations < 0.1 mg kg^{-1}, although some values reached up to 2.32 mg kg^{-1} (Table 9.1). The concentrations determined in the Gulf of Mexico seem to be associated with Cd levels in water and sediment reported during the 90s. Between 1983 and 1990, Cd presented a mean range from 0.001 ± 0.005 mg L^{-1} to 2.49 ± 1.60 mg L^{-1} in waters off the coast of Veracruz, higher than those recently reported in waters of the southern Gulf of Mexico, which do not exceed 0.005 mg L^{-1} (Benítez et al., 2012).

In the Pacific, Cd concentration in sharks reported in the literature since 2013 (Table 9.1) are in agreement with the lowest concentrations present in water ranging from 0.05 to 0.09 mg L^{-1}, depending on the time of year (Pérez-Moreno et al., 2016). These data suggest a relationship between the presence of this metal in water and its presence in muscle, although biomagnification processes cannot be ruled out.

A relationship with sampling periods is also observed concerning sediment reports, as cadmium levels in coastal areas of the Gulf of Mexico from 1983 to 1990 were higher: 0.02 ± 0.01 mg kg^{-1} dry weight (d.w.) in Tamaulipas and 1.64 ± 0.26 mg kg^{-1} d.w. in Veracruz (Villanueva and Botello, 1992), decreasing to 0.28 mg kg^{-1} d.w. in the south of the Gulf of Mexico (Benítez et al., 2012) in 2012. Furthermore, Cd levels in Pacific sharks reported since 2013 (except *R. typus*) are consistent with the lower values observed in more recent central Pacific sediments: no greater than 0.42 ± 0.23 mg kg^{-1} d.w. (Ramírez-Ayala et al., 2021).

Compared to other worldwide assessments, Cd concentrations revealed similarities and differences depending on the assessed species and geographical region. Two Mexican species (*C. limbatus* and *S. lewini*) have also been evaluated in Papua New Guinea and in the North Atlantic (only *S. lewini*). *Rhizoprionodon acutus* and *S. zygaena* have also been evaluated in Papua New Guinea and in the Ionian Sea in Italy, respectively (Table 9.2).

Concerning *C. limbatus*, Cd values observed in the Gulf of Mexico are at the lower limit of the range reported by Powell et al. (1981) and Powell and Powell (2001). These authors recorded concentrations between 0.01 and 2,500 mg kg^{-1} in muscle tissue. With regard to *S. lewini*, Cd values observed in the waters of the Gulf of Mexico were slightly above those reported by Powell et al. (1981). However, between 1977 and 1987, these authors observed a considerable increase in Cd concentrations in hammerhead sharks, reaching 370 mg kg^{-1} (Powell and Powell, 2001). Likewise, the values reported in Mexican waters for this species were lower than those reported in North Atlantic waters by Windom et al. (1973). Concerning the same genera found in Mexico, *R. acutus* from Papua New Guinea present values ranging from 0.01 to 330 mg kg^{-1}, well above the 0.04 and 0.08 mg kg^{-1} observed in *R. longurio* and *R. terraenovae* in Mexico, respectively. For *S. zygaena*, the concentration range observed in Papua New Guinea was of 0.02 to 0.03 mg kg^{-1}, similar to those observed in *S. lewini* from the Mexican Pacific coast (Table 9.2).

Another species that has been studied in other regions is *C. falciformis*, which had Cd levels of 0.098 mg kg^{-1} d.w. (0.04 mg kg^{-1} wet weight [w.w.]) in the North Atlantic (Windom et al., 1973), lower than the range observed for Mexico's central Pacific and Northwest coasts, reported as a minimum of 0.37 mg kg^{-1} (Table 9.1).

According to Powell et al. (1981) and Powell and Powell (2001), the high presence of Cd in sharks from Papua New Guinea is due to the significant contribution of mining waste discharges, which were significantly dumped decades ago and have contaminated the coastal waters. On the other hand, Windom et al. (1973) observed that Cd levels were higher in cartilaginous fish compared to bony fish in North Atlantic waters. No significant differences were noted between

Table 9.2   Cadmium concentrations (mean ± standard deviation; in mg kg^{-1} wet weight) in the muscle of Carcharhiniform sharks of the world

Species	Location	Cd	n	References
Carcharhinus limbatus	Papua New Guinea-W. Pacific	0.01–0.05 10.0–2500.0	88 1425	Powell et al., 1981; Powell and Powell, 2001
Carcharhinus falciformis	North Atlantic	0.24*	2	Windom et al., 1973
Galeorhinus australis	Australia-Tasmania	<0.05	6	Eustace, 1974
Galeorhinus galeus	Belgium-Celtic Sea	0.4	6	Domi et al., 2005
	Ireland	<0.02	2	Vas, 1991
Galeus melastomus	Belgium-Celtic Sea	0.7	7	Domi et al., 2005
	Israel-Mediterranean Sea	0.07	63	Hornung et al., 1993
	Scotland	0.08	7	Vas, 1991
Halaelurus bivis	Argentina-Bahia Blanca Estuary	0.15	73	Marcovecchio et al., 1991
Mustelus asterias	Belgium-Celtic Sea	<0.17	7	Domi et al., 2005
Mustelus schmitti	Argentina-Bahia Blanca Estuary	0.14	570	Marcovecchio et al., 1991
Prionace glauca	USA–North Atlantic	0.19 ± 0.15	5	Hauser-Davis et al., 2021
	Ecuador	0.04 ± 0.03	46[m]	Castro-Rendón et al., 2021
		0.04 ± 0.02	34[f]	
	England	0.45	5	Vas, 1991
	Peru	0.02 ± 0.01	19[m]	Reátegui-Quispe and Pariona-Velarde, 2019
		0.02 ± 0.01	6[f]	Reátegui-Quispe and Pariona-Velarde, 2019
Rhizoprionodon acutus	Papua New Guinea-W. Pacific	1.–0.33 10.0–330.0	65 314	Powell et al., 1981; Powell and Powell, 2001
Scyliorhinus canicula	Belgium-Celtic Sea	1.1	8	Domi et al., 2005
	England	1.08	2	Vas, 1991
	Ireland	0.78	13	
Scyliorhinus stellaris	Ireland	<0.02	1	Vas, 1991
Sphyrna lewini	Papua New Guinea-W. Pacific	0.01–0.03 10.0–370.0	21 98	Powell et al., 1981; Powell and Powell, 2001
	North Atlantic	0.098*	1	Windom et al., 1973
Sphyrna zygaena	Italy-Ionian Sea	0.03 ± 0.005 (0.02–0.03)	4	Storelli et al., 2003

m = males; f = females; n = sample size. * Original dry weight values transformed by 0.245 factor to wet weight values.

coastal and marine species, suggesting that the differences could be related to the size of the organisms and physiological aspects associated to this growth.

P. glauca is the species found in Mexico and is the most studied in other regions, mainly in the North Atlantic, Ecuador, Peru and England (Table 9.2). Cadmium concentrations ranged from 0.02 mg kg^{-1} in Peru to 0.45 mg kg^{-1} in England, lower than the 0.75 mg kg^{-1} detected on the coasts of Colima, Mexico (Table 9.1). Vas (1991) considered that the slight difference observed between the Cd concentrations in different sharks sampled from British coasts from contaminated and uncontaminated areas is the result of physiological metal regulation mechanisms, such as metallothionein-like proteins.

These proteins are rich in cysteine and thiolic groups and display a high affinity for metal cations, playing a fundamental role in metal regulation and intracellular chelation. As a result, metal ions are neutralized up to a certain limit, thus avoiding adverse effects (Bouquegneau et al., 2003; Hauser-Davis et al., 2021; Nunez-Nogueira et al., 2006). Recently, Hauser-Davis et al. (2021) studied the presence of metallothioneins in blue sharks, reporting that this molecule detoxifies various metals in muscle in addition to Cd, such as As, Cs, Pb, Se, Ti and Zn, while also detoxifying other elements, such as Cu and Hg, in the liver. Parallel to the defensive responses associated with metallothionein-like proteins, intracellular antioxidant responses associated with the production of reactive oxygen species (ROS), which are capable of inducing various deleterious alterations, have also served as markers to assess exposure to toxic metals in aquatic organisms. Thus, several proteins and enzymes such as glutathione–GSH (Hauser-Davis et al., 2021; Kehrig et al., 2016), superoxide dismutase–SOD, catalase–CAT, glutathione peroxidase–GPx (Das et al., 2019), as well as cellular endpoints, such as lipid peroxidation–LPO (Kumar and Singh, 2010) have been employed to determine the antioxidant efficiency against this type of pollutants.

Many Carcharhiniformes species have been evaluated concerning muscle tissue Cd levels (Table 9.2), indicating a concentration range from <0.02 mg kg^{-1} (nursehoung shark *Scyliorhinus stellaris* (Linnaeus 1758), Vas, 1991) to 1.1 mg kg^{-1} (*S. canicula*, Domi et al., 2005; Vas, 1991). These results indicate that, in general, Cd muscle levels in Carcharhiniformes from Mexico are within the range observed for this group of sharks globally, including other orders such as Heterodontiformes, Squaliformes and Squatiniformes (Table 9.3). As for Lamniformes, previous reports are noted for the Porbeagle shark [*Lamna nasus* (Bonnaterre 1788)], a close relative of *I. oxyrinchus* and *A. pelagicus*, from Canadian (Beckett and Freeman, 1974) and Irish waters (Vas, 1991). In both cases, Cd values were within the low-medium range observed in Lamniformes from Mexico, well below the maximum of 2.32 mg kg^{-1} reported in *A. pelagicus* from Colima, Mexico. Differences among species seem to be related to their geographic and water column (depth) distributions, and to a lesser degree with their diets, since these species share ecological niches and exhibit predatory demersal habits. They do, however, reach different depths, and *L. nasus* does not inhabit equatorial waters and feeds on smaller animals than *I. oxyrinchus*, which is cosmopolitan and distributed throughout temperate and tropical waters (Compagno, 1990). Furthermore, sharks that reach higher depths may be exposed to higher metal concentrations because of metal transformation, remineralization and bioavailability processes that occur at the benthic level (Mille et al., 2018). Thus, all the aforementioned factors may contribute to the observed varying Cd accumulation responses.

Table 9.3 Cadmium concentrations (mean ± standard deviation and range; in mg kg^{-1} wet weight) in muscle of sharks from different orders (non-Carcharhiniformes) worldwide

Order	Species	Location	Cd	n	References
Heterodontiformes	*Heterodontus portusjacksoni*	Australia–Sydney	0.005	2	Gibbs and Miskiewicz, 1995
Hexanchiformes	*Hexanchus griseus*	Israel–Mediterranean Sea	0.04	6	Hornung et al., 1993
	Notorhinchus spp	Argentina–Bahia Blanca Estuary	0.18	14	Marcovecchio et al., 1991
Lamniformes	*Lamna nasus*	Canada	0.02 ± 0.01	22	Becket and Freeman, 1974
		Ireland	0.79	1	Vas, 1991
Orectolobiformes	*Chiloscyllium plagiosum*	Hong Kong–Southern waters	0.01	35	Cornish et al., 2007
Squaliformes	*Centrophorous granulosus*	Israel–Mediterranean Sea	0.06	33	Hornung et al., 1993
	Centroscymnus crepidator	Australia-Tasmania	0.003	10	Turoczy et al., 2000
	Centroscymnus owstonii		0.013	11	
	Deania calcea		0.01	18	
	Etmopterus spinax	Israel–Mediterranean Sea	0.08	8	Hornung et al., 1993
		Scotland	0.25	1	Vas, 1991
	Scymnorhinus licha		<0.02	1	
	Somniosus rostratus	Israel–Mediterranean Sea	0.07	8	Horung et al., 1993
	Squalus acanthias	Belgium–Celtic Sea	<0.16	6	Domi et al., 2005
	Squalus mitsukurii	Japan	0.007	59	Taguchi et al., 1979
Squatiniformes	*Squatina guggenheim*	Brazil	0.03 ± 0.01[†]	9	Martins et al., 2021

n = sample size; [†]Original dry weight values transformed by 0.245 factor to wet weight values.

Finally, the Orectolobiformes group is represented by the whale shark *R. typus* in Mexican waters (Table 9.1) and by *Chiloscyllium plagiosum* [Anonymous (Bennett) 1830] (white-spotted bamboo shark) in the coasts of Hong Kong (Table 9.3). Cornish et al. (2007) reported concentrations of 0.01 mg kg^{-1} in the muscle of *C. plagiosum*, within the range of that observed in *R. typus* from the Mexican Pacific Northwest. However, a marked difference between the biology of the bamboo shark and the whale shark is noted, even though both belong to the same order. The bamboo shark is one of the smallest sharks (no larger than 95 cm, or 37 inches). It is nocturnal, presenting benthic and reef habits, and its diet is mainly based on small invertebrates and fish (Compagno, 1990) while the whale shark is recognized as the largest fish in the world, reaching sizes of up to 12 meters in length (472 inches), and displaying pelagic and planktonic feeding habits (Compagno, 1990). Thus, metal variability, in general, seems to be more related to the physiological accumulation and detoxification capacities (Vas, 1991), although other authors disagree, such as Adel et al. (2016), who consider that metal accumulation depends on shark trophic level, longevity, and predatory behavior. These differences seem to depend on the metal, as reported for Hg in some pelagic sharks, such as *I. oxyrinchus* and *A. vulpinus* (Teffer et al., 2014).

In this sense, the association between metal concentrations and shark length is also interesting. One of the main metal exposure routes is through food ingestion, which is directly affected by length and, indirectly, by age (Chouvelon et al., 2014). In general, this relationship is considered linear, allowing for interindividual variability to be explained and reducing the effects of deviations related to ecosystem aspects (Mille et al., 2018). However, in case of Cd, this type of correlation exhibits highly varied results among shark species. As presented in Table 9.4, three species of sharks sampled from Mexico exhibited correlations between Cd concentrations in muscle and size. Concerning liver, only two species exhibited positive correlations when comparing sexes (Table 9.4). It is interesting to note that *A. pelagicus*, *C. falciformis* and *R. typus* exhibited positive correlations for females only, where increasing size resulted in increasing Cd concentrations in muscle, while the opposite was observed for males (Alvaro-Berlanga et al., 2021; Pancaldi et al., 2021b; Terrazas-López et al., 2016). The reason for these contradictory results is not clear but may be due to the distinct maturation processes of each sex. Therefore, further studies in this regard are necessary. In contrast, male *A. pelagicus*, exhibited a positive correlation between Cd and size. The other eight species studied in Mexico exhibited significant correlations between these variables, or no previous reports in this regard are available. Considering that the total number of individuals analyzed so far is low, it is difficult to develop this type of correlation and determine the degree of influence that growth has on cadmium uptake in general. Thus, Cd monitoring in sharks should be expanded to cover both sexes and various maturity stages. This, however, is only achievable with the support of the fishing communities and specimen donation, thus allowing enough samples without needlessly sacrificing individuals from an already exhausting resource affected by overfishing, pollution, and alarming human actions, such as finning.

The Cd results observed in Mexico suggest a lower presence of this metal in national coastal areas compared to other areas of the globe. However, even at low levels, this element may represent a consumption risk, discussed later on in this chapter, especially for fishing populations that frequently ingest the product of their work in high amounts, resulting in higher exposure to this pollutant (see the health risk assessment section). Unfortunately, few species are also studied in other parts of the world regarding Cd in muscle, hindering a more comprehensive and detailed comparative analysis of Cd in sharks.

Table 9.4 Relationship between cadmium concentration in muscle or liver and body length in Mexican sharks

Species	Muscle	Liver	References
Alopias pelagicus	ND	(+)	Lara et al., 2020
	(+)[m]	(+)[m]	Álvaro-Berlanga et al., 2021
Carcharhinus falciformis	(+)	(+)	Vega Barba, 2018
	NC	NC	Álvaro-Berlanga et al., 2021
	(+)[f]	(+)[f]	Terrazas-López et al., 2016
	(−)[m]	(+)[m]	Terrazas-López et al., 2016
Carcharhinus leucas	ND	ND	Ruelas-Inzunza and Páez-Osuna, 2007
Carcharhinus limbatus	NC	NC	Mendoza-Díaz et al., 2013
	NC	ND	Núñez-Nogueira, 2005
Isurus oxyrinchus	ND	ND	Vélez-Alavez et al., 2013
Mustelus henlei	NC	NC	Pantoja-Echeverria et al., 2020
Prionace glauca	ND	ND	Galván-Magaña et al., 2011
	ND	ND	Rodríguez-Romero et al., 2019
	NC	NC	Álvaro-Berlanga et al., 2021
	ND	ND	Barrera-García et al., 2013
Rhincodon typus	ND	ND	Pancaldi et al., 2021a
	(+)[f]	ND[f]	Pancaldi et al., 2021b
	(−)[m]	ND[m]	Pancaldi et al., 2021b
Rhizoprionodon longurio	ND	ND	Frías-Espericueta et al., 2014
Rhizoprionodon terraenovae	NC	ND	Núñez-Nogueira, 2005
Sphyrna lewini	ND	ND	Ruelas-Inzunza and Páez-Osuna, 2007

(+) positive correlation. (−) negative correlation. f = females, m = males; NC = No correlation observed, ND = No data available.

CADMIUM IN LIVER

The presence of cadmium in liver has been reported in the same eleven shark species previously described for muscle samples. The total number of liver samples is of 233 individual sharks. The two species *Carcharhinus falciformis* and *Mustelus henlei* were the most assessed regarding liver samples, investigated in 51 and 52 individuals, respectively. Hammerhead sharks and bull sharks were the less sampled, with only one specimen each (Table 9.5).

Cadmium concentrations in the liver range from minimum values such as in *Isurus oxyrinchus* (0.0001–1.03 mg kg^{-1} w.w.; Vélez-Alavez et al., 2013) to 0.18–1.60 mg kg^{-1} w.w. in *C. falciformis*; Vega-Barba, 2018), respectively. In fact,

Table 9.5 Cadmium concentrations (mean ± standard deviation and range; in mg kg^{-1} wet weight) in the liver of sharks from Mexican waters

Order	Species	Location	Cd	n	References
Lamniformes	*Alopias pelagicus*	Baja California Sur	86.53 ± 56.44	34	Lara et al., 2020
	Isurus oxyrinchus	Baja California Sur	0.0001–1.03	20	Vélez-Álvarez et al., 2013
	Carcharhinus falciformis	Jalisco	253.9 ± 453.0*	43	Vega-Barba, 2018
		Baja California Sur	127.11 ± 0.21	8	Terrazas-López et al., 2016
	Carcharhinus leucas	Sinaloa	40.42 ± 5.39**†	1	Ruelas-Inzunza and Páez-Osuna, 2007
	Carcharhinus limbatus	Veracruz	0.015 ± 0.002†	4	Núñez-Nogueira, 2005
		Veracruz	0.083 ± 0.011†	12	Mendoza-Díaz et al., 2013
Carcharhiniformes	*Prionace glauca*	Baja California Sur	34.66 ± 29.61 (4.87–155.35)	35	Barrera-García et al., 2013
	Rhizoprionodon longurio	Sinaloa	0.40 ± 0.28	15	Frias-Espericueta et al., 2014
	Rhizoprionodon terraenovae	Veracruz	0.75 ± 1.28†	5	Núñez-Nogueira, 2005
	Mustelus henlei	Baja California Sur	0.23 ± 0.23	52	Pantoja-Echevarría et al., 2020
	Sphyrna lewini	Sinaloa	40.42 ± 27.44**†	1	Ruelas-Inzunza and Páez-Osuna, 2007
Orectolobiformes	*Rhincodon typus*	Baja California Sur	4.6–17.5	3	Pancaldi et al., 2019b; 2021b

n = sample size. *Mean ± sd calculated by adding the means from juveniles plus adults and standard deviation difference was calculated according to Williams (1993) for differences between two means. ** Mean value obtained from duplicated analyses of a sample. *** Original dry weight values transformed by a factor of 0.245 to wet weight values.

C. falciformis presented the highest values, with mean metal Cd values ranging from 1.86 ± 1.41 mg kg^{-1} w.w. in male juveniles to 749 ± 602 mg kg^{-1} w.w. in females (Vega-Barba, 2018). It is worth mentioning that *C. falciformis* juveniles presented the highest mean Cd concentrations, correlating with the fishing area.

Coastal ecosystems have been the most assessed, particularly the Gulf of California and Mid-Mexican Pacific. Reports of Cd contamination are frequent concerning mainly organisms from the coasts of Jalisco, including bivalve mollusks, crustaceans and fish, suggesting an important presence of this metal in the area due to urban coastal activities and volcanism (Ramírez-Ayala et al., 2020). The high presence of Cd in sharks is generally associated to the fact that these are larger predatory organisms and to Cd biomagnification throughout the food chain. However, in the particular case of Mexican Pacific sharks, some authors suggest a relationship between metal presence with upwelling processes in areas rich in phosphorite deposits, a mineral rich in calcium phosphates, which may contain metallic impurities such Cd (Secretaría de Economía, 2017). The Baja California Peninsula is also known for its geotectonic activity and seabed subduction (Álvaro-Berlanga et al., 2021; Lara et al., 2020; Pantoja-Echevarría et al., 2020) as well as the Central Pacific (Ramírez-Ayala et al., 2021).

Observing geographical distribution, Cd concentrations in *C. falciformis* from Barra de Navidad and Bahía de Tenacatita, Jalisco (Vega-Barba, 2018) are larger than those reported in Bahía de Todos los Santos, Baja California Sur (Terrazas-López et al., 2016). It is interesting that just two years later, Vega-Barba (2018) re-analyzed liver samples from *C. falciformis* sampled in Bahía de Todos los Santos and found mean levels to be twice as those reported by Terrazas-López et al. (2016) (Table 9.5). The reason for this increase in such a short period is not very clear, but it should be considered that *C. falciformis* is a pelagic species that moves along the entire Mexican Pacific coast (Compagno, 1984; Saldaña-Ruiz et al., 2019) and might be exposed to different metal concentrations for different amounts of times, but constant, which allows a greater exposure that could hypothetically explain its increase in a short period. This resembles with what was observed in muscle for *R. typus* with higher Cd values observed in Baja California (Table 9.1).

The differences between whale sharks and silky sharks may be influenced by the planktonic diet of the whale shark and Cd contributions in water enriched by tectonic activity, while silky sharks exhibit predatory feeding habits. Vega-Barba (2018) considers that high Cd concentrations in *C. falciformis* are due to its role as a coastal epipelagic predator, areas, where a greater abundance of prey is noted, unlike other pelagic or benthic species. Cephalopods may also be part of their diet (e.g., squid; Compagno, 1984), and as these animals, mainly squid, may accumulate high Cd levels, a high transfer capacity may be in place (Bustamante et al., 2002).

The highest Cd values were detected in *C. falciformis* liver samples. Four female samples from Bahía de Tenacatita displayed a mean concentration of 749 ± 602 mg kg^{-1} d.w. In contrast, a combination of juvenile males and females presented slightly lower values, with a mean of 457 ± 453 mg kg^{-1} (Vega-Barba, 2018). The authors, however, indicated the need to evaluate a more significant number of adult organisms to better define the possible effects of individual growth on metal accumulation, as Cd accumulation may be more associated with maturation or metabolic rates.

Comparing shark Cd values from different Mexican coasts, only two studies reporting Cd contamination in liver were observed from the state of Veracruz. Núñez-Nogueira (2005) reported a mean concentration of 3.08 ± 5.23 mg kg^{-1} d.w. (0.75 ± 1.28 mg kg^{-1} w.w.) in *R. terraenovae* liver (n = 5), while *C. limbatus* values where 0.06 ± 0.01 mg kg^{-1} d.w. (0.015 ± 0.002 mg kg^{-1} w.w.). Eight years later, Mendoza-Díaz et al. (2013) reported a mean Cd concentration of 0.083 ± 0.011 mg kg^{-1} w.w. in female *C. limbatus* from northern Veracruz, a 5.5-fold increase compared to 2005 (Table 9.5), suggesting a significant increase in Cd levels in the area. If true, this may be associated with the impact of the Deep-Water Horizon spill in 2010 off the coast of Louisiana, USA, as this element is one of the metallic trace elements present in the Macondo oil mixture (Nowell et al., 2013; Soto and Vázquez-Botello, 2013). The impact of the spill on the aquatic biota has been evident on North American coasts of the Gulf of Mexico (Incardona et al., 2014; Murawski et al., 2014; Prohaska et al., 2021; Wise et al., 2014) so some *C. limbatus* prey may have been exposed. However, no conclusive evidence in this regard has been reported to date.

The coastal habits displayed by *R. terraenovae* contrasts with *C. limbatus*, which inhabits more offshore waters. This may explain the higher concentrations in *R. terraenovae* livers, which is likely more exposed to continental contributions, as suggested previously for whale sharks (Pancaldi et al., 2019b, 2021b) and *C. falciformis* (Vega-Barba, 2018). An interesting aspect is that the *Rhizoprionodon* spp. genus comprises two species that inhabit Mexican waters, with *R. longurio* inhabiting Pacific coasts, and *R. terraenovae*, those of the Gulf of Mexico and the Caribbean. These two species are the only ones that have been analyzed on both coasts regarding Cd in liver so far (Table 9.5). In the Pacific, Frías-Espericueta et al. (2014) reported a mean concentration of 0.40 ± 0.28 mg kg^{-1} w.w. in *R. longurio* female individuals off the coast of Mazatlán, Sinaloa, similar to those observed sixteen years before (0.75 mg kg^{-1} w.w.) by Núñez-Nogueira (2005) in Veracruz. Because these studies were conducted almost ten years apart, it is difficult to determine Cd exposure levels over time between the Gulf of Mexico and the central Mexican Pacific, although, as mentioned previously, environmental concentration differences are noted comparing both coasts. Therefore, the similarity of the concentrations observed between both species suggests that the differences are not associated to sex, diet, or migration areas, contrary to what has been proposed for *Prionace glauca* by Castro-Rendón et al. (2021) when comparing different geographical areas.

Concerning differences to sharks from other areas, the blue shark *P. glauca* is one of the two species reported in both international waters and the Mexican Pacific, with thirty-five livers sampled from organisms from Punta Belcher, Baja California Sur (Barrera-García et al., 2013) (Table 9.5). In this case, the authors indicate Cd concentrations ranging from 4.87–155.35 mg kg^{-1} w.w. in juveniles, higher than those reported by Vas (1991) for adult organisms from the coasts of England, of 0.25 mg kg^{-1} w.w., which is 19-fold lower than the lowest concentration observed in Punta Belcher, Mexico (Tables 9.5 and 9.6). According to Barrera-García et al. (2013), this is associated to a greater Cd availability in Pacific waters compared to the coasts of England.

The other species also reported in international waters is *Sphyrna lewini*, with sightings in Trinidad and Tobago (Mohammed and Mohammed, 2017). The concentrations reported in ten sharks of this island ranged from 0.27 to 1.23 mg kg^{-1} w.w. (Table 9.6), while in Mexico, the observed value is higher, of 40.42 ± 27.44 mg kg^{-1} w.w. (Table 9.5) for one specimen sampled in 2007 from the Sinaloa coast. This suggests a significant difference concerning Cd contamination in both areas.

Table 9.6 Cadmium concentrations (mean ± standard deviation and range; mg kg^{-1} wet weight) in liver of sharks around the world.

Species	Sex	Location	Cd	n	References
Carcharhinus albimarginatus	Mix	Japan	0.26 ± 0.46	8	Endo et al., 2008
Carcharhinus plumbeus	Female	Japan	0.73	1	Endo et al., 2008
Carcharhinus porosus	NS	Trinidad and Tobago	0.61–4.05	12	Mohammed and Mohammed, 2017
Carcharhinus leucas	Female	Japan	2.97	1	Endo et al., 2008
Centrophorus granulosus	NS	Haifa, Israel	7.74	8	Hornung et al., 1993
Chiloscyllium plagiosum	Mix	Hong Kong	0.240 (0.040–1.09)	26	Cornish et al., 2007
Etmopterus spinax	NS	UK	1.76	1	Vas, 1991
	NS	Haifa, Israel	3.06	3	Hornung et al., 1993
Galeocerdo cuvier	Mix	Japan	0.15 ± 0.24	42	Endo et al., 2008
Galeorhinus galeus	NS	UK	<0.02	2	Vas, 1991
Galeus melastomus	NS	UK	0.07	7	Vas, 1991
	NS	Haifa, Israel	5.12	9	Hornung et al., 1993
Halaeulurus bivius	NS	Argentina	7.93 ± 1.78 (3.52–13.70)	73	Marcovecchio et al., 1991
Hexanchus griseus	NS	Haifa, Israel	1.06	6	Hornung et al., 1993
Mustelus schmitii	NS	Argentina	5.62 ± 1.65 (1.53–13.60	570	Marcovecchio et al., 1991
Notorhynchus sp	NS	Argentina	8.41–0.32 (5.31–9.16	14	Marcovecchio et al., 1991
Prionace glauca	NS	UK	0.25	5	Vas, 1991
	Males	USA North Atlantic	1.83 ± 1.34	8	Hauser-Davis et al., 2021
Scyliorhinus canicula	Mix	UK	<0.02	13	Vas, 1991
	NS	UK	0.93	2	Vas, 1991
Scyliorhinus stellaris	NS	UK	<0.02	1	Vas, 1991
Scymnorhinus licha	NS	UK	<0.02	1	Vas, 1991

Somniosus microcephalus	Mix	Canada	3.91 ± 0.436*	24	McMeans et al., 2007
Somniosus pacificus	Mix	USA-Alaska	2.64 ± 0.354*	14	McMeans et al., 2007
Sornniosus rostratus	Mix	Haifa, Israel	12.2	2	Hornung et al., 1993
Sphyrna lewini	NS	Trinidad and Tobago	0.27–1.23	10	Mohammed and Mohammed, 2017
Sphyrna zygaena	Males	Italy	19.77 ± 1.29 (18.43–21.00)	4	Storelli et al., 2003
Squatina guggenheim	NS	Brasil	0.05 ± 0.05** (0.006–0.176)**	9	Martins et al., 2021

*means ± standard error. ** Original dry weight values transformed by 0.245 factor to wet weight values. NS = not specified.

Cadmium accumulates at higher levels in shark liver compared to muscle worldwide (Tiktak et al., 2020), corroborating the results reported herein. Even though Cd values from Mexico are among the highest reported worldwide, Storelli et al. (2003) reported a range of high concentrations between 18.43 and 21.0 mg kg^{-1} w.w. in four *Sphyrna zygaena* liver samples from Italy, although still 12-fold less than the highest mean value reported in livers of Mexican sharks (*C. falciformis* from Jalisco, 253.9 mg kg^{-1} w.w.). The high presence of the metal in *C. falciformis* juveniles may be related to metabolic changes compared to adults. For example, Mull et al. (2012) point out that the presence of toxic metals, such as Hg, in the liver of the white shark *Carcharodon carcharias* may be due to more efficient detoxification mechanisms in liver than in other tissues, such as muscle, as well as difference between juveniles and adults, as different development stages involve changes in growth rates. Those authors highlight that juveniles use more energy for growth and muscle development, resulting in a more accelerated hepatic metabolism, allowing for more significant metal accumulation in this organ.

An important aspect to consider is that the liver is the largest organ in sharks, with an important surface-volume ratio and blood supply, ideal for metabolizing significant amounts of many molecules, including metals. Shark livers display three essential functions: to manufacture enzymes required in different metabolic processes, maintain lipid reserves (fat) for use when food is limited and, finally, to act as a buoyancy organ. Sharks lack swim bladders; thus they have developed long, bilobed livers that produce large amounts of oils that give them the necessary water buoyancy. Likewise, they have developed the ability to accumulate urea as a major organic osmolyte (Ballantyne, 2016). This urea retention capacity significantly determines the biochemistry and physiology of elasmobranchs and is associated to various liver enzymes (Kajimura et al., 2006). In this regard, Ji et al. (2020) reported interactions between urea and Cd, favoring the uptake of this metal within certain types of plant cells. Apparently, the presence of urea allows a greater Cd accumulation through increased affinity of Cd^{2+} ions with membrane transporters, thus allowing for increased cell permeability capacity, which, in turn, increases active Cd uptake and influx rate in plant roots (Ji et al., 2020), while urea acts as a ROS antioxidant (Barrera-García et al., 2013) derived from the presence of Cd. If this relationship exists in fish liver cells as well, the capability of livers to store urea would thus increase the capacity for hepatic Cd accumulation. Furthermore, the liver would also be receiving high irrigation. Of course, this potential relationship would act in conjunction with the antioxidant and chelating response of metallothionein-like proteins previously observed in shark livers (Hauser-Davis et al., 2021; Hauser-Davis, 2020; Hidalgo et al., 1985) and high lipid content with an affinity for metals (Terrazas-López et al., 2016), thus promoting an efficient metal retention capacity in this organ. Detailed histopathological and physiological studies are crucial to define the adverse effects that this non-essential metal may cause in shark liver tissue.

CADMIUM IN DIFFERENT SHARK TISSUES

Although the muscle and liver of sharks are the most studied tissues regarding metal contamination, some studies in Mexico included other tissues and organs, with other organs from *Rhincodon typus* and *Carcharhinus limbatus* specimens as the most analyzed (Mendoza-Díaz et al., 2013; Núñez-Nogueira, 2005; Pancaldi et al., 2019b) (Table 9.7). As for the blacktip shark *C. limbatus*, Núñez-Nogueira (2005) reported similar Cd concentrations in the brain (0.32 ± 0.43 mg kg^{-1}) as observed in *R. typus* (0.34 mg kg^{-1}). Whale sharks in Mexican waters exhibited the highest Cd concentrations in heart samples (1.87 mg kg^{-1}), followed by filtering parches (1.48 mg kg^{-1}) and gills (1.24 mg kg^{-1}) (Pancaldi et al., 2019b). These results suggest a very homogeneous distribution among tissues, although further studies on a higher number of specimens are required to better describe the body distribution patterns. Regarding the number of tissues analyzed by species, the whale shark *R. typus* is represented by ten different tissues, namely brain, gills, heart, kidney, muscle, filtering parches, stomach, testicles, epidermis and dermal denticles (Table 9.7). In general, for the rest of the species reported in Mexican waters, without considering muscle and liver, only four other tissues are reported, namely brain, gills, kidney and placenta. In whale sharks from Mexico, only heart, stomach and testicles have been analyzed to date (Table 9.7).

These results are interesting as, even though eleven species have been studied in Mexico, few tissues have been studied compared to those reported for other species worldwide, such as blackmouth catshark *Galeus melastomus* (Rafinesque 1810) in the waters of Israel and Scotland, where nine different tissues were assessed (Hornung et al., 1993; Vas, 1991). Other species such as *Galeorhinus galeus* (Linnaeus 1758), little sleeper shark *Somniosus rostratus* (Risso 1827), gulper shark *Centrophorus granulosus* (Bloch & Schneider 1801), and velvet belly *Etmopterus spinax* (Linnaeus 1758) have been studied in up to seven different tissues (Table 9.8). In Mexico, the only shark with a higher number of organs or tissues analyzed is *R. typus*, surpassing those studied in various species in other countries, such as England, Scotland, Israel and Trinidad and Tobago, for example (Table 9.8). These species have been investigated regarding from one up to six tissues, compared to the eight *R. typus* tissues from the Mexican Pacific (Table 9.7).

Table 9.7 Cadmium concentrations (mean ± standard deviation; in mg kg-1 wet weight) in different tissues of Mexican shark species.

Tissue	Species	State	Location	Cd	n	References
Brain	*Rhincodon typus*	Baja California Sur	Bahia La Paz	0.34	1	Pancaldi et al., 2019b
	Carcharhinus limbatus	Veracruz	Gulf of Mexico Coast	0.32 ± 0.43*	21	Núñez-Nogueira, 2005
	Rhizoprionodon terraenovae	Veracruz	Gulf of Mexico Coast	0.52 ± 0.39*	21	Núñez-Nogueira, 2005
Filtering parches	*Rhincodon typus*	Baja California Sur	Bahia La Paz	1.48	1	Pancaldi et al., 2019b
Gills	*Rhincodon typus*	Baja California Sur	Bahia La Paz	1.24	1	Pancaldi et al., 2019b
	Carcharhinus limbatus	Veracruz	Gulf of Mexico Coast	0.21 ± 0.26*	21	Núñez-Nogueira, 2005
	Rhizoprionodon terraenovae	Veracruz	Gulf of Mexico Coast	0.49 ± 0.24*	21	Núñez-Nogueira, 2005
Heart	*Rhincodon typus*	Baja California Sur	Bahia la Paz	1.87	1	Pancaldi et al., 2019b
Kidney	*Prionace glauca*	Baja California Sur	Bahia la Paz	0.68	1	Pancaldi et al., 2019b
	Isurus oxyrinchus	Baja California Sur	Punta Belcher	0.017 ± 7.28	35	Barrera-García et al., 2013
	Rhincodon typus	Baja California Sur	San Lázaro, Isla Magdalena	0.62	20	Vélez-Alavez et al., 2013
Placenta	*Rhizoprionodon longurio*	Sinaloa	Mazatlán	0.07 ± 0.06	15	Frías-Espericueta et al., 2014
Epidermis/ denticles	*Rhincodon typus*	Baja California Sur	Bahia de la Paz	0.001–0.51	133	Pancaldi et al., 2019b; 2021a
Stomach	*Rhincodon typus*	Baja California Sur	Bahia de la Paz	0.89	1	Pancaldi et al., 2019b
Testicle	*Rhincodon typus*	Baja California Sur	Bahia de la Paz	0.58	1	Pancaldi et al., 2019b

n = sample size. *Original dry weight values transformed by 0.245 factor to wet weight values.

Table 9.8 Cadmium concentrations (mg kg^{-1} wet weight) in different body tissues of sharks from around the world

Tissue	Species	Location	Cd	n	References
Bowels (intestines)	*Centrophorus granulosus*	Israel	1.68	23	Hornung et al., 1993
	Etmopterus spinax		2.33	8	
	Galeus melastomus		1.79	42	
	Hexanchus griseus		1.32	6	
	Somniosus rostratus		5.51	6	
Brain	*Centrophorus granulosus*	Israel	0.11	12	Hornung et al., 1993
	Somniosus rostratus		0.14	1	
Dorsal fin	*Carcharhinus porosus*	Trinidad and Tobago	2.9–12.1	12	Mohammed and Mohammed, 2017
	Sphyrna lewini		1.11–2.98	10	
Egg sac	*Somniosus rostratus*	Israel	0.16	2	Hornung et al., 1993
Epaxial muscle	*Carcharinus porosus*	Trinidad and Tobago	0.69–5.86	12	Mohammed and Mohammed, 2017
Gills	*Galeorhinus galeus*	Ireland	<0.02	2	Vas, 1991
	Galeus melastomus	Scotland	0.09	7	
	Prionace glauca	England	0.99	5	
	Scyliorhinus canicula	England	1.1	2	
Gonads	*Centrophorus granulosus*	Israel	0.13	13	Hornung et al., 1993
	Etmopterus spinax	Israel	0.17	5	
		Scotland	<0.02	1	Vas, 1991
	Galeorhinus galeus	Ireland	<0.02	2	
	Galeus melastomus	Israel	0.16	11	Hornung et al., 1993
		Scotland	0.06	7	Vas, 1991
	Scyliorhinus canicula	England	0.95	2	
		Ireland	<0.02	13	
	Scyliorhinus stellaris	Ireland	<0.02	1	
	Somniosus rostratus	Israel	0.1	4	Hornung et al., 1993
Heart	*Centrophorus granulosus*	Israel	1.2	5	Hornung et al., 1993
	Galeus melastomus		0.25	4	
		Scotland	0.13	7	Vas, 1991
	Somniosus rostratus	Israel	0.38	1	Hornung et al., 1993
Hypaxial muscle	*Carcharinus porosus*	Trinidad and Tobago	0.71–7.21	12	Mohammed and Mohammed, 2017

(Contd.)

Table 9.8 Cadmium concentrations (mg kg^{-1} wet weight) in different body tissues of sharks from around the world (*Contd.*)

Tissue	Species	Location	Cd	n	References
Jaw	*Etmopterus spinax*	Scotland	1.94	1	Vas, 1991
	Prionace glauca	England	0.55	5	
	Scymnorhinus licha	Scotland	1.7	1	
	Squatina squatina	England	<0.02	1	
Kidney	*Centrophorus granulosus*	Israel	1.21	27	Hornung et al., 1993
	Etmopterus spinax		1.37	6	
	Galeorhinus galeus	Ireland	<0.02	2	Vas, 1991
	Galeus melastomus	Israel	0.99	48	Hornung et al., 1993
		Scotland	0.25	7	Vas, 1991
	Hexanchus griseus	Israel	0.71	5	Hornung et al., 1993
	Scyliorhinus canicula	Ireland	<0.02	13	Vas, 1991
	Scyliorhinus stellaris		<0.02	1	
	Somniosus rostratus	Israel	1.91	8	Hornung et al., 1993
Pancreas	*Galeorhinus galeus*	Ireland	<0.02	2	Vas, 1991
Skin	*Centrophorus granulosus*	Israel	0.13	5	Hornung et al., 1993
	Etmopterus spinax	Scotland	0.38	1	Vas, 1991
	Galeorhinus galeus	Ireland	<0.02	2	Vas, 1991
	Galeus melastomus	Israel	1.14	2	Hornung et al., 1993
		Scotland	0.08	7	Vas, 1991
	Prionace glauca	England	0.97	5	
	Rhincodon typus	África	<0.008[†]	20	Boldrocchi et al., 2020
	Scyliorhinus canicula	England	1.07	2	Vas, 1991
		Ireland	0.95	13	
	Scyliorhinus stellaris	Ireland	<0.02	1	
	Scymnorhinus licha	Scotland	0.48	1	
Spleen	*Galeorhinus galeus*	Ireland	<0.02	2	Vas, 1991
	Galeus melastomus	Scotland	0.09	7	
	Scyliorhinus canicula	England	0.89	2	
		Ireland	<0.02	13	
	Scyliorhinus stellaris	Ireland	<0.02	1	

Tissue	Species	Location	Cd	n	References
Stomach	*Centrophorus granulosus*	Israel	0.11	3	Hornung et al., 1993
	Etmopterus spinax		0.91	2	
	Galeus melastomus		1.91	8	
	Somniosus rostratus		0.28	3	
Vertebrae	*Carcharhinus porosus*	Trinidad and Tobago	7.96–29.89	12	Mohammed and Mohammed, 2017
	Etmopterus spinax	Scotland	1.09	1	Vas, 1991
	Galeorhinus galeus	Ireland	0.03	2	
	Galeus melastomus	Scotland	0.54	7	
	Prionace glauca	England	<0.02	5	
	Scyliorhinus canicula	Ireland	1.57	13	
	Scymnorhinus licha	Scotland	0.57	1	
	Sphyrna lewini	Trínidad and Tobago	6.19–27.29	10	Mohammed and Mohammed, 2017
	Squalus acanthias	Ireland	0.14	12	Vas, 1991
	Lamna nasus	Ireland	0.13	1	

†Original dry weight values transformed by 0.245 factor to wet weight values.

Evaluating tissues individually, the brain has been studied in three Mexican species, with concentrations ranging from 0.34 mg kg^{-1} (*R. typus*) to 0.52 ± 0.39 mg kg^{-1} w.w. (*R. terraenovae*), higher than values reported in UK species, whose concentrations ranged from 0.21 to 1.24 mg kg^{-1} w.w. and in *C. granulosus* and *S. rostratus* (0.11–0.14 mg kg^{-1} w.w.) from Israel (Table 9.8). In gills, mean concentrations ranged from 0.84 ± 1.06 to 1.99 ± 0.98 mg kg^{-1} in the same three species from Mexican waters. Regarding cardiac tissue, only one *R. typus* has been analyzed, from Baja California Sur, with a Cd concentration of 1.87 mg kg^{-1} w.w., slightly higher than those reported in *G. melastomus*, *S. rostratus* and *C. granulosus*, which ranged from 0.13 to 1.2 mg kg^{-1} (Table 9.8).

The kidney is one of the most studied organs globally in sharks and one of the main Cd accumulators. International reports indicate Cd concentrations from <0.02 to 1.91 mg kg^{-1} (Table 9.8). The renal filtering of the blood is undoubtedly essential for the osmoregulation and homeostasis of different ions in blood. Thus, kidneys are an important metabolism site regarding Cd, generating cumulative processes. Catalase is one of the antioxidant enzymes reported as activated by metal exposure in shark kidneys (Barrera-García et al., 2013). In Mexico, concentrations ranged from 0.017 ± 7.28 mg kg^{-1} (*I. oxyrinchus*) to 0.62 mg kg^{-1} (*R. typus*) and 0.68 mg kg^{-1} (*P. glauca*) in this organ (Table 9.7). As observed for *P. glauca* and *R. typus*, Cd concentration in this organ is higher than that observed in *I. oxyrinchus* (Table 9.7). This pattern was also described by Barrera-García et al. (2013), who highlighted that, despite this pattern, mako sharks develop greater activity of antioxidant markers than blue sharks, which suggests variations possibly associated with differences in metabolic rates or sensitivities between sharks.

The skin has only been studied in the whale shark (*R. typus*) from the waters of the Baja California Peninsula in Mexico. Two studies reported Cd concentrations in the epidermis and dermal denticles (Pancaldi et al., 2019b, 2021a) ranging from 0.001 to 0.51 mg kg^{-1} w.w., which include the values reported for the same species (<0.008 mg kg^{-1} w.w.) in northeast Africa (Boldrocchi et al., 2020). Cadmium values in the skin of this species are not as dissimilar from those reported in the skin of other sharks such as *G. galeus* and *S. stellaris* in Ireland and *G. melastomus* in Scotland (Vas, 1991). The presence of higher or lower Cd concentrations in this tissue suggest deeper associations with shark distribution ranges instead of type of feeding, as noted for liver and muscle. This idea is supported by Cd skin differences observed in *G. melastomus* (1 mg kg^{-1} on the coasts of Israel and 0.008 mg kg^{-1} in Scotland). Concentrations have been reported as ranging from <0.02 mg kg^{-1} in *S. stellaris* in Irish waters to 1.07 mg kg^{-1} in lesser spotter dogfish *S. canicula* in British waters, indicating that, although they belong to the same genus, differences in metal availability in these species seems to vary with geographical areas (Table 9.8). On the other hand, intrinsic factors to each species can also be the cause for this differential accumulation pattern as, depending on their geographical location, different shark populations may display different growth rates and physiology.

Pancaldi et al. (2019b) reported a Cd concentration of 0.89 mg kg^{-1} w.w. in the stomach of a *R. typus* specimen (Table 9.8), agreeing with the range reported for sharks from Israel of 0.11 to 1.91 mg kg^{-1} w.w. (Table 9.8). Based on freshwater fish studies, Cd seems to follow intestinal-digestive tract acquisition routes similar to that observed in gills, through the calcium physiological mechanism (Klinck et al., 2009), particularly associated with divalent metal transporters or DMT1 (divalent metal transport-1), including cadmium ions from diets rich in this element (Cooper et al., 2006). The stomach appears to retain lower Cd concentrations than the intestine (Klinck et al., 2009), as observed in five shark species analyzed in Israel (Table 9.8). This tissue accumulation difference may also occur in *R. typus*, although no intestine samples have been assessed to date.

Studies reporting the presence of Cd in the placenta and testis of sharks are limited, with only two species from Mexican waters analyzed in this regard. The first report was carried out by Frías-Espericueta et al. (2014) for a *R. longurio* placenta from the coasts of Sinaloa, where fifteen sharks were analyzed, with a mean concentration of 0.07 ± 0.06 mg kg^{-1} w.w. Likewise, Pancaldi et al. (2019b) reported 0.58 mg kg^{-1} w.w. in whale shark testis from Baja California Sur, Mexico for the first time. Almost no reports of metals in these reproductive tissues in sharks are available, especially in the testis, which is curious as this organ is used as a model for study of cadmium sperm toxicity in sharks (McClusky, 2006; 2008).

Gills and kidneys are also important Cd uptake sites in sharks, and the scarcity of studies in these tissues in Mexican sharks represent an opportunity for future research in both coastal and pelagic species, complementing assessments of muscle and liver as target organs, independently of interest in these organs as meat or natural products use, such as those derived from liver oil.

SHARK MEAT HEALTH RISK ASSESSMENTS

Although some of the individual metal concentrations observed in Mexican shark muscles were low, mean metal values indicate excessive Cd concentrations compared to values considered safe for human consumption.

The criteria to determine the risk in consumption can be evaluated from three main approaches:

- Maximum allowed limit in meat for national (NMAL) or international consumption (IMAL)
- Maximum allowable intake limit per week of a certain contaminant in food (MWIR)
- Non-carcinogenic risk rate for consumption of contaminated meat (HQ).

In the first approach, the maximum level of cadmium allowed in fishery products, particularly for fish meat, is 0.5 mg kg^{-1}, according to the Mexican regulation (SSA, 2011). The European Union establishes a concentration of 1 mg kg^{-1} as the maximum permissible level of Cd (European Commission, 2014), applied herein for international comparisons.

The second approach involves the weekly metal intake rate or MWIR (mg kg^{-1} week^{-1}) compared with the provisional tolerable weekly intake or PTWI (7 mg kg^{-1} week^{-1}; Ruelas-Inzunza and Páez-Osuna, 2007). The MWIR is calculated by the following formula:

$$MWIR = \frac{C \times FWC}{W}$$

Where C is the metal concentration in fish meat, FWC is the amount of fish meat consumed per week (234 g week^{-1} in adults and 200 g week^{-1} in children) (Castro-Rendón et al., 2021), and W is human body weight (70 kg per adult or 15 kg per children) according to Man et al. (2014).

The third criterion is the hazard quotation (HQ). The HQ value determines the relationship between the level at which an adverse effect is not expected and the potential exposure to a substance (i.e., the potential for toxic effects caused by exposure to non-carcinogenic chemicals) according to the following two equations:

$$E = \frac{C \times I}{W} \quad \text{and} \quad HQ = \frac{E}{RfD}$$

First, the level of metal (E) consumption is calculated, considering the concentration of the metal in the fish meat per species (C) in mg kg^{-1} w.w., the *per capita* fish meat ingestion rate (I; g day^{-1}) and total adult body weight (W; 70 kg). The *per capita* fish meat ingestion rate was considered for the general Mexican population (36 g d^{-1}) and fishing communities (400 g d^{-1}).

Subsequently, to determine the HQ, the level of consumption of the pollutant (E) is divided by the ratio of the estimated daily exposure of each contaminant or a chronic reference dose (RfD) for the metal, also known as the oral exposure reference dose (Ramírez-Ayala et al., 2021). In this case, the RfD was established as 0.001 mg kg^{-1} person^{-1} for Cd (EPA, 2000). An HQ value greater than or equal to one establishes a risk or non-carcinogenic adverse effects for human health.

Table 9.9 displays the three risk assessment criteria for consumption, considering fish intake rates in Mexico per person for both adults and children. In general terms, four species showed maximum Cd values above the established limits, namely *A. pelagicus, C. falcisformis, R. typus* and *P. glauca* (Table 9.9), above 0.5 and 1 mg kg^{-1} in meat (NMAL and IMAL limits, respectively), also exceeding the values of the other two consumption risk indicators, the rate of intake per week (MWIR) and non-carcinogenic hazard rate (HQ). For this last criterion, it is interesting that almost all species exhibited non-carcinogenic risks. The exception was *I. oxyrinchus*, the only species with a high enough Cd value in meat below the legal limits.

Table 9.9 Maximum allowed cadmium limits in fish meat, maximum weekly intake rate per adults and children, and non-carcinogenic risk ratio consumption of the metal. The allowed limits are 0.5 mg kg^{-1} in fish meat in Mexico (SSA, 2011) and 1 mg kg^{-1} internationally (European Commission, 2014). The PTWI (provisional tolerable weekly intake of 7 mg week^{-1} kg^{-1}, WHO/OMS, 2016) was considered for the maximum weekly intake rates (MWIR). The non-carcinogenic risk rate (HQ) was calculated for the general Mexican population (HQN) and the fishing community (HQP)

Species	NMAL	IMAL	MWIR (adults)	MWIR (children)	HQN	HQP
Alopias pelagicus	>0.5	>0.0025	>7	>7	>1	>1
Carcharhinus falciformis	>0.5	>0.0025	>7	>7	>1	>1
Carcharhinus limbatus	0.086	>0.0025	2.03	3.99	>1	>1
Mustelus henlei	0.04	>0.0025	0.91	1.89	>1	>1
Rhizoprionodon longurio	0.01	>0.0025	0.21	0.49	>1	>1
Rhizoprionodon terraenovae	0.083	>0.0025	1.96	3.85	>1	>1
Sphyrna lewini	0.038	>0.0025	0.91	1.75	>1	>1
Rhincodon typus	>0.5	>0.0025	>7	>7	>1	>1
Isurus oxyrinchus	0.0001	>0.0025	0.002	0.005	0.36	0.57
Prionace glauca	>0.5	>0.0025	>7	>7	>1	>1

NMAL = national maximum allowed limit (SSA, 2011; 0.5 mg kg^{-1}), IMAL = international maximum allowed limit (European Commission, 2014; 0.0025 mg kg^{-1}), MWIR (metal weekly intake rate, mg week^{-1} kg^{-1}), HQN (calculated based on the national mean fish meat intake; 36 g day^{-1}), HQP (calculated based on fish meat consumption reported in fishing communities in México; 400 g day^{-1}). Adult body weight (70 kg), Children body weight (15 kg).

It is important to highlight that several of the reported cases represent only one sampled shark, as is the case for *C. leucas, S. lewini* and *R. typus* (Table 9.1), so it cannot be generalized that these fishery resources represent a health risk. For the rest of the species, more exhaustive monitoring seems necessary to allow for deeper assessments, as indicated by Vas (1991), who considered that the observed Cd accumulation patterns are a clear sign of the need for careful monitoring, especially regarding commercially exploited species.

Another critical aspect to consider comprises of culinary treatment effects on metal concentrations and bioavailability. A very recent study by Schmidt et al. (2021) evidenced that Cd concentrations decrease scallops between 37% and 53% after culinary treatments, with the metal transferred to boiling water. Bioavailability, however, was not modified. In the case of shark meat, concentrations do not vary significantly following culinary treatment, but bioavailability decreases in up to 41%. This suggests that frying, boiling or sauteing could help reduce Cd availability prior to ingestion. Likewise, the dietary use of natural antagonists for certain metals like Cd and Pb are highly recommended in the diets of communities with a high risk of dietary exposure. In this regard, Zhai et al. (2015) suggest a frequent consumption of essential elements, vitamins, vegetables, and fruits, to reduce the adverse effects derived from metal ingestion.

The results observed herein for Carcharhiniformes and Lamniformes mainly agree with the findings reported by Tiktak et al. (2020) in their systematic review on the possible risk of pollutants in elasmobranchs, such as skates, rays and sharks. They observed that these orders usually present the most significant pollutants accumulation. Likewise, values above the risk limits are consistent with what has recently been observed in *P. glauca* from the southern Pacific coasts, particularly for Cd and mercury in Ecuador (Castro-Rendón et al., 2021). On the other hand, Adel et al. (2016) did not observe risks regarding Cd for the consumption of shark meat (whitecheek shark *Carcharhinus dussumieri*). These differences may be related to biological aspects of the different species, such as growth rates, sizes, or metal bioavailability. In Mexican waters, availability seems to be higher because of the geotectonic activity of the Pacific coasts (Pantoja-Echevarría et al., 2020), as discussed previously.

Medium- and long-term monitoring programs, both regarding the presence of metals in sharks, sediments and water in coastal Mexican areas, are required to allow for more precise comparisons, defining not only the origin of the detected variability between species but also the origin or cause of the metal presence in fish, with respect to environmental levels. Ruelas-Inzunza and Páez-Osuna (2007) highlighted the importance of more studies that consider seasonal aspects, age and gonadal fish status to better understand variations in metal concentrations. Furthermore, Pastorelli et al. (2012) mentioned the importance of considering other factors such as nutritional status or certain health conditions (diseases and pregnancies), which could increase dietary Cd absorption, especially in more vulnerable groups, such as fishing communities.

It is important to highlight that all shark species analysed for Cd in Mexico are categorized as protected status by the IUCN (International Union of Conservation of Nature), five are listed in CITES (the Convention on International Trade in Endangered Species of Wild Fauna and Flora), and less than half of are regulated for trade (Table 9.10). On the other hand, the NOM-029-PESC-2006 (SAGARPA, 2006) comprises a legal instrument that provides regulations for exploiting elasmobranchs and promotes responsible fishing for sharks and rays in Mexican waters. Thus, further concerning metal concentrations may also aid in establishing appropriate protective measures for their conservation.

Table 9.10 Status of conservation of Mexican sharks involved in cadmium studies.

Species	IUCN	CITES
Alopias pelagicus	EN	II
Carcharhinus falciformis	VU	II
Carcharhinus leucas	VU	Unlisted
Carcharhinus limbatus	VU	Unlisted
Mustelus henlei	LC	Unlisted
Rhizoprionodon longurio	VU	Unlisted
Rhizoprionodon terraenovae	LC	Unlisted
Sphyrna lewini	CR	II
Rhincodon typus	EN, LD	II
Isurus oxyrinchus	EN	II
Prionace glauca	NT	Unlisted

IUCN, 2021. The IUCN Red List of Threatened Species. Versión 2021-3. <https://www.iucnredlist.org> ISSN2307-8235
CR = Critically Endangered; EN = Endangered; LC = Least Concern; LD = Largely Deplected; NT = Near Threatened; VU = Vulnerable
CITES: https://cites.org/sites/default/files/esp/app/2021/S-Appendices-2021-06-22.pdf

CONCLUSIONS

Despite increasing scientific studies on metals in sharks, only eleven species have been studied in Mexico in the last 24 years concerning Cd, mostly in muscle, liver, and skin, and *C. falciformis* display a seemingly higher accumulation capacity than other reported species. The target organs of most significant interest are undoubtedly the liver and the muscle, so it becomes essential to include analysis of more organs and tissues, which will allow for further understanding on medium- and long-term metabolic and physiological effects. It is likely that the Cd accumulation capacity of these animals is related to urea retention physiology, the synthesis of metallothioneins and specific antioxidant enzymes, which regulate cellular Cd levels.

From a geographic perspective, studies have focused mostly on the coasts of the central Pacific and the Baja California Peninsula, with few studies in Veracruz in the Gulf of Mexico and no reports of Cd in sharks from the South Pacific and the Mexican Caribbean waters. Regarding concentrations, Cd levels are diverse and, in many cases, comparable with similar or the same species from other parts of the world. However, higher Cd values may also reflect local environmental conditions related to shark population characteristics and natural and anthropogenic metal contributions, especially in the Pacific.

Risk assessments regarding the consumption of shark meat indicates that four of the eleven analyzed species presented higher Cd levels in muscle, several above maximum permissible Mexican values. Concerning international guidelines, most species exceeded the intake limit for adults and children, and all species presented Cd concentrations at levels that may generate non-carcinogenic effects. Although a small number of samples have been assessed and a low number of studies carried out, these findings are an alert to carry out more extensive and permanent monitoring programs, which provide information that may aid in establishing measures for practical conservation efforts. Furthermore, this will also ensure higher consumer shark product quality and also motivate for more extensive protection efforts, as these organisms are also under conservation or protection statuses, both in Mexico and internationally.

REFERENCES

Adel, M., Oliveri Conti, G., Dadar, M., Mahjoub, M., Copat, C. and Ferrante, M. 2016. Heavy metal concentrations in edible muscle of whitecheek shark, *Carcharhinus dussumieri* (elasmobranchii, chondrichthyes) from the Persian Gulf: A food safety issue. Food Chem. Toxicol. 97: 135–140. https://doi.org/10.1016/j.fcCasret.2016.09.002

Álvaro-Berlanga, S., Calatayud-Pavía, C.E., Cruz-Ramírez, A., Soto-Jiménez, M.F. and Liñán-Cabello, M.A. 2021. Trace elements in muscle tissue of three commercial shark species: *Prionace glauca*, *Carcharhinus falciformis*, and *Alopias pelagicus* off the Manzanillo, Colima coast, Mexico. Environ. Sci. Pollut. Res. 28(18): 22679–22692.

ATSDR. 2012. Toxicological profile of cadmium.

Ballantyne, J.S. 2016. Some of the most interesting things we know, and don't know, about the biochemistry and physiology of elasmobranch fishes (sharks, skates and rays). Comp. Biochem. Physiol. B. Biochem. Mol. Biol. 199: 21–28. https://doi.org/10.1016/j.cbpb.2016.03.005

Barrera-García, A., O'Hara, T., Galván-Magaña, F., Méndez-Rodríguez, L.C., Castellini, J.M. and Zenteno-Savín, T. 2013. Trace elements and oxidative stress indicators in the liver and kidney of the blue shark (*Prionace glauca*). Comp. Biochem. Phys. A. Mol. Integr. Physiology 165(4): 483–490.

Beckett, J.S. and Freeman, H.C. 1974. Mercury in swordfish and other pelagic species from the western Atlantic Ocean. Microfiche Report, COM-75-50075, 1974: 154–159

Benítez, J.A., Vidal, J., Brichieri-Colombi, T. and Delgado-Estrella, A. 2012. Monitoring ecosystem health of the Terminos Lagoon region using heavy metals as environmental indicators. pp. 349–358. *In*: Brebbia, C.A. and Chon, T.S. (eds). Enviromental Impact. Southampton, UK.

Bevacqua, L., Reinero, F.R., Becerril-García, E.E., Elorriaga-Verplancken, F.R., Juaristi-Videgaray, D. Micarelli, P., et al. 2021. Trace elements and isotopes analyses on historical samples of white sharks from the Mediterranean Sea. Eur. Zool. J. 88(1): 132–141.

Bhat, S.A., Hassan, T. and Majid, S. 2019. Heavy metal toxicity and their harmful effects on living organism—A review. Int. J. Med. Sci. Diag. Res. 3(1): 106–122.

Boldrocchi, G., Monticelli, D., Butti, L., Omar, M. and Bettinetti, R. 2020. First concurrent assessment of elemental- and organic-contaminant loads in skin biopsies of whale sharks from Djibouti. Sci. Total. Environ. 722: 137841.

Bouquegneau, J-M., Pillet, S., Das, K. and Debacker, V. 2003. Heavy metals in marine mammals. pp. 135–167. *In*: Vos, J.V., Bossart, G.D., Fournier, M. and O'Shea, T. (eds). Toxicology of Marine Mammals. Taylor & Francis.

Bryan, C.E., Christopher, S.J., Balmer, B.C. and Wells, R.S. 2007. Establishing baseline levels of trace elements in blood and skin of bottlenose dolphins in Sarasota Bay, Florida: Implications for non-invasive monitoring. Sci. Total Environ. 388(1–3): 325–342.

Bustamante, P., Cosson, R.P., Gallien, I., Caurant, F. and Miramand, P. 2002. Cadmium detoxification processes in the digestive gland of cephalopods in relation to accumulated cadmium concentrations. Mar. Environ. Res. 53: 227–241.

Carreón-Zapiain, M.T., Tavares, R., Favela-Lara, S. and Oñate-González, E.C. 2020. Ecological risk assessment with integrated genetic data for three commercially important shark species in the Mexican Pacific. Reg. Stud. Mar. Sci. 39: 101431.

Castillo-Géniz, J.L., Márquez-Farias, J.F., Rodriguez De La Cruz, M.C., Cortés, E. and Cid del Prado, A. 1998. The mexican artisanal shark fishery in the Gulf of Mexico: Towards a regulated fishery. Mar. Freshw. Res. 49(7): 611–620.

Castillo-Géniz, J.L. and Tovar-Ávila, J. 2021. Tiburones mexicanos de importancia pesquera en la CITES. p. 95. *In*: Tovar-Ávila, J. and Castillo-Géniz, J.L. (eds). Instituto Nacional de Pesca, México, D.F.

Castro, J.I. 1993. The shark nursery of Bulls Bay, South Carolina, with a review of the shark nurseries of the southeastern coast of the United States. Environ. Biol. Fishes 38(1): 37–48.

Castro-Rendón, R.D., Calle-Morán, M.D., García-Arévalo, I., Ordiano-Flores, A. and Galván-Magaña, F. 2021. Mercury and Cadmium concentrations in muscle tissue of the Blue Shark (*Prionace glauca*) in the central eastern pacific ocean. Biol. Trace Elem. Res. J. 200: 3400–3411. https://doi.org/10.1007/s12011-021-02932-7

Chouvelon, T., Caurant, F., Cherel, Y., Simon-Bouhet, B., Spitz, J. and Bustamante, P. 2014. Species-and size-related patterns in stable isotopes and mercury concentrations in fish help refine marine ecosystem indicators and provide evidence for distinct management units for hake in the Northeast Atlantic. ICES J. of Mar. Sci. 71: 1073–1087. https://doi.org/10.1093/icesjms/fst199

Cid, A., Castillo, J.L., Soriano, S.R., Ramírez, C., Márquez, J.F. and Tovar, J.R. 2000. Tiburón. pp. 179–258. *In*: Sustentabilidad y Pesca Responsable en México: Evaluación y Manejo, 1999–2000. Instituto Nacional de Pesca México (INAPESCA). CINVESTAV Unidad Mérida, Yucatán, México.

Cliff, G.M. and Dudley, S.F.J. 1991. Sharks caught in the protective gill nets off Natal, South Africa. 4. The bull shark *Carcharhinus leucas* valenciennes. S. Afr. J. Mar. Sci. 10(1): 253–270.

Compagno, L.J.V. 1984. Sharks of the world. An annotated and illustrated catalogue of shark species known to date. FAO Species Catalogue. Vol. 4. Part 2—Carcharhiniformes. 1st edición. FAO Fisheries Synopsis Rome, Italy, 125(4): 251–655.

Compagno, L.J.V. 1990. Alternative life-history styles of cartilaginous fishes in time and space. Environ. Biol. Fishes. 28: 33–75.

CONAPESCA (Comisión Nacional de Acuacultura y Pesca). 2018. Anuario estadístico de acuacultura y pesca 2018. pp. 300. SAGARPA, México City.

Consales, G. and Marsili, L. 2021. Assessment of the conservation status of Chondrichthyans: underestimation of the pollution threat. Eur. Zool. J. 88(1): 165–180.

Cooper, C.A., Handy, R.D. and Bury, N.R. 2006. The effects of dietary iron concentration on gastrointestinal and branchial assimilation of both iron and cadmium in zebrafish (*Danio rerio*). Aquat. Toxicol. 79: 167–175.

Cornish, A.S., Ng, W.C., Ho, V.C.M., Wong, H.L., Lam, J.C.W., Lam, P.K.S., et al. 2007. Trace metals and organochlorines in the bamboo shark *Chiloscyllium plagiosum* from the southern waters of Hong Kong, China. Sci. Total Environ. 376(1–3): 335–345. https://doi.org/10.1016/j.scitotenv.2007.01.070

Corrill, L.S. and Huff, J.E. 1976. Occurrence, physiologic effects, and toxicity of heavy metals—arsenic, cadmium, lead, mercury, and zinc–in marine biota: an annotated literature collection. Environ. Health Perspec 18: 181–217.

Das, S., Tseng, L.C., Chou, C., Wang, L., Souissi, S. and Hwang, J.S. 2019. Effects of cadmium exposure on antioxidant enzymes and histological changes in the mud shrimp *Austinogebia edulis* (Crustacea: Decapoda). Environ. Sci. Pollut. Res. 26(8): 7752–7762. https://doi.org/10.1007/s11356-018-04113-x

De Borhegyim, S.F. 1961. Shark teeth, stingray spines, and shark fisihing in ancient Mexico and Central America. Southwestern J. Anthropol. 17(3): 273–296.

De Marco, S.G., Botté, S.E. and Marcovecchio, J.E. 2006. Mercury distribution in abiotic and biological compartments within several estuarine systems from Argentina: 1980–2005 period. Chemosphere 65(2): 213–223.

Del Moral-Flores, L.F., Morrone, J.J., Alcocer Durand, J., Espinosa-Pérez, H. and Pérez-Ponce De León, G. 2015. Lista patrón de los tiburones, rayas y quimeras (Chondrichthyes, Elasmobranchii, Holocephali) de México. Arx. Misc. Zool. 13: 47–163.

Di Marzio, A., Lambertucci, S., Fernandez, A.J. and Martínez-López, E. 2019. From Mexico to the Beagle Channel: A review of metal and metalloid pollution studies on wildlife species in Latin America. Environ. Res. 176: 108462.

Domi, N., Bouquegneau, J.M. and Das, K. 2005. Feeding ecology of five commercial shark species of the Celtic Sea through stable isotope and trace metal analysis. Mar. Environ. Res. 60(5): 551–569. https://doi.org/10.1016/j.marenvres.2005.03.001

Dutton, J. and Venuti, V.M. 2019. Comparison of maternal and embryonic trace element concentrations in common thresher shark (*Alopias vulpinus*) muscle tissue. Bull. Environ. Contam. Toxicol. 103(3): 380–384.

Ebert, D., Dando, M. and Fowler, S. 2021. Sharks of the World: A Complete Guide. Princeton University Press.

Endo, T., Haraguchi, K. and Sakata, M. 2002. Mercury and selenium concentrations in the internal organs of toothed whales and dolphins marketed for human consumption in Japan. Sci. Tot. Environ. 300(1–3): 15–22.

Endo, T., Hisamichi, Y., Haraguchi, K., Kato, Y., Ohta, C. and Koga, N. 2008. Hg, Zn and Cu levels in the muscle and liver of tiger sharks (*Galeocerdo cuvier*) from the coast of Ishigaki Island, Japan: Relationship between metal concentrations and body length. Mar. Pollut. Bull. 56: 1774–1780.

EPA (Environmental Protection Agency). 2000. Integrated Risk Information System (IRIS) on Cadmium. 65(8): 1863–1865. National Center for Environmental Assessment, Office of Research and Development, Washington, DC.

European Commission Regulation (EU). 2014. No 488/2014 of 12 May 2014. Amending Regulation (EC). No 1881/2006 as regards maximum levels of cadmium in foodstuffs. L138/75-L138/79

Eustace, I.J. 1974. Zinc, Cadmium, Copper and Manganese in species of finfish and shellfish caught in the derwent estuary, Tasmania. Aust. J. Mar. Freshw. Res. 25(2): 209–220.

Faroon, O., Ashizawa, A., Wright, S., Tucker, P., Jenkins, K., Ingerman, L., et al. 2012. Toxicological Profile for Cadmium. ATSDR (US). Atlanta, GA, USA.

Fernández, J.I., Álvarez-Torres, P., Arreguín-Sánchez, F., López-Lemus, L.G., Ponce, G., Díaz-de León, A. 2011. Coastal fisheries of Mexico. pp. 231–284. *In*: Salas, S., Chuenpagdee, R., Charles, A. and Seijo, J.C. (eds). Coastal Fisheries of Latin America and the Caribbean. FAO Fisheries and Aquaculture Technical Paper. Rome, Italy.

Filippini, T., Torres, D., Lopes, C., Carvalho, C., Moreira, P., Naska, A., et al. 2020. Cadmium exposure and risk of breast cancer: A dose-response meta-analysis of cohort studies. Environ. Int. 142: 105879. https://doi.org/10.1016/j.envint.2020.105879

Flores, H.D. and Ramos, M.J. 2004. Las pesquerías Artesanales en el Golfo de México. In: Manejo Costero en México. pp. 540–550. Centro EPOMEX, Universidad Autónoma de Campeche.

Flores, J. and Albert, L.A. 2005. Environmental lead in Mexico, 1990–2002. Rev. Environ. Contam. Toxicol. 181: 37–109.

Flores-Galván, M.A., Daesslé, L.W., Arellano-García, E., Torres-Bugarín, O., Macías-Zamora, J.V. and Ruiz-Campos, G. 2020. Genotoxicity in fishes environmentally exposed to As, Se, Hg, Pb, Cr and toxaphene in the lower Colorado River basin, at Mexicali valley, Baja California, México. Ecotoxicol. 29(4): 493–502.

Flores-Hernandez, D., Ramos-Miranda, J., 2004. Las pesquerías Artesanales en el Golfo de Mexico. In: El manejo costero en Mexico. Campeche, Mexico: EPOMEXUniversidad Autonoma de Campeche, SEMARNAT, Campeche.

Frías-Espericueta, M.G., Cardenas-Nava, N.G., Márquez-Farías, J.F., Osuna-López, J.I., Muy-Rangel, M.D., Rubio-Carrasco, W., et al. 2014. Cadmium, copper, lead and zinc concentrations in female and embryonic pacific sharpnose shark (*Rhizoprionodon longurio*) tissues. Bull. Environ. Contam. Toxicol. 93(5): 532–535. https://doi.org/10.1007/s00128-014-1360-0

Fricke, R., Eschmeyer, W.N. and Van der Laan, R. (eds) 2022. Eschmeyer's Catalog of Fishes: Genera, Species, References. (http://researcharchive.calacademy.org/research/ichthyology/catalog/fishcatmain.asp). Electronic version accessed 08 Agosto 2022.

Galván-Magaña, F., Escobar-Sánchez, O. and Carrera-Fernández, M. 2011. Embryonic bicephaly in the blue shark, *Prionace glauca*, from the Mexican Pacific Ocean. Mar. Biodivers. Rec. 4.

García-Hernández, J., Cadena-Cárdenas, L., Betancourt-Lozano, M., García-de La Parra, L.M., García-Rico, L. and Márquez-Farías, F. 2007. Total mercury content found in edible tissues of top predator fish from the Gulf of California, Mexico. Toxicol. Environ. Chem. 89(3): 507–522.

Gibbs, J.P. and Miskiewicz, A.G. 1995. Heavy metals in fish near a major primary treatment sewage plant outfall. Mar. Pollut. Bull. 30: 667–674.

Gilbert, J.M., Reichelt-Brushett, A.J., Butcher, P.A., McGrath, S.P., Peddemors, V.M., Bowling, A.C., et al. 2015. Metal and metalloid concentrations in the tissues of dusky *Carcharhinus obscurus*, sandbar *C. plumbeus* and white *Carcharodon carcharias* sharks from south-eastern Australian waters, and the implications for human consumption. Mar. Pollut. Bull. 92(1–2): 186–194.

Gilman, H. and Leeper, R.W. 1951. Organometallic compounds of lead, tin and germanium. J. Org. Chem. 16(3): 466–475.

Gutiérrez-Zavala, R.M. and Cabrera-Mancilla, E. 2019. Especies marinas de valor comercial en el estado de Guerrero. México: Instituto Nacional de Pesca y Acuacultura.

Hauser-Davis, R.A. 2020. The current knowledge gap on metallothionein mediated metal-detoxification in Elasmobranchs. PeerJ. 8: e10293. https://doi.org/10.7717/peerj.10293

Hauser-Davis, R.A., Rocha, R.C.C., Saint'Pierre, T.D. and Adams, D.H. 2021. Metal concentrations and metallothionein metal detoxification in blue sharks, *Prionace glauca* L. from the Western North Atlantic Ocean. J. Trace. Elem. Med. Biol. 68: 126813.

Haynes, D. and Michalek-Wagner, K. 2000. Water quality in the great barrier reef world heritage area: Past perspectives, current issues and new research directions. Mar. Pollut. Bull. 41(7–12): 428–434.

Heithaus, M., Frid, A., Vaudo, J., Worm, B. and Wirsing, A. 2010. Unraveling the ecological importance of elasmobranchs. pp. 611–637. *In*: Sharks and their relatives II: Biodiversity, Adaptive Physiology, and Conservation, CRC Press.

Helfman, G., Collette, B.B., Facey, D.E. and Bowen, B.W. 2009. The Diversity of Fishes: Biology, Evolution, and Ecology, 2nd Ed. Wiley-Blackwell, Hoboken, New Jersey.

Hidalgo, J., Tort, L. and Flos, R. 1985. Cd-, Zn-, Cu-binding protein in the elasmobranch *Scyliorhinus canicula*. Comp. Biochem. Physiol. C. Toxicol. Pharmacol. 81: 159–165.

Hornung, H., Krom, M.D., Cohen, Y. and Bernhard, M. 1993. Trace metal content in deep-water sharks from the eastern Mediterranean Sea Mar. Biol. 115.

Hurtado-Banda, R., Gomez-Alvarez, A., Márquez-Farías, J.F., Cordoba-Figueroa, M., Navarro-García, G. and Medina-Juárez, L.Á. 2012. Total mercury in liver and muscle tissue of two coastal sharks from the Northwest of Mexico. Bull. Environ. Contam. Toxicol. 88(6): 971–975.

Incardona, J.P., Gardner, L.D., Linbo, T.L., Brown, T.L., Esbaugh, A.J., Mager, E.M., et al. 2014. Deepwater Horizon crude oil impacts the developing hearts of large predatory pelagic fish. Proc. Natl. Acad. Sci. 111(15): E1510–518.

Jaishankar, M., Tseten, T., Anbalagan, N., Mathew, B.B. and Beeregowda, K.N. 2014. Toxicity, mechanism and health effects of some heavy metals. Interdiscip. Toxicol. 7(2): 60–72.

Ji, S., Gao, L., Chen, W., Su, J. and Shen, Y. 2020. Urea application enhances cadmium uptake and accumulation in Italian ryegrass. Environ. Sci. Pollut. Res. 27: 34421–34433.

Jonathan, M.P., Aurioles-Gamboa, D., Campos Villegas, L.E., Bohórquez-Herrera, J., Hernández-Camacho, C.J. and Sujitha, S.B. 2015. Metal concentrations in demersal fish species from Santa Maria Bay, Baja California Sur, Mexico (Pacific coast). Mar. Pollut. Bull. 99(1–2): 356–361.

Kajimura, M., Walsh, P.J., Mommsen, T.P. and Wood, C.M. 2006. The Dogfish Shark (*Squalus acanthias*) Increases both hepatic and extrahepatic ornithine urea cycle enzyme activities for nitrogen conservation after feeding. Physiol. Biochem. Zool. 79: 602–613.

Kehrig, H., Hauser-Davis, R., Seixas, T., Pinheiro, A. and APM, D.B. 2016. Mercury species, selenium, metallothioneins and glutathione in two dolphins from the southeastern Brazilian coast: mercury detoxification and physiological differences in diving capacity. Environ. Pollut. 213: 785–792.

Klaine, S.J., Alvarez, P.J.J., Batley, G.E., Fernandes, T.F., Handy, R.D., Lyon, D.Y. et al. 2008. Nanomaterials in the environment: Behaviour, fate, bioavailablity, and effects. Environ. Toxicol. Chem. 27(9): 1825–1851.

Klinck, J.S., Ng, T.Y. and Wood, C.M. 2009. Cadmium accumulation and in vitro analysis of calcium and cadmium transport functions in the gastro-intestinal tract of trout following chronic dietary cadmium and calcium feeding. Comp. Biochem. Physiol. C. Toxicol. Pharmacol. 150: 349–360.

Kumar, P. and Singh, A. 2010. Cadmium toxicity in fish: An overview. GERF Bulletin of Biosciences 1(1): 41–47. www.gerfbb.com

Kyne, P.M. and Simpfendorfer, C.A. 2007. A collation and summarization of available data on deepwater chondrichthyans: Biodiversity, life history and fisheries. pp. 1–137. *In*: Report by the IUCN SSC Shark Specialist Group for the Marine Conservation Biology Institute.

Lambert, M., Leven, B.A. and Green, R.M. 2000. New methods of cleaning up heavy metal in soils and water. Environmental Science and Technology Briefs for Citizens. 7(4): 133–163.

Lara, A., Galván-Magaña, F., Elorriaga-Verplancken, F., Marmolejo-Rodríguez, A.J., Gonzalez-Armas, R., Arreola-Mendoza, L., et al. 2020. Bioaccumulation and trophic transfer of potentially toxic elements in the pelagic thresher shark *Alopias pelagicus* in Baja California Sur., Mexico. Mar. Pollut. Bull. 156: 111192.

Lopes, T.O.M., Passos, L.S., Vieira, L.V., Pinto, E., Dorr, F., Scherer, R., et al. 2020. Metals, arsenic, pesticides, and microcystins in tilapia (*Oreochromis niloticus*) from aquaculture parks in Brazil. Environl. Sci. Poll. Res. 27(16): 20187–20200.

Man, Y.B., Wu, S.C. and Wong, M.H. 2014. Shark fin, a symbol of wealth and good fortune may pose health risks: the case of mercury. Environ. Geochem. Health. 36: 1015–1027.

Marcovecchio, J.E., Moreno, V.J. and Pérez, A. 1991. Metal accumulation in tissues of sharks from the Bahía Blanca estuary, Argentina. Mar. Environ. Res. 31: 263–274.

Martínez-Ortíz, J. and García-Domínguez, M. 2013. Chondrichthyes of Ecuador Field Guide. Chimaeras, Sharks and Rays. Manta: Ministry of Agriculture, livestock, Aquaculture and Fisheries (MAGAP)/Vice Ministry of Aquaculture and Fisheries (VMAP)/ Undersecretariat of Fisheries Resources (SRP).

Martins, M.F., Costa, P.G., Gadig, O.B.F. and Bianchini, A. 2021. Metal contamination in threatened elasmobranchs from an impacted urban coast. Sci. Total. Environ. 757: 143803.

Mathews, T. and Fisher, N.S. 2009. Dominance of dietary intake of metals in marine elasmobranch and teleost fish. Sci. Total Environ. 407(18): 5156–5161. https://doi. org/10.1016/j.scitotenv.2009.06.003

Matulik, A.G., Kerstetter, D.W., Hammerschlag, N., Divoll, T., Hammerschmidt, C.R. and Evers, D.C. 2017. Bioaccumulation and biomagnification of mercury and methylmercury in four sympatric coastal sharks in a protected subtropical lagoon. Mar. Pollut. Bull. 116(1–2): 357–364.

Maz-Courrau, A., López-Vera, C., Galván-Magaña, F., Escobar-Sánchez, O., Rosíles-Martínez, R. and Sanjuán-Muñoz, A. 2012. Bioaccumulation and biomagnification of total mercury in four exploited shark species in the Baja California Peninsula, Mexico. Bull. Environ. Cont. Toxicol. 88(2): 129–134.

McClusky, L.M. 2006. Stage-dependency of apoptosis and the blood-testis barrier in the dogfish shark (*Squalus acanthias*): Cadmium-induced changes as assessed by vital fluorescence techniques. Cell Tissue Res. 325: 541–553.

McClusky, L.M. 2008. Cadmium accumulation and binding characteristics in intact Sertoli/ germ cell units, and associated effects on stage-specific functions in vitro: insights from a shark testis model. J. Appl. Toxicol. 28.

McKinney, M.A., Dean, K., Hussey, N.E., Cliff, G., Wintner, S.P., Dudley, S.F.J., et al. 2016. Global versus local causes and health implications of high mercury concentrations in sharks from the east coast of South Africa. Sci. Total Environ. 541: 176–183.

McMeans, B.C., Borgå, K., Bechtol, W.R., Higginbotham, D. and Fisk, A.T. 2007. Essential and non-essential element concentrations in two sleeper shark species collected in arctic waters. Environ. Pollut. 148: 281–290.

Medina-Morales, S.A., Corro-Espinosa, D., Escobar-Sánchez, O., Delgado-Alvarez, C.G., Ruelas-Inzunza, J., Frías-Espericueta, M.G., et al. 2020. Mercury (Hg) and selenium (Se) content in the shark *Mustelus henlei* (Triakidae) in the northern Mexican Pacific. Environ. Sci. Poll. Res. Int. 27(14): 16774–16783.

Mendoza-Díaz, F., Serrano, A., Cuervo-López, L., López-Jiménez, A., Galindo, J.A. and Basañez-Muñoz, A. 2013. Concentración de Hg, Pb, Cd, Cr y As en hígado de *Carcharhinus limbatus* (Carcharhiniformes: Carcharhinidae) capturado en Veracruz, México. Rev. Biol. Trop. 61(2): 821–828.

Mille, T., Cresson, P., Chouvelon, T., Bustamante, P., Brach-Papa, C. and Sandrine, B., et al. 2018. Trace metal concentrations in the muscle of seven marine species: Comparison between the Gulf of Lions (North-West Mediterranean Sea) and the Bay of Biscay (North-East Atlantic Ocean). Mar. Pollut. Bull. 135: 9–16. https://doi.org/10.1016/j.marpolbul.2018.05.051

Mohammed, A. and Mohammed, T. 2017. Mercury, arsenic, cadmium and lead in two commercial shark species (*Sphyrna lewini* and *Carcharinus porosus*) in Trinidad and Tobago. Mar. Pollut. Bull. 119: 214–218.

Monteiro-Neto, C., Vinhas Itavato, R. and De souza Morales, L.E. 2003. Concentrations of heavy metals in *Sotalia fluviatilis* (Cetacea: Delphinidae) off the coast of Ceará, northeast Brazil. Environ. Poll. 123: 319–324.

Moore, A.B.M., Bolam, T., Lyons, B.P. and Ellis, J.R. 2015. Concentrations of trace elements in a rare and threatened coastal shark from the Arabian Gulf (smoothtooth blacktip *Carcharhinus leiodon*). Mar. Pollut. Bull. 100(2): 646–650.

Morais, S., García, e. Costa, F., Pererira, M. and de, L. 2012. Heavy metals and human health. pp. 227–245. *In*: Oosthuizen, J. (ed.). Environmental Health—Emerging Issues and Practice. Rijeka, Croatia: InTech.

Mull, C., Blasius, M., O'Sullivan, J. and Lowe, C. 2012. Metals, Trace elements, and organochlorine contaminants in muscle and liver tissue of juvenile white sharks (*Carcharodon carcharias*) from the Southern California Bight. pp. 59–75. *In*: Global Perspectives on the Biology and Life History of Great White Sharks. CRC Press, Boca Raton, Florida.

Murawski, S.A., Hogarth, W.T., Peebles, E.B. and Barbeiri, L. 2014. Prevalence of external skin lesions and polycyclic aromatic hydrocarbon concentrations in Gulf of Mexico fishes, Post-Deepwater Horizon. Trans. Am. Fish. Soc. 143: 1084–1097.

Murillo-Cisneros, D.A., Zenteno-Savín, T., Harley, J., Cyr, A., Hernández-Almaraz, P., Gaxiola-Robles, R., et al. 2021. Mercury concentrations in Baja California Sur fish: Dietary exposure assessment. Chemosphere 267: 129233.

Norris, S.B., Reistad, N.A. and Rumbold, D.G. 2021. Mercury in neonatal and juvenile blacktip sharks (*Carcharhinus limbatus*). Part II: Effects assessment. Ecotoxicology 30(2): 311–322.

Nowell, L.H., Ludtke, A.S., Mueller, D.K. and Scott, J.C. 2013. Organic contaminants, trace and major elements, and nutrients in water and sediment sampled in response to the Deepwater Horizon oil spill. pp. 96. U.S. Geological Survey Open-File Report 2012–5228.

Núñez Nogueira, G., Bautista-Ordoñez and Janitzio Rósiles-Martínez, R. 1998. Concentración y distribución de mercurio en tejidos del cazón *(Rhizopreonodon terraenovae)* del Golfo México. Veterinaria México 29(1): 15–21.

Núñez-Nogueira, G. 2005. Concentration of essential and non-essential metals in two shark species commonly caught in Mexican (Gulf of Mexico) coastline. pp. 451–473. *In*: Vázquez-Botello, A., Rendón-von Osten, J., Gold-Bouchot, G. and Agraz-Hernández, C. (eds). Golfo de México Contaminación e Impacto Ambiental: Diágnostico y Tendencias. México: Universidad Autónoma de Campeche, Universidad Nacional Autónoma de México, Instituto Nacional de Ecología.

Nunez-Nogueira, G., Mouneyrac, C., Amiard, J.C. and Rainbow, P.S. 2006. Subcellular distribution of zinc and cadmium in the hepatopancreas and gills of the decapod crustacean *Penaeus indicus*. Mar. Biol. 150: 197–211.

Núñez-Nogueira, G., Pérez-López, A. and Santos-Córdova, J.M. 2019. As, Cr, Hg, Pb, and Cd concentrations and bioaccumulation in the dugong *Dugong dugong* and manatee *Trichechus manatus*: A review of body burdens and distribution. International Journal of Environ. Res. Public Health, 16(3): 404.

Pacyna, J. and Pacyna, E. 2001. An assessment of global and regional emissions of trace metals to the atmosphere from anthropogenic sources worldwide. Env. Rev. 9(4): 269–298.

Páez-Osuna, F., Álvarez-Borrego, S., Ruiz-Fernández, A.C., García-Hernández, J., Jara-Marini, M.E., Bergés-Tiznado, M.E. et al. 2017. Environmental status of the Gulf of California: A pollution review. Earth-Sci. Rev. 166: 181–205.

Pancaldi, F., Galván-Magaña, F., González-Armas, R., Soto-Jimenez, M.F., Whitehead, D.A., O'Hara, T., et al. 2019a. Mercury and selenium in the filter–feeding whale shark (*Rhincodon typus*) from two areas of the Gulf of California, Mexico. Mar. Pollut. Bull. 146: 955–961.

Pancaldi, F., Páez-Osuna, F., Soto-Jiménez, M.F., González-Armas, R., O'Hara, T., Marmolejo-Rodríguez, A.J., et al. 2019b. Trace Elements in tissues of whale sharks (*Rhincodon typus*) stranded in the Gulf of California, Mexico. Bull. Environ. Cont. Toxicol. 103(4): 515–520.

Pancaldi, F., Páez-Osuna, F., Marmolejo-Rodríguez, A.J., Whitehead, D.A., González-Armas, R., Soto-Jiménez, M.F., et al. 2021a. Variation of essential and non-essential trace elements in whale shark epidermis associated to two different feeding areas of the Gulf of California. Environ. Sci. Pollut. Res. 28: 36803–36816.

Pancaldi, F., Marmolejo-Rodríguez, A.J., Soto-Jiménez, M.F., Murillo-Cisneros, D.A,. Becerril-García, E.E., Whitehead, D.A., et al. 2021b. Trace elements in the whale shark *Rhincodon typus* liver: An indicator of the health status of the ecosystem base (plankton). Lat. Am. J. Aquat. Res. 49(2): 359–364.

Pantoja-Echevarría, L.M., Marmolejo-Rodríguez, A.J., Galván-Magaña, F., Arreola-Mendoza, L., Tripp-Valdéz, A., Verplancken, F.E., et al. 2020. Bioaccumulation and trophic transfer of Cd in commercially sought brown smoothhound *Mustelus henlei* in the western coast of Baja California Sur, Mexico. Mar. Pollut. Bull. 151: 110879.

Pastorelli, A.A., Baldini, M., Stacchini, P., Baldini, G., Morelli, S., Sagratella, E., et al. 2012. Human exposure to lead, cadmium and mercury through fish and seafood product consumption in Italy: a pilot evaluation. Food Addit Contam, Part A, Chem. Anal. Control Expo. Risk Assess. 29: 1913–1921.

Pérez-Moreno, V., Ramos-López, M.Á., Zavala-Gómez, C.E. and Rico Rodríguez, M,Á. 2016. Heavy metals in seawater along the mexican pacific coast. Interciencia 41(6): 419–423. http://www.redalyc.org/articulo.oa?id=33945816008

Pinheiro, M.A., Silva, P.P.G., De Almeida Duarte, L.F., A.A. and Zanotto, F.P. 2012. Accumulation of six metals in the mangrove crab *Ucides cordatus* (Crustacea: Ucididae) and its food source, the red mangrove *Rhizophora mangle* (Angiosperma: Rhizophoraceae). Ecotoxicol. Environ Saf. 81: 114–121.

Powell, J.H. and Powell, R.E. 2001. Trace elements in fish overlying subaqueous tailing in the tropical West Pacific. Water Air Soil Pollut. 125: 81–104.

Powell, J.H., Powell, R.E. and Fielder, D.R. 1981. Trace element concentrations in tropical marine fish at Bougainville Island, Papua New Guinea. Water Air Soil Pollut. 16(2): 143–158. https://doi.org/10.1007/BF01046851

Priede, I.G., Froese, R., Bailey, D.M., Bergstad, O.A., Collins, M.A., Dyb, J.E. et al. 2006. The absence of sharks from abyssal regions of the world's oceans. Proc. Royal. Soc. B: Biol. Sci. 273(1592): 1435–1441.

Prohaska, B.K., Talwar, B.S. and Grubbs, R.D. 2021. Blood biochemical status of deep-sea sharks following longline capture in the Gulf of Mexico. Conserv. Physiol. 9: 1–9.

Queen, M.A.J., Bright, K., Delphine, S.M. and Udhaya, P.A. 2020. Spectroscopic investigation of supramolecular organometallic compound L-threonine cadmium acetate monohydrate. Spectrochim Acta A: Mol. Biomol. Spectrosc. 228: 117802.

Ramírez-Ayala, E., Arguello-Pérez, M.A., Tintos-Gómez, A., Pérez, J.A.M., Díaz-Gómez, J.A., Pérez-Rodríguez, R.Y., et al. 2021. Heavy metals in sediment and fish from two coastal lagoons of the Mexican Central Pacific. Lat. Am. J. Aquat. Res. 49(5): 818–827.

Ramírez-Ayala, E., Arguello-Pérez, M.A., Tintos-Gómez, A., Pérez-Rodríguez, R., Díaz-Gómez, J.A., Borja-Gómez, I. et al. 2020. Review of the biomonitoring of persistent, bioaccumulative, and toxic substances in aquatic ecosystems of Mexico: 2001–2016. Lat. Am. J. Aquat. Res. 48(5): 705–738.

Reátegui-Quispe, A. and Pariona-Velarde, D. 2019. Determinación de plomo, cadmio, mercurio y Bases Volátiles Nitrogenadas Totales (NBVT) en el músculo de tiburón azul *Prionace glauca* procedente de la zona sur del Perú. Rev. Biol. Mar. Oceanogr. 54(3): 336–342.

Roberts, S.B., Lane, T.W. and Morel, F.M.M. 1997. Carbonic anhydrase in the marine diatom *Thalassiosira weissflogii* (Bacillariophyceae). J. Phycol. 33(5): 845–850.

Rodriguez-Romero, J., Simeón-De La Cruz, A., Ochoa-Díaz, M.R. and Monsalvo-Spencer, P. 2019. New report of malformations in blue shark embryos (*Prionace glauca*) from the western coast of Baja California Sur, Mexico. J. Mar. Biolog. Assoc. U.K. 99(2): 497–502.

Ruelas-Inzunza, J. and Páez-Osuna, F. 2007. Essential and toxic metals in nine fish species for human consumption from two coastal lagoons in the Eastern Gulf of California. J. Environ. Sci., Health A Tox Hazard Subst. Environ. Eng. 42: 1411–1416.

Ruelas-Inzunza, J., Hernández-Osunam, J. and Páez-Osuna, F. 2011a. Total and organic mercury in ten fish species for human consumption from the Mexican Pacific. Bull. Environ. Contam. Toxicol. 86(6): 679–683.

Ruelas-Inzunza, J., Páez-Osuna, F., Ruiz-Fernández, A.C. and Zamora-Arellano, N. 2011b. Health risk associated to dietary intake of mercury in selected coastal areas of Mexico. Bull. Environ. Contam. Toxicol. 86(2): 180–188.

Ruelas-Inzunza, J., Delgado-Alvarez, C., Frías-Espericueta, M. and Páez-Osuna, F. 2013. Mercury in the atmospheric and coastal environments of Mexico. Rev. Environ. Contam. Toxicol. 226: 65–99.

Ruelas-Inzunza, J., Amezcua, F., Coiraton, C. and Páez-Osuna, F. 2020. Cadmium, mercury, and selenium in muscle of the scalloped hammerhead *Sphyrna lewini* from the tropical Eastern Pacific: Variation with age, molar ratios and human health risk. Chemosphere 242: 125180.

SAGARPA (Secretaría de Agricultura, Ganadería, Desarrollo Rural, Pesca y Alimentación). Norma Oficial Mexicana NOM-029-PESC-2006. Pesca responsable de tiburones y rayas. Especificaciones para su aprovechamiento.

Saldaña-Ruiz, L.E., García-Rodríguez, E., Pérez-Jiménez, J.C., Tovar-Ávila. J. and Rivera-Téllez, E. 2019. Biodiversity and conservation of sharks in Pacific Mexico. Adv. Mar. Biol. 83: 11–60.

Schmidt, L., Novo, D.L.R., Druzian, G.T., Landero, J.A., Caruso, J., Mesko, M.F., et al. 2021. Influence of culinary treatment on the concentration and on the bioavailability of cadmium, chromium, copper, and lead in seafood. J. Trace Elem. Med. Biol. 65: 126717.

Secretaría de Economía. 2017. Perfil de mercado de la maca. Dirección General de Desarrollo Minero, 4–5.

Seijo, J.C., Caddy, J.F., Arzápalo, W.W. and Cuevas, A.J. 2013. Considerations for an ecosystem approach to fisheries management in the southern Gulf of Mexico. pp. 319–336. *In*: Gulf of Mexico Origin, Waters, and Biota: Vol. 4, Ecosystem-Based Management. Texas A&M University Press, Project Muse.

Seixas, T.G., Kehrig, H.A., Di Beneditto, A.P.M., Souza, C.M.M., Malm, O. and Moreira, I. 2009. Essential (Se, Cu) and non-essential (Ag, Hg, Cd) elements: What are their relationships in liver of *Sotalia guianensis* (Cetacea, Delphinidae)? Mar. Pollut. Bull. 58(4): 629–634.

Shipley, O.N., Lee, C.S., Fisher, N.S., Sternlicht, J.K., Kattan, S., Staaterman, E.R., et al. 2021. Metal concentrations in coastal sharks from The Bahamas with a focus on the Caribbean Reef shark. Sci. Rep. 11(1): 1–11.

Simpson, W.R. 1978. A critical review of Cadmium in the marine enviroment. Prog. Oceanogr. 10(1): 1–70.

Soto, L.A. and Vázquez-Botello, A. 2013. Legal Issues and Scientific Constraints in the Environmental Assessment of the Deepwater Horizon Oil Spill in Mexico Exclusive Economic Zone (EEZ) in the Gulf of Mexico. Int. J. Geosci. 04: 39–45.

Souza-Araujo, J., Souza-Junior, O.G., Guimarães-Costa, A., Hussey, N.E., Lima, M.O. and Giarrizzo, T. 2021. The consumption of shark meat in the Amazon region and its implications for human health and the marine ecosystem. Chemosphere 265: 129132.

SSA (Secretaría de Salud). 2011. Norma Oficial Mexicana NOM-242-SSA1-2009, Productos y servicios. Productos de la pesca frescos, refrigerados, congelados y procesados. Especificaciones sanitarias y métodos de prueba. DOF (Diario Oficial de la Federación).

Stevens, J. and McLoughlin, K. 1991. Distribution, size and sex composition, reproductive biology and diet of sharks from Northern Australia. Mar. Freshw. Res. 42(2): 151–199.

Storelli, M.M., Ceci, E., Storelli, A. and Marcotrigiano, G.O. 2003. Polychlorinated biphenyl, heavy metal and methylmercury residues in hammerhead sharks: Contaminant status and assessment. Mar. Pollut. Bull. 46: 1035–1039.

Surgiewicz, J., 2012. Ocena zagrozeń organicznymi zwiazkami metali w przemysłowych procesach produkcji i przetwarzania polichlorku winylu—Assessment of hazards posed by metallo-organic compounds in industrial production and processing of polyvinyl chloride. Medycyna. Pracy. 63(4): 419–429.

Taguchi, M.K., Toda, Y.S. and Shimizu, M. 1979. Study of metal contents of elasmobranch fishes. 1. Metal concentration in the muscle tissues of a dogfish, *Squalus mitsukurii*. Mar. Environ. Res. 2: 239–249.

Teffer, A.K., Staudinger, M.D., Taylor, D.L. and Juanes, F. 2014. Trophic influences on mercury accumulation in top pelagic predators from offshore New England waters of the northwest atlantic ocean. Mar. Environ. Res. 101(1): 124–134. https://doi.org/10.1016/j.marenvres.2014.09.008

Terrazas-López, R., Arreola-Mendoza, L., Galván-Magaña, F., Anguiano-Zamora, M., Sujitha, S.B. and Jonathan, M.P. 2016. Cadmium concentration in liver and muscle of silky shark (*Carcharhinus falciformis*) in the tip of Baja California south, México. Mar. Pollut. Bull. 107(1): 389–392.

Tiktak, G.P., Butcher, D., Lawrence, P.J., Norrey, J., Bradley, L., Shaw, K. et al. 2020. Are concentrations of pollutants in sharks, rays and skates (*Elasmobranchii*) a cause for concern? A systematic review. Mar. Pollut. Bull. 160: 111701.

Turoczy, N.J., Laurenson, L.J.B., Allinson, G., Nishikawa, M., Lambert, D.F., Smith, C., et al. 2000. Observations on metal concentrations in three species of shark (*Deania calcea, Centroscymnus crepidater*, and *Centroscymnus owstoni*) from Southeastern Australian waters. J. Agric. Food. Chem. 48: 4357–4364.

Unión Europea. 2009. Metales Pesados: Contenidos Máximos en Metales Pesados en Productos Alimenticios.

Vas, P. 1991. Trace metal levels in sharks from British and Atlantic waters. Mar. Pollut. Bull. 22: 67–72.

Vega-Barba, C. 2018. Elementos potencialmente tóxicos en el tiburón piloto *Carcharhinus falciformis* y sus presas en la costa sur de Jalisco. Instituto Poltécnico Nacional.

Vélez-Alavez, M., Labrada-Martagón, V., Méndez-Rodriguez, L.C., Galván-Magaña, F. and Zenteno-Savín, T. 2013. Oxidative stress indicators and trace element concentrations in tissues of mako shark (*Isurus oxyrinchus*). Comp. Biochem. Phys. A. Mol. Int. Phys. 165(4): 508–514.

Villanueva, S. and Botello, A.V. 1992. Metales pesados en la zona costera del Golfo de México y Caribe Mexicano: Una revisión. Rev. Int. Contam. Amb. 8(1): 47–61.

Vosylienė, M.Z. 2007. Review of the methods for acute and chronic toxicity assessment of single substances, effluents and industrial waters. Act. Zoo. Lit. 17(1): 3–15.

Weigmann, S. 2016. Annotated checklist of the living sharks, batoids and chimaeras (Chondrichthyes) of the world, with a focus on biogeographical diversity. J. Fish Biol. 88(3): 837–1037.

Windom, H., Stickney, R., Smith, R., White, D. and Taylor, F. 1973. Arsenic, Cadmium, Copper, Mercury, and Zinc in some species of North Atlantic Finfish. J. Fish. Res. Board Can. 30(2): 275–279.

Wise, J.P., Wise, J.T.F., Wise, C.F., Wise, S.S., Gianios, C., Xie, H., et al. 2014. Concentrations of the genotoxic metals, chromium and nickel, in whales, tar balls, oil slicks, and released oil from the Gulf of Mexico in the immediate aftermath of the deepwater horizon oil crisis: Is genotoxic metal exposure part of the deepwater horizon legacy? Environ. Sci. Tech. 48(5): 2997–3006.

Wourms, J.P. and Demski, L.S. 1993. The reproduction and development of sharks, skates, rays and ratfishes: introduction, history, overview, and future prospects. pp. 7–21. *In*: Wourms, J.P. and Demski, S. (eds). The Reproduction and Development of Sharks, Skates, Rays and Ratfishes. Springer, Dordrecht.

Xu, Y., Feng, L., Jeffrey, P.D., Shi, Y. and Morel, F.M.M. 2008 Structure and metal exchange in the cadmium carbonic anhydrase of marine diatoms. Nature 452(7183): 56–61.

Yu, X., Khan, S., Khan, A., Tang, Y., Nunes, L.M., Yan, J., et al. 2020. Methyl mercury concentrations in seafood collected from Zhoushan Islands, Zhejiang, China, and their potential health risk for the fishing community: Capsule: Methyl mercury in seafood causes potential health risk. Environ. Int. 137: 105420.

Zhai, Q., Narbad, A. and Chen, W. 2015. Dietary strategies for the treatment of cadmium and lead toxicity. Nutrients 7: 552–571.

Chapter 10

A Review on Cadmium, Mercury, and Lead Loads in *Pontoporia blainvillei*, the Most Endangered Dolphin Species from the Southwest Atlantic

Leila Soledade Lemos*[1] and Rachel Ann Hauser-Davis[2]

[1] Institute of Environment, Florida International University,
North Miami, United States.
Email: *leslemos@hotmail.com*

[2] Oswaldo Cruz Foundation, Rio de Janeiro, Brazil
Email: *rachel.hauser.davis@gmail.com*

INTRODUCTION

Historically, human activities have critically impacted the marine ecosystem. Habitat degradation, over exploitation of marine resources, and pollution are just some of the alarming issues that significantly threaten marine biota. Chemical pollution is of particular concern, with many pollutants displaying a tendency towards persistence, bioaccumulation and toxicity (EPA, 2021), comprising a significant negative stressor for aquatic ecosystems. The main classes of chemical

---

*Corresponding author: leslemos@hotmail.com

contaminants in these ecosystems (i.e., hydrocarbons, pesticides, polychlorinated, polybrominated and perfluorinated compounds, and metals) are characterized by their environmental persistence, bioavailability, tendency to bioaccumulate along the trophic chain and potential toxic effects (Ali et al., 2019; Chen et al., 2014; Cui et al., 2020; Cullen et al., 2019; Pulster et al., 2020), mainly affecting top predators (Ali and Khan, 2019).

Metals, in particular, although naturally occurring compounds, are introduced in significant amounts into aquatic environments, even more so in coastal environments, through different anthropogenic activities including mineral extraction, pesticide and fertilizer use and industrial, agricultural, and domestic waste disposals. Essential metals such as copper (Cu), selenium (Se), and zinc (Zn) play important roles in biological systems, participating in many metabolic and biochemical processes (Saad et al., 2016). Although these metals are essential at low doses, they may become toxic at high concentrations (Zoroddu et al., 2019). Non-essential metals such as cadmium (Cd), mercury (Hg), and lead (Pb), however, may provoke toxic effects at even extremely low doses (ATSDR, 2019; Chowdhury and Chandra, 1987).

These elements are three of the most toxic metals to living organisms (ATSDR, 2019), implicated as the cause of both deleterious sublethal effects, such as oxidative stress, and endpoint mutagenic, carcinogenic and neurotoxic effects, as well as hormonal cycle alterations and immunosuppressant consequences (ATSDR, 1999; 2012; 2020). They are also capable of accumulating in target tissues, such as reproductive organs, paving the way for decreased fertility and populational declines.

Marine mammal populations have undergone populational declines worldwide due to multicausal factors such as hunting, fisheries entanglements, climate change effects and chemical pollution. However, although chemical pollution assessments have been carried out for decades, significant knowledge gaps concerning several marine mammal species are still noted, due to inherent difficulties in sampling marine mammals, and infrequent stranding events. This is the case for the most endangered dolphin species from the Southwest Atlantic, the Franciscana dolphin (*Pontoporia blainvillei*). In this context, this chapter provides a comprehensive assessment on the current information on Cd, Hg, and Pb concentrations in this species.

FRANCISCANA DOLPHINS

Franciscana dolphins (*Pontoporia blainvillei*, Gervais and D'Orbigny 1844) are small odontocete cetaceans of the Pontoporiidae family found in shallow waters (<30 m) in both marine and estuarine areas (Crespo, 1998; Danilewicz et al., 2009; Di Beneditto et al., 2011; Di Beneditto and Ramos, 2014). The distribution range of the species occurs from Itaúnas, Espírito Santo, Brazil (18° 25′ S, 30° 42′ W; Siciliano, 1994) to Golfo Nuevo, Argentina (42° 35′ S, 64° 48′ W; Bastida et al., 2007). However, this distribution is not continuous, with two gaps (or hiatuses) noted along the Brazilian coast, the first between Macaé (22° 25′S) and Ilha Grande (23° S), in the state of Rio de Janeiro, and the second between Regência, Espírito Santo (19° 40′ S), and Barra do Itabapoana, Rio de Janeiro (21°18′ S). An offshore hiatus between 30 and 50 m has also been previously described (Zerbini et al., 2010). The species' geographic range has been categorized into four Franciscana Management Areas (FMAs; Figure 10.1) based on genotypic, phenotypic, life history, and distributional data to improve management and conservation efforts (Secchi et al., 2003). The FMA I corresponds to the Rio de Janeiro and Espírito Santo states, in Brazil, the FMA II ranges from São Paulo to Santa Catarina, also in Brazil, the FMA III comprises coastal waters off Rio Grande do Sul, in both Brazil, and Uruguay, and the FMA IV includes coastal waters off Argentina. New genetic and morphological data, however, have indicated some degree of differentiation between individuals in FMA I, suggesting reproductive isolation and prompting the subdivision of the area into two distinct management units, the (FMA Ia in Espírito Santo and FMA Ib in Rio de Janeiro; Cunha et al., 2014).

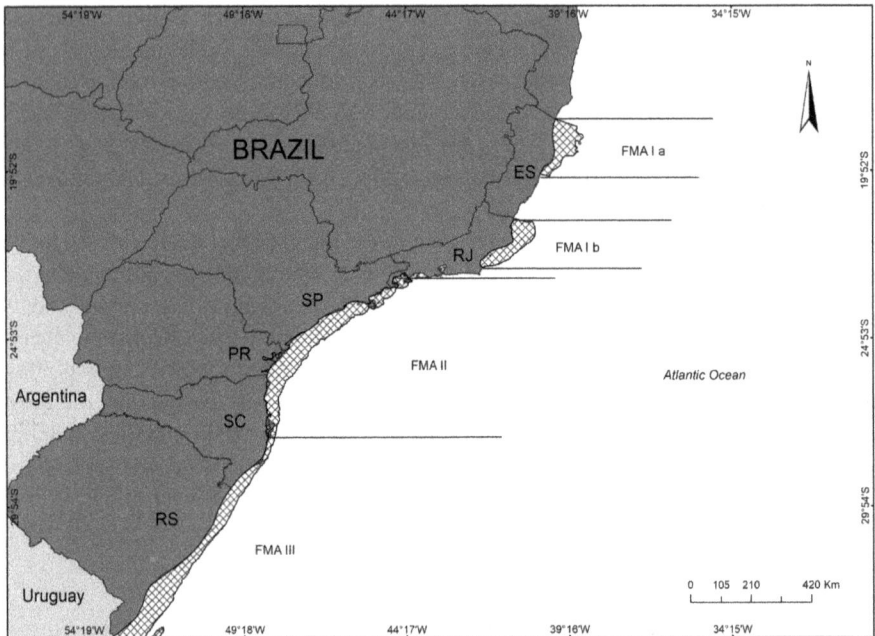

Figure 10.1   Franciscana dolphin (*Pontoporia blainvillei*) geographic range divided into four Franciscana Management Areas (FMAs).

The FMA I contain the smallest population among all FMAs. Aerial surveys conducted along the FMA Ia detected no individuals, suggesting that species abundance in this area is very small (Danilewicz et al., 2012). This hypothesis is further supported by Cunha et al. (2014), who identified very low genetic diversity in the population. At the FMA Ib, the population was estimated at 2,000 individuals (CV = 0.46; Sucunza, 2020), increasing to 6,800 individuals at the FMA II (2008–2009; CV = 0.26; Sucunza, 2020) similar to the population of the state of Rio Grande do Sul, Brazil (included in the FMA III), also estimated at 6,800 individuals (2004; CV = 0.32; Danilewicz et al., 2010). A population estimate in the other FMA III portion (Uruguay) has yet to be conducted, now comprising a research priority in this FMA. Finally, the FMA IV has an estimated population of 15,000 individuals (2003–2004; CV = 0.42; Crespo et al., 2009).

Due to its coastal habits, Franciscana dolphins inhabit areas subject to intense anthropogenic contamination sources, such as domestic and industrial sewage and chemical run-off. In fact, this species is one of the most anthropogenically impacted cetacean species in the Southwest Atlantic Ocean (Secchi and Wang, 2002), classified as "Vulnerable" by the International Union for Conservation of Nature (IUCN; Zerbini et al., 2017) and as "Critically Endangered" by the Brazilian government (Ministério do Meio Ambiente, 2014). This classification is based on a projected decline of over 30% of the species population over three generations (Taylor et al., 2007), mainly due to incidental mortality in gillnet fisheries, reported since the early 1940s (Van Erp, 1969). However, this decline rate is probably underestimated. The causes of the inferred population declines have not ceased and are, in fact, likely increasing, comprising growing fishery activities and consequent bycatches, prey abundance decreases, lack of mitigation actions (Zerbini et al., 2017) in addition to increasing environmental contamination. Therefore, the species conservation is considered a priority.

Concerning environmental contamination, Franciscana dolphins are considered excellent sentinel species due to their high site fidelity (Bordino et al., 2008; Cremer and Simões-Lopes, 2008). Some studies have assessed contaminant loads in the species, including persistent organic pollutants (Lailson-Brito et al., 2011; Lavandier et al., 2016; Leonel et al., 2010), metals and metalloids (Baptista et al., 2016; Kehrig et al., 2016; Lemos et al., 2013; Seixas et al., 2008). These assessments are, however, still scarce, making further monitoring studies for this species crucial.

FRANCISCANA DOLPHINS AND METAL EXPOSURE

In order to better discuss Franciscana Cd, Pb and Hg loads reported in the literature, all values were converted to a dry weight basis when reported as wet basis, by assuming a moisture content of 69.73% in liver samples, 72.21% in muscle samples, and 77.32% in kidney samples (based on Yang and Miyazaki, 2003). Compiled assessments on Franciscana dolphin Cd, Hg and Pb loads are displayed in Tables 10.1 through 10.4 for liver, muscle, kidney, and brain samples, respectively. Each metal is discussed separately for all tissues.

Lead, Mercury and Cadmium in the Aquatic Environment

Table 10.1 Mean cadmium (Cd), mercury (Hg), and lead (Pb; µg.g^{-1}, dry weight) concentrations ± standard deviation and ranges in **liver** samples of Franciscana dolphins (*Pontoporia blainvillei*) from the South American coast

Study	Local	n	Cd	Hg	Pb
Marcovecchio et al. (1990)	Buenos Aires, Argentina	2	10.90 ± 4.62	12.55 ± 5.29	NA
Gerpe et al. (2002)	Buenos Aires, Argentina	1[a] 1[b]	ND 7.50	ND 29.04	NA
Lailson-Brito et al. (2002)	Rio de Janeiro, Brazil	17	1.42 ± 1.42	17.74 ± 36.01	NA
Dorneles et al. (2007)	Rio Grande do Sul, Brazil	44	1.92 0.13 – 13.68	NA	NA
Seixas et al. (2007)	Rio de Janeiro, Brazil Rio Grande do Sul, Brazil	18 13	0.6 1.5	2.6 10.7	NA
de Carvalho et al. (2008)	Rio de Janeiro, Brazil	7	0.36 0.16 – 0.56	3.73 0.99 – 8.92	NA
Moreira et al. (2009)	Rio Grande do Sul, Brazil Rio de Janeiro, Brazil	31	NA	5.98 0.83 – 51.65	NA
Panebianco et al. (2012)	Buenos Aires, Argentina	18[c] 6[d]	4.59 ± 3.47 1.02 – 10.31 11.83 ± 11.23 2.94 – 32.87	NA	1.88 ± 2.61 ND – 6.44 1.62 ± 1.95 ND – 9.85
Lemos et al. (2013)	Rio de Janeiro, Brazil	1	0.53	1.10	NA
Kehrig et al. (2016)	Rio de Janeiro, Brazil	7[c] 4[d]	NA	1.65 ± 0.66 5.29 ± 3.31	NA
Romero et al. (2016)	Buenos Aires, Argentina	2[a] 9[e] 22[f] 9[g]	NA	2.54 – 5.04 1.29 ± 0.79 3.62 ± 3.46 26.88 ± 22.25	NA
Romero et al. (2017)	Buenos Aires, Argentina	*Estuarine:* 1[a] 7[e] 8[f] 5[g] *Marine:* 1[a] 2[e] 14[f] 4[g]	NA	NA	0.057 0.088 ± 0.052 0.070 ± 0.031 0.055 ± 0.003 ND ND 0.054 ± 0.011 0.059 ± 0.01

Legend: ND: non-detectable (below the limit of detection); NA: non-applicable; [a] fetus; [b] mother; [c] immature individuals; [d] mature individuals, [e] calf, [f] juvenile, [g] adult. All data are expressed in dry weight. Wet weight basis concentration was converted to dry weight, assuming a moisture content of 69.73% in liver samples (Yang and Miyazaki, 2003).

Table 10.2 Mean cadmium (Cd), mercury (Hg), and lead (Pb; μg.g^{-1}, dry weight) concentrations ± standard deviation and ranges in **muscle** samples of franciscana dolphins (*Pontoporia blainvillei*) from the South American coast

Study	Local	n	Cd	Hg	Pb
Marcovecchio et al. (1990)	Buenos Aires, Argentina	2	0.36 ± 0.36	10.79 ± 4.32	NA
Gerpe et al. (2002)	Buenos Aires, Argentina	1[a] 1[b]	ND 4.14	ND 7.20	NA
de Carvalho et al. (2008)	Rio de Janeiro, Brazil	7	0.76 0.36–1.15	0.61 0.22–0.97	NA
Panebianco et al. (2013)	Buenos Aires, Argentina	36	ND	NA	ND
Kehrig et al. (2016)	Rio de Janeiro, Brazil	7[c] 4[d]	NA	1.08 ± 0.14 1.80 ± 0.07	NA
Baptista et al. (2016)	Rio de Janeiro, Brazil	16	NA	1.92 ± 0.96	NA
Romero et al. (2016)	Buenos Aires, Argentina	2[a] 9[e] 22[f] 9[g]	NA	1.32–2.93 1.03 ± 1.09 1.49 ± 0.72 3.10 ± 1.96	NA

Legend: ND: non-detectable (below the limit of detection); NA: non-applicable; [a] fetus; [b] mother; [c] immature individuals; [d] mature individuals, [e] calf, [f] juvenile, [g] adult. All data are expressed in dry weight. Wet weight basis concentration was converted to dry weight, assuming a moisture content of 72.21% in muscle samples (Yang and Miyazaki, 2003).

Table 10.3 Mean cadmium (Cd), mercury (Hg), and lead (Pb; μg.g^{-1}, dry weight) concentrations ± standard deviation and ranges in **kidney** samples of franciscana dolphins (*Pontoporia blainvillei*) from the South American coast

Study	Local	n	Cd	Hg	Pb
Marcovecchio et al. (1990)	Buenos Aires, Argentina	2	43.65 ± 17.20	8.38 ± 3.09	NA
Gerpe et al. (2002)	Buenos Aires, Argentina	1[a] 1[b]	ND 28.17	ND 7.63	NA
Lailson-Brito et al. (2002)	Rio de Janeiro, Brazil	15	1.32 ± 1.41	6.00 ± 4.76	NA
Seixas et al. (2007)	Rio de Janeiro, Brazil Rio Grande do Sul, Brazil	18 13	3.4 5.5	1.4 1.7	NA
Moreira et al. (2009)	Rio Grande do Sul, Brazil Rio de Janeiro, Brazil	31	NA	1.52 0.45–5.11	NA
Panebianco et al. (2011)	Buenos Aires, Argentina	37	25.22 ± 33.33 0.88–131.61	NA	NA

Study	Local	n	Cd	Hg	Pb
Romero et al. (2016)	Buenos Aires, Argentina	2[a] 9[c] 22[d] 9[e]	NA	1.10–2.54 0.76 ± 0.72 1.35 ± 1.11 3.78 ± 4.35	NA
Romero et al. (2017)	Buenos Aires, Argentina	*Estuarine:* 1[a] 7[e] 8[f] 5[g] *Marine:* 1[a] 2[e] 14[f] 4[g]	NA	NA	0.058 0.079 ± 0.027 0.056 ± 0.003 0.054 ± 0.006 0.056 0.058 0.054 ± 0.004 0.036 ± 0.026

Legend: ND: non-detectable (below the limit of detection); NA: non-applicable; [a] fetus; [b] mother; [c] immature individuals; [d] mature individuals, [e] calf, [f] juvenile, [g] adult. All data are expressed in dry weight. Wet weight basis concentration was converted to dry weight, assuming a moisture content of 77.32% in kidney samples (Yang and Miyazaki, 2003).

Table 10.4 Mean cadmium (Cd), mercury (Hg), and lead (Pb; $\mu g.g^{-1}$, dry weight) concentrations ± standard deviation and ranges in **brain** samples of franciscana dolphins (*Pontoporia blainvillei*) from the South American coast

Study	Local	n	Cd	Hg	Pb
Romero et al. (2016)	Buenos Aires, Argentina	2[a] 9[b] 22[c] 9[d]	NA	1.16 0.16 ± 0.05 0.39 ± 0.17 0.38 ± 3.85	NA
Romero et al. (2017)	Buenos Aires, Argentina	*Estuarine:* 1[a] 7[b] 8[c] 5[d] *Marine:* 1[a] 2[b] 14[c] 4[d]	NA	NA	NA 0.02 ± 0.01 0.023 ± 0.023 ND ND 0.041 0.026 ± 0.027 0.043 ± 0.035

Legend: ND: non-detectable (below the limit of detection); NA: non-applicable; [a] fetus; [b] calf, [c] juvenile, [d] adult.

CADMIUM

Cadmium concentrations were usually higher in kidney samples compared to other matrices. This is expected, as kidneys are a critical Cd toxicity target (Satarug, 2018). For example, Franciscana dolphins from Buenos Aires,

Argentina, displayed a higher mean concentration of cadmium in kidney samples (43.65 µg.g^{-1}) compared to liver (10.90 µg.g^{-1}) and muscle samples (0.36 µg.g^{-1}; Marcovecchio et al., 1990). In the same location, Gerpe et al. (2002) reported the concentrations of 28.17 µg.g^{-1} in kidney, 7.50 µg.g^{-1} in liver, and 4.14 µg.g^{-1} in muscle in a mother while its fetus displayed non-detectable (ND) Cd concentrations in the three tissues. Not only the concentrations in the tissues had the same trend as previously reported for Cd, but also non-transference or very small (i.e., ND) transference via mother-fetus was noted for this element. Later, in the same location, Panebianco et al. (2011) reported the mean Cd concentration of 25.22 µg.g^{-1} (range: 0.88–131.61 µg.g^{-1}) in kidney samples, while Panebianco et al. (2012) reported a mean Cd concentration of 4.59 µg.g^{-1} (range: 1.02–10.31 µg.g^{-1}) in livers of immature individuals and 11.83 µg.g^{-1} (range: 2.94–32.87 µg.g^{-1}) in mature individuals. Panebianco et al. (2013) were not able to detect Cd (< 1.90 µg.g^{-1}) concentrations in both kidney and skin samples of Franciscana dolphins from the same location.

In Rio Grande do Sul, Brazil, the mean Cd concentration in liver samples of Franciscana dolphins (n = 44; Table 10.1) was 1.92 µg.g^{-1} (range: 0.13–13.68 µg.g^{-1}; Dorneles et al., 2007). In Rio de Janeiro, Lailson-Brito et al. (2002) reported a mean Cd concentration of 1.42 µg.g^{-1} in liver and 1.32 µg.g^{-1} in kidney of franciscana dolphins. In 2007, Seixas et al. (2007) analyzed Cd concentrations in Franciscana dolphins from Rio Grande do Sul (RS) and Rio de Janeiro (RJ) states, finding higher concentrations in both liver and kidney samples from the first state (RS: 5.5 µg.g^{-1} in kidney and 1.5 µg.g^{-1} in liver; RJ: 3.4 µg.g^{-1} in kidney and 0.6 µg.g^{-1} in liver). Still in Rio de Janeiro, de Carvalho et al. (2008) reported the mean concentration of 0.36 (range: 0.16–0.56 µg.g^{-1}) in liver and 0.76 (range: 0.36–1.15 µg.g^{-1}) in muscle, as opposed to the trends found in other studies. Lemos et al. (2013) reported the concentration of 0.53 µg.g^{-1} in liver of a single immature individual, which is in the same range of the other studies.

Overall, Cd concentrations in Franciscana dolphins from Brazil were in the same range as the concentrations found in Argentina, indicating similar exposures to this toxic element. Unfortunately, no studies to date have been conducted in Uruguay.

MERCURY

Mercury concentrations were relatively higher in liver samples. Marcovecchio et al. (1990) reported concentrations of 12.55 µg.g^{-1} in livers, 10.79 µg.g^{-1} in muscles and 8.38 µg.g^{-1} in kidneys of Franciscana dolphins from Buenos Aires, Argentina. Later, in the same area, Gerpe et al. (2002) reported ND concentrations of Hg in liver, muscle and kidney tissues of a Franciscana fetus while its mother displayed a higher Hg concentration in liver (29.04 µg.g^{-1}) compared to kidney (7.63 µg.g^{-1}) and muscle (7.20 µg.g^{-1}) samples. In 2016, Romero et al. (2016) assessed Hg concentrations in liver, muscle, kidney, and brain samples of Franciscana dolphins from Buenos Aires, Argentina, of varied age groups. Liver samples had the highest Hg concentrations in all age groups, while the other matrices varied. Fetuses displayed higher concentrations in liver, followed by muscle, brain, and kidney, while calves and juveniles displayed higher concentrations in liver, followed by muscle, kidney, and brain. The trend in adults was liver followed by kidney, muscle, and finally brain. Interestingly, Hg concentrations were higher in all matrices in fetuses compared to calves, indicating maternal transfer of this metal. Another important factor to highlight is the higher concentration in fetuses' brains (1.16 $\mu g.g^{-1}$) compared to the other age groups (0.16, 0.39, and 0.38 $\mu g.g^{-1}$ in calves, juveniles, and adults, respectively). Previous studies in humans have described the potential of the methylmercury (formed from inorganic Hg) to cross the placenta and the blood-brain barrier and affect the fetus without causing severe damage to the mother (Harada, 1978). Cases of cerebral palsy and mental deficiency have also been reported in human fetuses, due to the extreme neurotoxicity of this metal (Branco et al., 2021). This may be the cause for the relatively high concentrations of Hg in the fetuses' brains and should be further investigated, since only one study has assessed this matrix in the species.

In Rio de Janeiro, Brazil, Lailson-Brito et al. (2002) reported mean Hg concentrations of 17.74 $\mu g.g^{-1}$ in liver and 6.00 $\mu g.g^{-1}$ in kidney samples of Franciscana dolphins. Seixas et al. (2007) also compared Hg concentrations in both liver and kidney samples of Franciscanas from Rio Grande do Sul and Rio de Janeiro, finding higher concentrations in liver (RJ: 2.6 $\mu g.g^{-1}$; RS: 10.7 $\mu g.g^{-1}$) compared to kidney (RJ: 1.4 $\mu g.g^{-1}$; RS: 1.7 $\mu g.g^{-1}$). Carvalho et al. (2008) reported the mean of 3.73 $\mu g.g^{-1}$ (range: 0.99–8.92 $\mu g.g^{-1}$) in liver, and 0.61 $\mu g.g^{-1}$ (range: 0.22–0.97 $\mu g.g^{-1}$) in muscle samples. Moreira et al. (2009) assessed Hg concentrations in Franciscana dolphins from Rio Grande do Sul and Rio de Janeiro and described higher concentrations in liver (5.98 $\mu g.g^{-1}$) compared to kidney samples (1.52 $\mu g.g^{-1}$), which is in agreement with previous studies. It is also important to highlight that the Hg concentrations in liver samples from Rio Grande do Sul were greater than Rio de Janeiro. Later, in Rio de Janeiro, Lemos et al. (2013) reported the concentration of 1.10 $\mu g.g^{-1}$ in liver of an immature individual, while Kehrig et al. (2016) reported the mean concentrations of 1.65 $\mu g.g^{-1}$ in liver of immature individuals and 5.29 $\mu g.g^{-1}$ of mature individuals. In muscle samples, the concentrations were relatively lower: 1.08 $\mu g.g^{-1}$ in immature individuals and 1.80 $\mu g.g^{-1}$ in mature individuals. In the same year and location, Baptista et al. (2016) reported the mean concentration of 1.92 $\mu g.g^{-1}$ in Franciscana muscle samples.

LEAD

Panebianco et al. (2012) reported a mean Pb concentration of 1.88 $\mu g.g^{-1}$ (range: ND – 6.44 $\mu g.g^{-1}$) in livers of immature individuals and 1.62 $\mu g.g^{-1}$ (range: ND – 9.85 $\mu g.g^{-1}$) in mature individuals of Franciscana dolphins from Buenos Aires, Argentina. Later, however, Panebianco et al. (2013) was not able to detect Pb (< 5.34 $\mu g.g^{-1}$) concentrations in kidney and skin samples of Franciscana dolphins from the same location. The latest study was conducted by Romero et al. (2017), who then analyzed lead concentrations in liver, kidney, and brain samples of estuarine and marine Franciscana dolphins, also from Buenos Aires. In liver, estuarine fetus, calves, and juveniles displayed higher concentrations than the marine individuals, with calves displaying the highest mean Pb concentration (0.088 $\mu g.g^{-1}$; Table 10.1). In kidney, all age groups exhibited higher mean concentrations compared to the marine specimens (Table 10.3). Finally, in the brain, the opposite has been observed in which marine individuals displayed relatively higher concentrations in calves, juveniles, and adults. No other studies to date have been conducted on Pb concentrations in Franciscana dolphins, revealing the scarcity of these assessments and the importance to further expand these baseline data. In the case of Pb, several target organs are noted, including the nervous system, hematopoietic system, cardiovascular system, kidneys, reproductive system, and fetuses (ATSDR, 2021), demonstrating the importance of Pb biomonitoring efforts.

DIFFERENTIAL ORGAN CONTAMINATION

Concerning metal organ differences, overall, liver metal loads are expected to be higher in both low contamination scenarios and in cases where the exposed organism can cope with metal loads through metal homeostasis and detoxifying processes. The liver is, in fact, the main detoxifying organ of the body, as it displays the capacity to convert lipophilic compounds into more water-soluble substances, which are then efficiently eliminated from the body via the excretion system (Grant, 1991). The liver also houses high concentrations of the most efficient metal detoxifying factor: the metal-binding proteins called metallothioneins, the main metal detoxification route in mammals. In the aforementioned studies, Franciscana dolphin livers displayed, overall, the highest metal concentrations, especially concerning Hg.

On the other hand, muscle tissue exhibits low metabolic rates, so higher muscle metal concentrations in muscle compared to liver indicate bioaccumulation processes due to an overwhelmed hepatic detoxification system (Marcovecchio et al., 1991). When this occurs, severe deleterious cellular and physiological defenses effects are expected. In our review, we did not observe cases of higher Cd, Hg and Pb concentrations in muscle compared to liver, indicating efficient detoxification efforts by the liver, although the presence of some metal loads in muscle indicates possible accumulation processes.

Metal toxicity also affect the kidneys due to its capacity of filter, reabsorb and concentrate ions (Lentini et al., 2017). The degree of damage will depend on the metal, dose, and exposure time, with Cd, Hg, and Pb being known metals implicated in kidney toxicity. Relatively high levels of cadmium were observed in kidneys of Franciscana dolphins in the studies discussed herein (e.g., Marcovecchio et al., 1990), which can be explained by the Franciscana dolphin diet that is composed by juvenile teleost fish, crustaceans, and small loliginid squids (Rodríguez et al., 2002). Cephalopods in general are known to accumulate more Cd than other marine invertebrates, thus a frequent intake of this prey would reflect a higher accumulation of this metal (Bustamante et al., 1998).

Franciscana dolphin brains accumulated the lowest levels of Hg and Pb compared to the other discussed organs. However, even the smallest concentrations in brain may cause deleterious effects in an individual, including metal-induced neurotoxicity, oxidative stress, disruptive mitochondrial function, and impaired activity of various enzymes, which could lead to permanent injuries like severe neurological disorders (Chen et al., 2016). Therefore, due to the scarcity of assessments in brain samples and a better understanding of the kinetics of these metals in odontocete species, especially in Franciscana dolphins, future studies should expand their assessments to this matrix.

Furthermore, it is also important to note that metals undergo internal subcellular compartmentalization following organism exposure, which in turn significantly alters their bioavailability and deleterious biochemical effects. Many metal accumulation aspects, such as toxicity, tolerance and trophic transfer, can be understood by examining subcellular metal partitioning (Wallace et al., 2003). Assessments in this regard, however, are still significantly lacking for marine mammals, and only one study is available for the Franciscana dolphin, concerning titanium (Monteiro et al., 2020). Some studies concerning subcellular Cd, Hg, and Pb partitioning and detoxification are available for other dolphins (i.e., *Sotalia guianensis*, *Stenella coeruleoalba*, *Steno bredanensis*, and *Tursiops truncatus*, see Hauser-Davis et al., 2020 for further details), indicating differential metal-detoxification mechanisms for these elements, with some degree of detoxification in different organs carried out by metallothioneins. These kinds of appraisals are gradually creating a new field of research on metal effect assessments and bioavailability studies in marine mammals, of significant value in environmental contamination mitigation and conservation efforts. Thus, it is paramount that subcellular partitioning assessments be carried out for this endangered species.

CONCLUSIONS

Based on the discussed background information, samples often analyzed for metal concentrations in Franciscana dolphins comprise muscle, liver, and kidney, with only some studies analyzing brain and skin samples. Given the importance of the metal transference through the placenta and blood-brain barrier from mother to fetus and potential negative effects in the offspring, it is crucial to further expand assessments of brain samples for a better understanding on the mother-fetus transference and later detoxification and excretion by the individuals, as well as to find potential causal links with individual deaths. Other types of matrices such as blubber, urine and fecal samples should also be evaluated to monitor threatened wild populations.

Another important factor in this review is the absence of assessments in Franciscana dolphins along the coast of Uruguay, demonstrating the importance of further monitoring studies for the species in this country as well as expanding monitoring efforts in other areas of the species' distribution.

Future studies should focus on filling current research gaps, which include the trophic transfer of pollutants in ecosystems used by Franciscana dolphins, the determination of subcellular metal partitioning and contaminant-specific biomarkers to assess exposure and effects in the species.

Franciscana dolphin populations are already under significant distress due to population declines caused by incidental bycatch, negative interactions with watercraft, potential absence of adequate prey, and chemical contamination to name a few. This indicates substantial concerns for the future of the species, clearly indicating that overall health monitoring efforts are paramount.

REFERENCES

Ali, H. and Khan, E. 2019. Trophic transfer, bioaccumulation, and biomagnification of non-essential hazardous heavy metals and metalloids in food chains/webs—Concepts and implications for wildlife and human health. Hum. Ecol. Risk Assess. An Int. J. 25: 1353–1376. https://doi.org/10.1080/10807039.2018.1469398

Ali, H., Khan, E. and Ilahi, I. 2019. Environmental Chemistry and Ecotoxicology of Hazardous Heavy Metals: Environmental Persistence, Toxicity, and Bioaccumulation. J. Chem. 1–14. https://doi.org/10.1155/2019/6730305

ATSDR. 1999. Toxicological profile for mercury. Atlanta, Georgia.

ATSDR. 2012. Toxicological profile for cadmium. Atlanta, Georgia.

ATSDR. 2019. Substance Priority List [WWW Document]. URL https://www.atsdr.cdc.gov spl/index.html (accessed 3.20.22).

ATSDR. 2020. Toxicological profile for lead. Atlanta, Georgia.

ATSDR. 2021. What Are Possible Health Effects from Lead Exposure? [WWW Document]. URL www.atsdr.cdc.gov/csem/leadtoxicity/physiological_effects.html

Bastida, R., Rodríguez, D., Secchi, E. and da Silva, V. 2007. Mamíferos Acuáticos de Sudamérica y Antártida. Buenos Aires.

Baptista, G., Kehrig, H.A., Di Beneditto, A.P.M., Hauser-Davis, R.A., Almeida, M.G., Rezende, C.E., et al. 2016. Mercury, selenium and stable isotopes in four small cetaceans from the Southeastern Brazilian coast: Influence of feeding strategy. Environ. Pollut. 218: 1298–1307.

Bordino, P., Wells, R.S. and Stamper, M.A. 2008. Satellite tracking of Franciscana Dolphins *Pontoporia blainvillei* in Argentina: preliminary information on ranging, diving and social patterns. *In*: IWC Scientific Committee Meeting. IWC, Santiago, Chile. p. SC60/SM14.

Branco, V., Aschner, M. and Carvalho, C. 2021. Neurotoxicity of mercury: An old issue with contemporary significance. Adv. Neurotoxicol. 5: 239-262. https://doi.org/10.1016/bs.ant.2021.01.001

Bustamante, P., Caurant, F., Fowler, S. and Miramand, P. 1998. Cephalopods as a vector for the transfer of cadmium to top marine predators in the north-east Atlantic Ocean. Sci. Total Environ. 220: 71–80. https://doi.org/10.1016/S0048-9697(98)00250-2

Chen, J., Chen, L., Liu, D. and Zhang, G. 2014. Organochlorine pesticide contamination in marine organisms of Yantai coast, northern Yellow Sea of China. Environ. Monit. Assess. 186: 1561–1568. https://doi.org/10.1007/s10661-013-3473-z

Chen, P., Miah, M.R. and Aschner, M. 2016. Metals and Neurodegeneration. F1000Research 5: 366. https://doi.org/10.12688/f1000research.7431.1

Chowdhury, B.A. and Chandra, R.K. 1987. Biological and health implications of toxic heavy metal and essential trace element interactions. Prog. Food Nutr. Sci. 11: 55–113.

Cremer, M. and Simões-Lopes, P.C. 2008. Distribution, abundance and density estimates of franciscanas, *Pontoporia blainvillei* (Cetacea: Pontoporidae), in Babitonga bay, Southern Brazil. Rev. Bras. Zool. 25: 397–402.

Crespo, E.A. 1998. Informe del Tercer Taller para la Coordinación de la Investigación y la Conservación de la franciscana (*Pontoporia blainvillei*) en el Atlántico Sudoccidental. Rep. to Conv. Migr. Species (UNEP), Bonn, Ger. 23.

Crespo, E.A., Pedraza, S.N., Grandi, M.F., Dans, S.L. and Garaffo, G.V. 2009. Abundance and distribution of endangered Franciscana dolphins in Argentine waters and conservation implications. Mar. Mammal Sci. 26: 17–35. https://doi.org/10.1111/j.1748-7692.2009.00313.x

Cui, D., Li, X. and Quinete, N. 2020. Occurrence, fate, sources and toxicity of PFAS: What we know so far in Florida and major gaps. TrAC Trends Anal. Chem. 130: 115976. https://doi.org/10.1016/j.trac.2020.115976

Cullen, J.A., Marshall, C.D. and Hala, D. 2019. Integration of multi-tissue PAH and PCB burdens with biomarker activity in three coastal shark species from the northwestern Gulf of Mexico. Sci. Total Environ. 650: 1158–1172. https://doi.org/10.1016/j.scitotenv.2018.09.128

Cunha, H.A., Medeiros, B.V., Barbosa, L.A., Cremer, M.J., Marigo, J., Lailson-Brito, J. et al. 2014. Population structure of the endangered franciscana dolphin (Pontoporia blainvillei): reassessing management units. PLoS One 9: e85633. https://doi.org/10.1371/journal.pone.0085633

Danilewicz, D., Secchi, E.R., Ott, P.H., Moreno, I.B., Bassoi, M. and Borges-Martins, M. 2009. Habitat use patterns of franciscana dolphins (*Pontoporia blainvillei*) off southern Brazil in relation to water depth. J. Mar. Biol. Assoc. U.K. 89: 943–949. https://doi.org/10.1017/S002531540900054X

Danilewicz, D., Moreno, I.B., Ott, P.H., Tavares, M., Azevedo, A.F., Secchi, E.R. et al. 2010. Abundance estimate for a threatened population of franciscana dolphins in southern coastal Brazil: uncertainties and management implications. J. Mar. Biol. Assoc. U.K. 90: 1649–1657. https://doi.org/10.1017/S0025315409991482

Danilewicz, D., Zerbini, A.N., Andriolo, A., Secchi, E.R., Sucunza, F., Ferreira, E., et al. 2012. Abundance and distribution of an isolated population of franciscana dolphins (*Pontoporia blainvillei*) in southeastern Brazil: red alert for FMA I? (No. Paper SC/64/SM17 presented to the IWC Scientific Committee). Panama City.

de Carvalho, C.E.V., Di Beneditto, A.P.M., Souza, C.M.M., Ramos, R.M.A. and Rezende, C.E. 2008. Heavy metal distribution in two cetacean species from Rio de Janeiro State, south-eastern Brazil. J. Mar. Biol. Assoc. U.K. 88: 1117–1120. https://doi.org/10.1017/S0025315408000325

Di Beneditto, A.P.M., de Souza, C.M.M., Kehrig, H.A. and Rezende, C.E. 2011. Use of multiple tools to assess the feeding preference of coastal dolphins. Mar. Biol. 158: 2209–2217. https://doi.org/10.1007/s00227-011-1726-3

Di Beneditto, A.P.M. and Ramos, R.M.A., 2014. Marine debris ingestion by coastal dolphins: What drives differences between sympatric species? Mar. Pollut. Bull. 83: 298–301. https://doi.org/10.1016/j.marpolbul.2014.03.057

Dorneles, P.R., Lailson-Brito, J., Secchi, E.R., Bassoi, M., Lozinsky, C.P.C., Torres, J.P.M., et al. 2007. Cadmium concentrations in franciscana dolphin (*Pontoporia blainvillei*) from south brazilian coast. Brazilian J. Oceanogr. 55: 179–186. https://doi.org/10.1590/S1679-87592007000300002

EPA. 2021. EPA's Report on the Environment (ROE).

Gerpe, M., Rodríguez, D., Moreno, V.J., Bastida, R.O. and Moreno, J.E. 2002. Accumulation of heavy metals in the franciscana (*Pontoporia blainvillei*) from Buenos Aires Province, Argentina. Lat. Am. J. Aquat. Mamm. 1. https://doi.org/10.5597/lajam00013

Grant, D.M. 1991. Detoxification pathways in the liver. J. Inherit. Metab. Dis. 14: 421–430. https://doi.org/10.1007/BF01797915

Harada, M. 1978. Congenital Minamata disease: Intrauterine methylmercury poisoning. Teratology 18: 285–288. https://doi.org/10.1002/tera.1420180216

Hauser-Davis, R.A., Figueiredo, L., Lemos, L., de Moura, J.F., Rocha, R.C.C., Saint'Pierre, T. et al. 2020. Subcellular Cadmium, Lead and Mercury Compartmentalization in Guiana Dolphins (Sotalia guianensis) From Southeastern Brazil. Front. Mar. Sci. 7: 1–8. https://doi.org/10.3389/fmars.2020.584195

Kehrig, H.A., Hauser-Davis, R.A., Seixas, T.G., Pinheiro, A.B. and Di Beneditto, A.P.M. 2016. Mercury species, selenium, metallothioneins and glutathione in two dolphins from the southeastern Brazilian coast: Mercury detoxification and physiological differences in diving capacity. Environ. Pollut. 213: 785–792.

Lailson-Brito, J., Azeredo, M.A.A., Malm, O., Ramos, R.A., Di Beneditto, A.P.M. and Saldanha, M.F.C. 2002. Trace metals in liver and kidney of the franciscana (*Pontoporia blainvillei*) from the northern coast of Rio de Janeiro State, Brazil. Lat. Am. J. Aquat. Mamm. 1. https://doi.org/10.5597/lajam00014

Lailson-Brito, J., Dorneles, P.R., Azevedo-Silva, C.E., Azevedo, A., de, F., Vidal, L.G., et al. 2011. Organochlorine concentrations in franciscana dolphins, *Pontoporia blainvillei*, from Brazilian waters. Chemosphere 84: 882–887.

Lavandier, R., Areas, J., Quinete, N., de Moura, J.F., Taniguchi, S., Montone, R., et al. 2016. PCB and PBDE levels in a highly threatened dolphin species from the Southeastern Brazilian coast. Environ. Pollut. 208: 442–449.

Lemos, L.S., de Moura, J.F., Hauser-Davis, R.A., de Campos, R.C. and Siciliano, S. 2013. Small cetaceans found stranded or accidentally captured in southeastern Brazil: Bioindicators of essential and non-essential trace elements in the environment. Ecotoxicol. Environ. Saf. 97: 166–175. https://doi.org/10.1016/j.ecoenv.2013.07.025

Lentini, P., Zanoli, L., Granata, A., Signorelli, S.S., Castellino, P. and Dellaquila, R. 2017. Kidney and heavy metals–The role of environmental exposure. Mol. Med. Rep. 15: 3413–3419. https://doi.org/10.3892/mmr.2017.6389

Leonel, J., Sericano, J.L., Fillmann, G., Secchi, E. and Montone, R.C. 2010. Long-term trends of polychlorinated biphenyls and chlorinated pesticides in franciscana dolphin (*Pontoporia blainvillei*) from Southern Brazil. Mar. Pollut. Bull. 60: 412–418.

Marcovecchio, J.E., Moreno, V.J., Bastida, R.O., Gerpe, M.S. and Rodríguez, D.H. 1990. Tissue distribution of heavy metals in small cetaceans from the Southwestern Atlantic Ocean. Mar. Pollut. Bull. 21: 299–304. https://doi.org/10.1016/0025-326X(90)90595-Y

Marcovecchio, J.E., Moreno, V.J. and Pérez, A. 1991. Metal accumulation in tissues of sharks from the Bahía Blanca estuary, Argentina. Mar. Environ. Res. 31: 263–274. https://doi.org/10.1016/0141-1136(91)90016-2

Ministério do Meio Ambiente. 2014. Lista nacional oficial de espécies da fauna ameaçada de extinção–Mamíferos, aves, répteis, anfíbios e invertebrados terrestres, Portaria MMA no. 444 de 17 de dezembro de 2014 [National list of the species of the fauna threatened of extinction - Mammals, Brasil].

Monteiro, F., Lemos, L.S., de Moura, J.F., Rocha, R.C.C., Moreira, I., Di Beneditto, A.P.M., et al. 2020. Total and subcellular Ti distribution and detoxification processes in

Pontoporia blainvillei and Steno bredanensis dolphins from Southeastern Brazil. Mar. Pollut. Bull. 153: 110975. https://doi.org/10.1016/j.marpolbul.2020.110975

Moreira, I., Seixas, T.G., Kehrig, H.A., Fillmann, G., Beneditto, A.P. Di, Souza, C.M. et al. 2009. Selenium and Mercury (Total and Organic) in Tissues of a CoastalSmall Cetacean, Pontoporia blainvillei. J. Coast. Res. 866–870.

Panebianco, M.V., Botte, S.E., Negri, M.F., Marcovecchio, J.E. and Cappppozzo, H.L. 2012. Heavy Metals in Liver of the Franciscana Dolphin, Pontoporia blainvillei, from the Southern Coast of Buenos Aires, Argentina. J. Brazilian Soc. Ecotoxicol. 7: 33–41. https://doi.org/10.5132/jbse.2012.01.006

Panebianco, M.V., Negri, M.F., Botte, S.E., Marcovecchio, J.E. and Cappozzo, H.L. 2011. Metales pesados en el rinon del delfin franciscana, Pontoporia blainvillei (Cetacea: Pontoporiidae) y su relacion con parametros biologicos. Lat. Am. J. Aquat. Res. 39: 526–533. https://doi.org/10.3856/vol39-issue3-fulltext-12

Panebianco, M.V, Negri, M.F., Botte, S.E., Marcovecchio, J.E. and Cappozzo, H.L. 2013. Essential and non-essential heavy metals in skin and muscle tissues of franciscana dolphins (Pontoporia blainvillei) from the southern Argentina coast. Chem. Ecol. 29: 511–518. https://doi.org/10.1080/02757540.2013.810727

Pulster, E.L., Gracia, A., Armenteros, M., Toro-Farmer, G., Snyder, S.M., Carr, B.E., et al. 2020. A first comprehensive baseline of hydrocarbon pollution in Gulf of Mexico fishes. Sci. Rep. 10: 6437. https://doi.org/10.1038/s41598-020-62944-6

Rodríguez, D., Rivero, L. and Bastida, R. 2002. Feeding ecology of the franciscana (*Pontoporia blainvillei*) in marine and estuarine waters of Argentina. Lat. Am. J. Aquat. Mamm. 1. https://doi.org/10.5597/lajam00012

Romero, M.B., Polizzi, P., Chiodi, L., Das, K. and Gerpe, M. 2016. The role of metallothioneins, selenium and transfer to offspring in mercury detoxification in Franciscana dolphins (*Pontoporia blainvillei*). Mar. Pollut. Bull. 109: 650–654. https://doi.org/10.1016/j.marpolbul.2016.05.012

Romero, M.B., Polizzi, P., Chiodi, L., Robles, A., Das, K. and Gerpe, M. 2017. Metals as chemical tracers to discriminate ecological populations of threatened Franciscana dolphins (*Pontoporia blainvillei*) from Argentina. Environ. Sci. Pollut. Res. 24: 3940–3950. https://doi.org/10.1007/s11356-016-7970-9

Saad, A.A.A., El-Sikaily, A. and Kassem, H. 2016. Essential, non-essential metals and human health. Blue Biotechnology Journal 3(4): 447–495.

Satarug, S. 2018. Dietary Cadmium Intake and Its Effects on Kidneys. Toxics 6: 15. https://doi.org/10.3390/toxics6010015

Secchi, E.R. and Wang, J.Y. 2002. Assessment of the conservation status of a franciscana (*Pontoporia blainvillei*) stock in the Franciscana Management Area III following the IUCN Red List Process. Lat. Am. J. Aquat. Mamm. 1: 183–190. https://doi.org/10.5597/lajam00023

Secchi, E.R., Danilewicz, D. and Ott, P.H. 2003. Applying the phylogeographic concept to identify franciscana dolphin stocks: Implications to meet management objectives. J. CETACEAN RES. Manag 5: 61–68.

Seixas, T., Kehrig, H., Fillmann, G., Di Beneditto, A., Souza, C., Secchi, E., et al. 2007. Ecological and biological determinants of trace elements accumulation in liver and kidney of *Pontoporia blainvillei*. Sci. Total Environ. 385: 208–220. https://doi.org/10.1016/j.scitotenv.2007.06.045

Seixas, T.G., Kehrig, H. do A., Costa, M., Fillmann, G., Beneditto, A.P.M., et al. 2008. Total mercury, organic mercury and selenium in liver and kidney of a South American coastal dolphin. Environ. Pollut. 154: 98–106.

Siciliano, S. 1994. Review of small cetaceans and fishery interactions in coastal waters of Brazil. Rep. Int. Whal. Commn. (Special Isssue) 15: 241–250.

Sucunza, F. 2020. Estimativas de densidade e abundância da toninha (Pontoporia blainvillei) a partir de levantamentos aéreos: produção e aplicação de fatores de correção. Universidade Federal de Juiz de Fora.

Taylor, B.L., Chivers, S.J., Larese, J. and Perrin, W.F. 2007. Generation length and percent mature estimates for IUCN assessments of cetaceans. Administrative Report LJ-07-01. National Marine Fisheries Service, Southwest Fisheries Science Center. https://aquadocs.org/bitstream/handle/1834/41281/LJ-07-01.pdf?sequence=1&isAllowed=y

Van Erp, I. 1969. In quest of the La Plata dolphin. Pacific Discov. 22: 18–24.

Wallace, W., Lee, B. and Luoma, S. 2003. Subcellular compartmentalization of Cd and Zn in two bivalves. I. Significance of metal-sensitive fractions (MSF) and biologically detoxified metal (BDM). Mar. Ecol. Prog. Ser. 249: 183–197. https://doi.org/10.3354/meps249183

Yang, J. and Miyazaki, N. 2003. Moisture content in Dall's porpoise (*Phocoenoides dalli*) tissues: A reference base for conversion factors between dry and wet weight trace element concentrations in cetaceans. Environ. Pollut. 121: 345–347. https://doi.org/10.1016/S0269-7491(02)00239-7

Zerbini, A.N., Secchi, E., Crespo, E., Danilewicz, D. and Reeves, R. 2017. *Pontoporia blainvillei* [WWW Document]. IUCN Red List Threat. Species 2017 e.T17978A123792204. URL https://dx.doi.org/10.2305/IUCN.UK.2017-3.RLTS.T17978A50371075.en. (accessed 5.15.21).

Zerbini, A.N., Secchi, E.R., Danilewicz, D., Andriolo, A., Laake, J.L. and Azevedo, A. 2010. Abundance and distribution of the franciscana (*Pontoporia blainvillei*) in the Franciscana Management Area II (southeastern and southern Brazil). Scientific Committee Paper SC/62/SM7.

Zoroddu, M.A., Aaseth, J., Crisponi, G., Medici, S., Peana, M. and Nurchi, V.M. 2019. The essential metals for humans: A brief overview. J. Inorg. Biochem. 195: 120–129. https://doi.org/10.1016/j.jinorgbio.2019.03.013

Index

For Product Safety Concerns and Information please contact our EU
representative GPSR@taylorandfrancis.com
Taylor & Francis Verlag GmbH, Kaufingerstraße 24, 80331 München, Germany